This book comes with access to more content online.

Quiz yourself, track your progress, and improve your grade!

Register your book or ebook at
www.dummies.com/go/getaccess.

Select your product, and then follow the prompts
to validate your purchase.

You'll receive an email with your PIN and instructions.

Anatomy & Physiology

ALL-IN-ONE

by Erin Odya

Anatomy & Physiology All-in-One For Dummies®

Published by: **John Wiley & Sons, Inc.**, 111 River Street, Hoboken, NJ 07030-5774, www.wiley.com

Copyright © 2023 by John Wiley & Sons, Inc., Hoboken, New Jersey

Published simultaneously in Canada

For general information on our other products and services, please contact our Customer Care Department within the U.S. at 877-762-2974, outside the U.S. at 317-572-3993, or fax 317-572-4002. For technical support, please visit https://hub.wiley.com/community/support/dummies.

Wiley publishes in a variety of print and electronic formats and by print-on-demand. Some material included with standard print versions of this book may not be included in e-books or in print-on-demand. If this book refers to media such as a CD or DVD that is not included in the version you purchased, you may download this material at http://booksupport.wiley.com. For more information about Wiley products, visit www.wiley.com.

Library of Congress Control Number: 2023930263

ISBN 978-1-394-15365-7; ISBN 978-1-394-15366-4 (ebk); ISBN 978-1-394-15367-1 (ebk)

SKY10064249_010924

Contents at a Glance

Table of Contents

Introduction

Whether your aim is to become a physical therapist or a pharmacist, a doctor or an acupuncturist, a nutritionist or a personal trainer, a registered nurse or a paramedic, a parent or simply a healthy human being, your efforts have to be based on a good understanding of anatomy and physiology. But knowing that the knee bone connects to the thigh bone (or does it?) is just the tip of the iceberg.

The human body is a miraculous biological machine capable of growing, interacting with the world, and even reproducing despite numerous environmental odds stacked against it. Understanding how the body's interlaced systems accomplish these feats requires a close look at everything from chemistry to structural mechanics.

Early anatomists relied on dissections to study the human body, which is why the Greek word *anatomia* means "to cut up or dissect." Anatomical references have been found in Egypt dating back to 1600 BC, but it was the Greeks — Hippocrates in particular — who first dissected bodies for medical study around 420 BC. That's why more than two millennia later, we still use words based on Greek and Latin roots to identify anatomical structures.

That's also part of the reason why studying anatomy and physiology feels like learning a foreign language. Truth be told, you *are* working with a foreign language, but it's the language of you and the one body you're ever going to have.

About This Book

Anatomy and Physiology All-In-One For Dummies, 1st Edition, combines the idea of a simplified textbook with the practice opportunities of a workbook. It isn't meant to replace a proper anatomy and physiology (A&P) textbook, and it's certainly not meant to replace going to an actual class. It's designed to be a supplement to your ongoing education and a study aid for prepping for exams. Sometimes, a slightly different presentation of a fact or of the relationship between facts can lead to a small "Aha!" moment; then technical details in your more comprehensive resources become easier to master. Consider reading the relevant chapter before class. That way, when your instructor covers the content, it'll be more likely to stick.

Your coursework might cover things in a different order from the one I've chosen for this book. As an instructor, I've had years of trial and error in the order in which I teach topics. I've found the flow of this book to be the most successful for my students, and it does differ from the traditional order, particularly when you reach Unit 4, "Traveling the World," which covers the fluids that circulate in your body. I welcome you to take full advantage of the table of contents and the index to find the material as it's addressed in your class.

Whatever you do, don't feel obligated to go through this book in the order I've chosen. But please do answer the practice questions and check the answers as you go. They are placed strategically, and in addition to providing the correct answers, I clarify why the right answer is the right answer and why the other answers are incorrect.

The goals of this book are to be informal but not unscientific, brief but not sketchy, and information-rich but accessible to readers at many levels. I've tried to present a light but serious survey of human anatomy and physiology that you can enjoy merely for the love of learning.

Foolish Assumptions

In writing this book, I had to make some assumptions about you, the reader — the most important of which is that you're not actually a dummy! If any of the following applies, this book is for you:

>> **Formal student:** You're enrolled in an intro A&P course for credit or in a similar course tied to a particular certification or credential. You need to pass an exam or otherwise demonstrate understanding and retention of terminology, identification, and the processes of human anatomy and physiology.

>> **Informal student:** You're not enrolled in a course, but you're interested in gaining some background knowledge or are reviewing content on human anatomy and physiology because it's important to you for personal or professional reasons.

>> **Casual reader:** High five to you for maintaining your love of learning. Your teachers would be so proud! Here you are with a book on your hands and a little time to spend reading it . . . and it's all about you! I sincerely hope that you enjoy this book.

Because this book is an all-in-one text, I had to limit the exposition of the topics so that I could include lots of practice questions to keep you guessing. (Believe me, I could go on forever about this A&P stuff!) The final assumption I made is that if you've picked up this book to help you perform well in a course, you have access to a formal textbook. Particularly with the labeling of anatomical structures, I cover the main players, but you'll likely be asked to identify more or to recognize more details about them.

Icons Used in This Book

Throughout this book, you'll find symbols in the margins that highlight critical ideas and information. Here's what they mean.

TIP

The Tip icon gives you juicy tidbits about how best to remember tricky terms or concepts. It also highlights helpful strategies for memory and understanding.

REMEMBER

The Remember icon highlights key material that you should pay extra attention to so you can keep everything straight. It's often included to cue you to make a connection with content on other body systems.

WARNING

This icon — otherwise known as the Warning icon — points out areas and topics where common pitfalls can lead you astray.

TECHNICAL STUFF

The Technical Stuff icon flags extra information that takes your understanding of anatomy or physiology to a slightly deeper level, but the text isn't essential for understanding the organ system under discussion.

PRACTICE

This icon draws your attention to the review questions included in particular places to check your understanding. Answers are included before the end of chapter quiz. You should complete all the questions as best you can, from memory (don't go back and look up the answers); then check and correct your answers after you complete the section. It's also important that you give yourself a brain break before continuing to the next section.

Beyond the Book

Within this book, you may note that some web addresses break across two lines of text. If you're reading this book in print and want to visit one of these web pages, simply key in the web address exactly as it's noted in the text, pretending that the line break doesn't exist. If you're reading this book as an e-book, you've got it easy: Just click the web address to be taken directly to the web page.

In addition to the material in the print or e-book you're reading right now, this product comes with some access-anywhere goodies on the web. Although it's important to study each anatomical system in detail, it's also helpful to know how to decipher unfamiliar anatomical terms the first time you see them. Check out the free Cheat Sheet by going to www.dummies.com and typing *"Anatomy and Physiology All in One Cheat Sheet"* in the search box.

You also get access to an online database of questions with even more practice for you. This database contains an interactive quiz for each chapter, allowing you to hone your new knowledge even more!

To gain access to the database, all you have to do is register. Follow these simple steps:

1. **Register your book at** www.dummies.com/go/getaccess **to get your personal identification number (PIN).**

2. **Choose your product from the drop-down menu on that page.**

3. **Follow the prompts to validate your product.**

4. **Check your email for a confirmation message that includes your PIN and instructions for logging in.**

If you don't receive this email within two hours, please check your spam folder before contacting Wiley at https://support.wiley.com or by phone at (877) 762-2974.

Now you're ready to go! You can come back to the practice material as often as you want. Simply log on with the username and password you created during your initial login; you don't need to enter the access code a second time. Your registration is good for one year from the day you activate your PIN.

Where to Go from Here

If you purchased this book and are already partway through an A&P class, check the table of contents, and zoom ahead to whichever segment your instructor is covering. When you have a few spare minutes, review the chapters that address topics your class has already covered— an excellent way to prep for a midterm or final exam.

If you haven't yet started a class, you have the freedom to start wherever you like (although I strongly recommend that you begin with Chapter 1) and then proceed onward and upward through the glorious machine that is the human body!

When you've finished, find a nice home on your bookshelf for this book. The unique, hand-drawn color plates alone make it worth the space it takes up. The book will be right there for you when you need a quick refresher, and it'll be a great reference after a doctor's appointment or the next time you wonder what the heck they're talking about in a drug commercial.

1

Locating Physiology on the Web of Knowledge

In This Unit . . .

Chapter **1**

Learning a New Language: How to Speak Fluent A&P

Human *anatomy* is the study of our bodies' structures; *physiology* is how they work. It makes sense, then, to learn the two in tandem. But before we can dive into the body systems and their intricate structures, you must learn to speak the language of the science.

A Word about Jargon

Why does science have so many funny words? Why can't scientists just say what they mean, in plain English? Good question! The answer is a great place to start this chapter.

Scientists need to be able to communicate with others in their field, just like in any other career. They develop vocabularies of technical terminology and other forms of jargon so they can better communicate with colleagues. It's important that both the scientist sending the information and the one receiving it use the same words to refer to the same phenomenon. They say what they mean (most of them, most of the time, to the best of their ability), but what they mean can't be said in the English language that people use to talk about routine daily matters. As a result, we get jargon. And while it seems like an unnecessary obstacle on your course of learning something new, jargon is a good and needed thing.

Consider this. We're sitting on a porch, enjoying a cool evening, and I say to you, "Hey! You got any tea?" You would likely offer whatever variety of hot tea you have (or kindly say that you don't have any). Nothing too confusing there, right? If I were to ask my son (or any teenager, really) the same question, I wouldn't be getting a refreshing beverage; I'd be getting the latest gossip. You know what I meant because the word *tea* means the same thing to us. But to my son, it's like jargon.

To understand anatomy and physiology, you must know and use the same terminology as those in the field. The jargon can be overwhelming at first, but understanding the reason for it and taking the time to learn it before diving into the complicated content will make your learning experience less painful.

A great way to start is to take a moment to look at just how these complicated terms are created. The words sound foreign to us because they're derived from Greek and Latin. Unlike other languages, those haven't changed in centuries. *Mono* has always meant *one*, but I don't even know what *cool* means anymore (except that my son says I'm not it!). In science, anatomy included, things are named by putting the relevant word roots together. Let's look at the word *dermatitis*. The root *derm-* means *skin*, and *-itis* means *inflammation*. Put them together, and you get skin inflammation, commonly referred to as a rash. Table 1-1 gets you going with a list of commonly used word roots in anatomy and physiology, matched up with the body system in which you'll likely encounter them.

TIP

Every time you come across an anatomical or physiological term that's new to you, see whether you recognize any parts of it. Using this knowledge, go as far as you can in guessing the meaning of the whole term. After studying Table 1-1 and the other vocabulary lists in this chapter, you should be able to make some pretty good guesses!

Table 1-1   Technical Anatomical Word Fragments

Body System	Root or Word Fragment	Meaning
Skeletal	os-, oste-, arth-	bone, joint
Muscular	myo-, sarco-	muscle, striated muscle
Integument	derm-	skin
Nervous	neur-	nerve
Endocrine	aden-, estr-	gland, steroid
Cardiovascular	card-, angi-, hema-, vaso-	heart (muscle), vessel, blood vessels
Respiratory	pulmon-, bronch-	lung, windpipe
Digestive	gastr-, enter-, dent-, hepat-	stomach, intestine, teeth, liver
Urinary	ren-, neph-, ur-	kidney, urinary
Lymphatic	lymph-, leuk-, -itis	lymph, white, inflammation
Reproductive	andr-, uter-	male, uterine

Looking at the Body from the Proper Perspective

Remember that story about your sister's friend's boyfriend's uncle who went in to have a foot amputated, only to awaken from surgery to find that the surgeon removed the wrong one? This story highlights the need for a consistent perspective to go with the jargon. Terms that indicate direction make no sense if you're looking at the body the wrong way. You likely know your right from your left, but ignoring perspective can get you all mixed up. This section shows you the anatomical position, planes, regions, and cavities, as well as the main membranes that line the body and divide it into major sections.

Getting in position

Stop reading for a minute and do the following: Stand up straight. Look forward. Let your arms hang down at your sides and turn your palms so they're facing forward. You are now in *anatomical position* (see Figure 1-1). Any time the position of one structure is compared with another, the relationship comes from the body in this position.

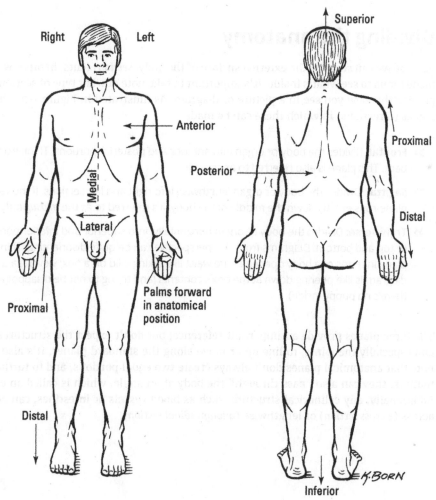

Right

Left

Anterior

Medial

Lateral

Proximal

Distal

Palms forward
in anatomical
position

Superior

Proximal

Posterior

Distal

Inferior

FIGURE 1-1:
The standard
anatomical
position.

Illustration by Kathryn Born, MA

Unless you're told otherwise, any reference to location (diagram or description) assumes this position. Using anatomical position as the standard removes confusion.

We also use directional terms to describe the location of structures. It helps to learn them as their opposing pairs to minimize confusion. The most commonly used terms are

>> **Anterior/posterior:** in front of/behind

>> **Ventral/dorsal:** toward the stomach/back (in humans, these mean the same as anterior/posterior)

>> **Superior/inferior:** above/below

>> **Medial/lateral:** closer to/farther from the midline (also used with rotation)

>> **Superficial/deep:** closer to/farther from the body surface

>> **Proximal/distal:** closer to/farther from the attachment point (used for appendages)

Right and *left* are also used quite often, but be careful! They refer to the patient's right and left, not yours.

Dividing the anatomy

Because we can see only the external surface of the body, *sections* (cuts, in other words) must be made for us to see what's inside. It's important to take note of what type of section was made to provide the view you see in a picture or diagram. As illustrated in Figure 1-2, There are three planes (directions) in which these can be made:

>> **Frontal:** Divides the body or organ into anterior and posterior portions. Think front and back. This plane is also referred to as *coronal*.

>> **Sagittal:** Divides the body or organ lengthwise into right and left sections. If the vertical plane runs exactly down the middle of the body, it's referred to as the *midsagittal plane*.

>> **Transverse:** Divides the body or organ horizontally, into superior and inferior portions, or top and bottom. Diagrams from this perspective can be quite disorienting. Think of this plane in terms of a body opened at the waist like taking a lid off a box; viewing a transverse diagram is like peering down at the box's contents (and trying to not be disappointed that there's no puppy inside).

The three planes provide an important reference; but don't expect the structures of the body, and especially the joints, to line up or move along the standard planes. It's also important to note that anatomical planes don't always create two equal portions, and to further complicate matters, they can also "pass through" the body at an angle, which is called an *oblique section*. Additionally, any cylindrical structure, such as blood vessels or intestines, can be cut straight across (a *cross section*) or lengthwise (a *longitudinal section*).

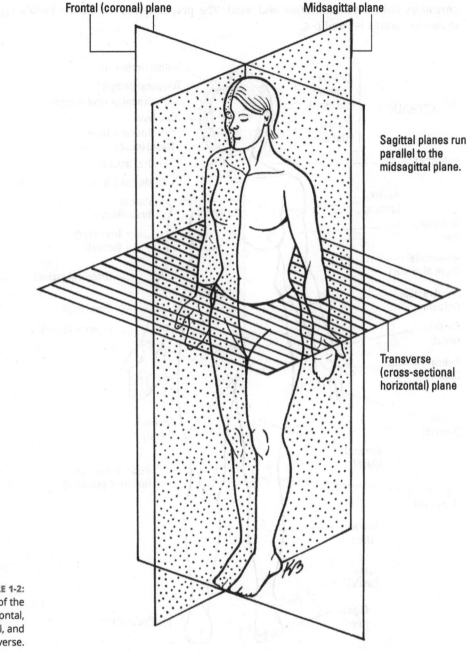

Frontal (coronal) plane

Midsagittal plane

Sagittal planes run parallel to the midsagittal plane.

Transverse (cross-sectional horizontal) plane

FIGURE 1-2: Planes of the body: frontal, sagittal, and transverse.

Illustration by Kathryn Born, MA

Mapping the regions

The anatomical planes orient you to the human body, but the *regions* compartmentalize it. Just like on a map, a region refers to a certain area. The body is divided into two major portions: axial and appendicular. The *axial body* runs right down the center (axis) and consists of everything except the limbs, meaning the head, neck, thorax (chest and back), abdomen, and pelvis. The *appendicular body* consists of appendages, otherwise known as *upper and lower extremities* (which you call arms and legs). The proper terms for all the body's regions are shown in Figures 1-3 and 1-4.

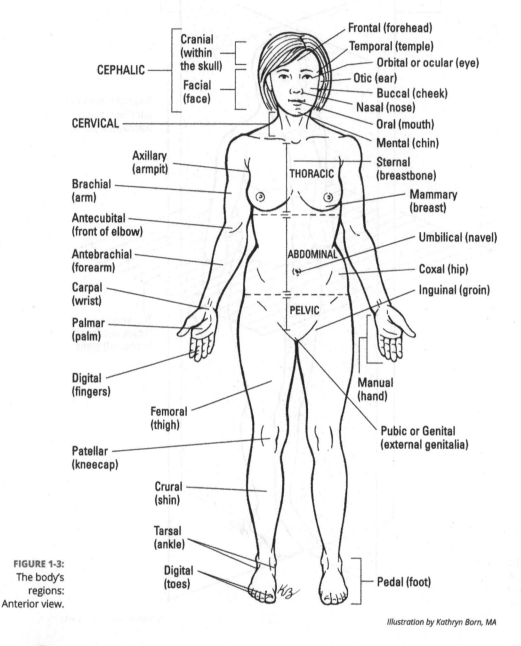

CEPHALIC — Cranial (within the skull)
Facial (face)

CERVICAL

Axillary (armpit)
Brachial (arm)
Antecubital (front of elbow)
Antebrachial (forearm)
Carpal (wrist)
Palmar (palm)
Digital (fingers)
Femoral (thigh)
Patellar (kneecap)
Crural (shin)
Tarsal (ankle)
Digital (toes)

Frontal (forehead)
Temporal (temple)
Orbital or ocular (eye)
Otic (ear)
Buccal (cheek)
Nasal (nose)
Oral (mouth)
Mental (chin)

THORACIC
Sternal (breastbone)
Mammary (breast)

ABDOMINAL
Umbilical (navel)
Coxal (hip)
Inguinal (groin)

PELVIC

Manual (hand)

Pubic or Genital (external genitalia)

Pedal (foot)

FIGURE 1-3: The body's regions: Anterior view.

Illustration by Kathryn Born, MA

WARNING It's helpful to practice these terms both visually (on a diagram) and descriptively, especially since you'll see them popping up throughout this book.

Additionally, the abdomen is divided into quadrants and regions. The midsagittal plane and a transverse plane intersect at an imaginary axis passing through the body at the *umbilicus* (navel or belly button). This axis divides the abdomen into *quadrants* (four sections). Putting an imaginary cross on the abdomen creates the right upper quadrant, left upper quadrant, right lower quadrant, and left lower quadrant. Physicians take note of these areas when a patient describes symptoms of abdominal pain.

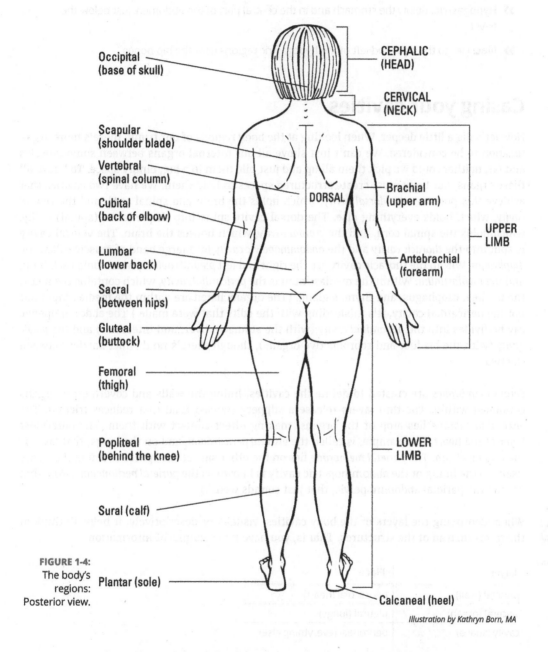

Occipital
(base of skull)

CEPHALIC
(HEAD)

CERVICAL
(NECK)

Scapular
(shoulder blade)

Vertebral
(spinal column)

Brachial
(upper arm)

DORSAL

Cubital
(back of elbow)

UPPER
LIMB

Lumbar
(lower back)

Antebrachial
(forearm)

Sacral
(between hips)

Gluteal
(buttock)

Femoral
(thigh)

Popliteal
(behind the knee)

LOWER
LIMB

Sural (calf)

FIGURE 1-4:
The body's
regions:
Posterior view.

Plantar (sole)

Calcaneal (heel)

Illustration by Kathryn Born, MA

The regions of the abdominopelvic cavity include the following:

>> **Epigastric:** The central part of the abdomen, just above the navel.

>> **Hypochondriac:** Doesn't moan about every little ache and illness, but lies to the right and left of the epigastric region and just below the cartilage of the rib cage. (*Chondral* means *cartilage,* and *hypo* means *below.*)

>> **Umbilical:** The area around the umbilicus.

>> **Lumbar:** Forms the region of the lower back to the right and left of the umbilical region.

>> **Hypogastric:** Below the stomach and in the central part of the abdomen, just below the navel.

>> **Iliac:** Lies to the right and left of the hypogastric regions near the hip bones.

Casing your cavities

Now let's dig a little deeper. When looking at the body regions of our trunk, there's more organization to be considered. We can't just shove all our internal organs between some muscles and fat; neither could we pick them all up and just pile them in a big, empty space. To house all these organs, our body must create structured spaces to hold them. We have two cavities that achieve this purpose: the *dorsal cavity*, which holds the brain and spinal cord, and the *ventral cavity*, which holds everything else. The dorsal cavity splits into the *vertebral* (spinal) *cavity*, which holds the spinal cord, and the *cranial cavity*, which houses the brain. The ventral cavity is split into the *thoracic cavity* and the *abdominopelvic cavity* by a large band of muscle called the *diaphragm*. Within the thoracic cavity are the right and left *pleural cavities*, which hold each lung, and the *mediastinum*. Within the mediastinum is the *pericardial cavity*, which contains the heart, the trachea, esophagus, and thymus gland. (The organs listed are within the mediastinum but not the pericardial cavity, it's misleading with the edits that were made.) The abdominopelvic cavity divides into the *abdominal cavity* (with the stomach, liver, and intestines) and the *pelvic cavity* (with the bladder and reproductive organs), though there's no distinct barrier between the two.

Serous membranes are created to define the cavities, lining the walls and covering the organs contained within. The thin layers release a slippery (*serous*) fluid that reduces friction. The *visceral membrane* lies atop of the organs, making direct contact with them. The outermost layer of the heart, for example, is called the *visceral pericardium*, and on the lungs, that layer is the *visceral pleura*. The *parietal membrane* lies on the other side of the spaces or lining the cavity itself. So the lining of the abdominopelvic cavity is known as the *parietal peritoneum*. (Note that it's not the parietal abdominopelvic, that just sounds weird.)

When identifying the layers of the body cavities, visually or descriptively, it helps to think of the terms instead of the structures. That is, you have two columns of information:

Layer	Place
parietal (wall)	pericardial (heart)
visceral (organ)	pleural (lungs)
cavity (space)	peritoneal (everything else)

Everything is identified by choosing one word from each column. That's much easier than remembering nine structures.

Getting Organized

When you think about learning anatomy and physiology, you likely think about humans as *organisms* — living beings that function as a whole. Then you might recall learning in grade school about the *levels of organization* of our bodies:

organism ← organ systems ← organs ← tissues ← cells ← molecule

"Wait," you might say. "I thought there were only five levels of organization, so why are there six here? And why are the arrows backward?"

Traditionally, scientists have ended the levels with cells — the smallest things that can be considered to be alive. It doesn't take a long swim in the pool of physiology, though, to see that our bodies' actions are carried out by macromolecules, such as proteins, or even just ions (Chapter 2 discusses what those things are) — hence, the six levels. Sometimes, you'll see atoms added as the seventh level.

About those backward arrows: You often see the levels of organization listed in the opposite order. A group of cells working together to achieve the same goal is a tissue, a group of tissues working together creates an organ, and so on. In this section, though, I start with you — your whole self. Then I'll break you down and build you back up as the book progresses.

Level 1: The organism level

This level is the real you; the good, the bad, and the ugly (I mean, beautiful). The entire context of anatomy and physiology is looking at how all the structures support you on the organism level, keeping you the best version of yourself that you can be!

Level 2: The organ system level

Human anatomists and physiologists have divided the human body into *organ systems* — groups of organs that work together to meet a major physiological need. The digestive system, for example, is the organ system responsible for obtaining energy from the environment. Realize, though, that this isn't a classification system for your organs. That is, each organ isn't placed in a single system; the systems are defined by the function they carry out. Organs can "belong" to more than one system. The pancreas, for example, produces enzymes that are vital to the breakdown of our food (digestive system), as well as hormones to maintain the balance of the body's many chemicals (endocrine system).

The chapter structure of this book is based on the definition of organ systems.

Level 3: The organ level

An *organ* is a group of tissues (at least two types) assembled to perform a specialized physiological function. The stomach, for example, is an organ that has the function of breaking down food, and the function of the biceps muscle is to move your forearm. This level is where the labeling fun begins. If only it were also where it ends!

Level 4: The tissue level

A *tissue* is a structure made of many cells — sometimes many different kinds of cells — that performs a specific function. Tissues are divided into four categories:

>> **Connective tissue** serves to support body parts and bind them together. Tissues as different as bone and fat are classified as connective tissue.

>> **Epithelial tissue (epithelium)** lines and covers organs, and also carries out absorption and secretion. The outer layer of the skin is made up of epithelial tissue.

>> **Muscle tissue** — surprise! — is found in the muscles (which allow your body parts to move), in the walls of hollow organs (such as intestines) to help move their contents along, and in the heart to move blood throughout the body. (Find out more about muscles in Chapter 7.)

>> **Nervous tissue** carries out communication within out body. Nerves and the brain are made of nervous tissue. (I talk about the nervous system in Chapter 8.)

Level 5: The cellular level

If you examine a sample of any human tissue under a microscope, you see cells, possibly millions of them. All living things are made of cells. The work of the body actually occurs within the cells. Your whole heart beats to push blood around your body because the muscle cells that create its walls shorten, causing the heart to contract.

Level 6: The molecular level

Though you can't see *molecules* without an incredibly powerful microscope, the creation and interaction of them "do our physiology." Those heart muscle cells that shorten to cause contraction do so because the molecules (proteins) inside the cell pull on each other, causing them to overlap. Even the familiar organelles of a cell are just organized collections of molecules.

Following that logic, that's all any of us are – just a bunch of molecules hanging out. That concept kind of blows your mind — or, well, your molecules — doesn't it?

That was a lot of new terms for just the first chapter! Take a moment now to see how they're sticking.

Whaddya Know? Chapter 1 Quiz

Quiz time! Complete each problem to test your knowledge on the various topics covered in this chapter. You can find the solutions and explanations in the next section.

1. Fill in Table 1-2 with the common and proper terms of your body's regions.

Table 1-2   The Body's Regions

Proper Term	Common Term	Proper Term	Common Term
	ankle		shin
Coxal			shoulder
Popliteal		Frontal	
	upper arm		forearm
	eye	Mental	
Plantar		Carpal	
	calf	Axillary	
	head		foot
	hand	Otic	
Femoral		Sternal	
Lumbar		Antecubital	
	elbow	Bucchal	
	neck		fingers/toes

For questions 2–11, label the body cavities illustrated in Figure 1-5.

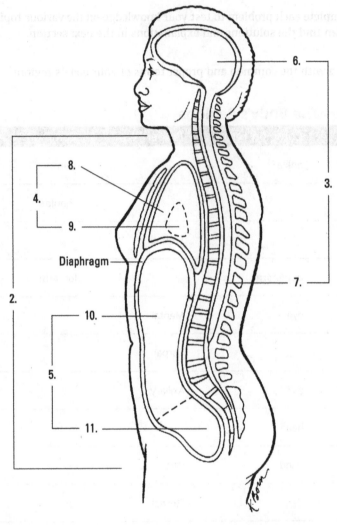

FIGURE 1-5:
Body cavities.

Illustration by Kathryn Born, MA

2 _____   7 _____

3 _____   8 _____

4 _____   9 _____

5 _____   10 _____

6 _____   11 _____

For questions 12–15, identify the statement as true or false. When the statement is false, identify the error.

12 The cephalic region is considered to be part of the appendicular body.

_____

13 The heart is located within the mediastinum, which is an area within the thoracic cavity.

_____

14 An organ, by definition, has to be made up of at least two different tissues.

_____

15 The terms *right* and *left* are dictated by the observer's perspective when the body is in anatomical position.

_____

16 Which of the following correctly matches the section term with the description? Choose all that apply:

(a) Frontal: separates anterior from posterior

(b) Sagittal: separates left from right

(c) Longitudinal: separates lengthwise

(d) Transverse: separates superior from inferior

(e) Oblique: separates at an angle

17 Why don't we stop at cells when breaking down our bodies' levels of organization?

_____

_____

_____

For questions 18–21, identify the abdominal regions by writing the corresponding letter from Figure 1-6.

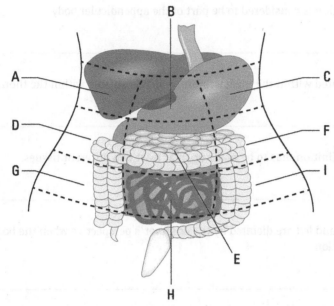

FIGURE 1-6: Abdominal regions.

18 _____ Left lumbar

19 _____ Hypogastric

20 _____ Iliac

21 _____ Hypochondriac

To complete questions 22–26, fill in the correct directional terms.

22 The nose is _____ to the eyes.

23 The rib cage is _____ to the lungs.

24 The fingers are _____ wrist.

25 The back is _____ to the chest.

26 The shoulders are _____ to the knees.

For questions 27–32, match the descriptions to identify the membranes that create the body's cavities:

(a) Parietal pericardium

(b) Parietal peritoneum

(c) Parietal pleura

(d) Visceral pericardium

(e) Visceral peritoneum

(f) Visceral pleura

27 _____ The outermost layer encasing the heart

28 _____ The membrane that lies on the surface of the liver

29 _____ The surface of the heart

30 _____ The lining of the thoracic cavity

31 _____ The membrane that makes direct contact with the lungs

32 _____ The layer that lines the abdominopelvic cavity

Answers to Chapter 1 Quiz

(1) Here is the completed table:

Proper Term	Common Term	Proper Term	Common Term
Tarsal	ankle	**Crural**	shin
Coxal	**hip**	**Scapular**	shoulder
Popliteal	**back of knee**	Frontal	**forehead**
Brachial	upper arm	**Antebrachial**	forearm
Orbital	eye	Mental	**chin**
Plantar	**sole/bottom of foot**	Carpal	**wrist**
Sural	calf	Axillary	**armpit**
Cephalic	head	**Pedal**	foot
Manual	hand	Otic	**ear**
Femoral	**thigh**	Sternal	**breastbone/sternum**
Lumbar	**lower back**	Antecubital	**inner elbow**
Cubital	elbow	Bucchal	**cheek**
Cervical	neck	**Digital**	fingers/toes

(2) **Ventral.**

(3) **Dorsal.**

(4) **Thoracic.**

(5) **Abdominopelvic or peritoneal.**

(6) **Cranial.**

(7) **Spinal or vertebral.**

(8) **Pleural.**

(9) **Pericardial.** It's within the thoracic cavity but not within the pleural cavity.

(10) **Abdominal.**

(11) **Pelvic.** There is not a structure that clearly marks the border between the abdominal and pelvic cavities.

(12) **False.** It's part of the axial skeleton. Appendicular is appendages, or arms and legs.

(13) **True.** The mediastinum contains everything within the thoracic cavity that isn't inside the pleural cavity.

(14) **True.**

(15) **False.** Although anatomical position is important here, it's the patient's right and left, not yours.

(16) **All the statements are accurate.** You didn't question yourself for that reason, did you?

TIP

Don't question your knowledge or your gut when you notice a pattern like this one, or like answering (a) five times in a row. Odds are that the test writer either didn't pay attention to patterns or created one on purpose to mess with you, as I just did.

(17) **It's true that cells are the smallest structures considered to be alive. Some organisms are themselves single-cell organisms (like bacteria). Obviously, we aren't single-celled, but that's why you've previously learned that cells are the smallest level of organization. In the context of physiology, however, the cells aren't what's changing to carry out functions; it's the molecules contained inside of them. That's the reason for the sixth level, which is molecules; they carry out the work.**

(18) **F.**

(19) **H.**

(20) **G or I.** G is the right iliac, and I is the left.

(21) **A or C.** A is the right hypochondriac, and C is the left.

(22) **Medial.**

(23) **Superficial.** Not anterior! The ribs aren't in front of the lungs; they surround them.

(24) **Distal.**

(25) **Posterior or dorsal.**

(26) **Superior.** Though the shoulders and the knees are both on appendages, they're not on the SAME one, so you wouldn't use proximal/distal.

(27) **(a) Parietal pericardium.**

(28) **(e) Visceral peritoneum.**

(29) **(d) Visceral pericardium.**

(30) **(c) Parietal pleura.**

(31) **(f) Visceral pleura.**

(32) **(b) Parietal peritoneum.**

Don't memorize all nine terms (cavities included), memorize the naming system. The space is always the cavity, and the visceral layer always makes direct contact with an organ. The pattern holds true everywhere except the brain and spinal cord; they're special.

TIP

IN THIS CHAPTER

» **Getting to the heart of all matter: Atoms**

» **Learning about building with your blocks**

» **Making sense of metabolism**

» **Exploring our body's energy**

Chapter **2**

Better Living through Chemistry

I can hear your cries of alarm. You thought you were getting ready to learn about the knee bone connecting to the thigh bone. How in the heck does that involve (horrors!) *chemistry?* As much as you may not want to admit it, chemistry — particularly *organic chemistry,* the branch of the field that focuses on carbon-based molecules — is a crucial starting point for understanding how the human body works.

There is, however, no agreed-upon baseline of chemistry understanding you need to fully grasp the body's physiological processes. Different instructors will hold you to different standards. The expectation, though, is that students enrolled in an anatomy and physiology (A&P) course have had at least a high-school-level course in chemistry. In this chapter, I walk you through what I think is essential background knowledge for success in A&P courses, based on my own experience teaching them.

TIP

When it comes to chemistry, what matters most is knowing that the things happen, not understanding how or why they happen. Don't let the details bog you down. Should you find yourself in a course whose instructor expects a lot more from you in terms of chemistry knowledge, I highly recommend checking out *Chemistry For Dummies,* 2nd Edition, by John T. Moore (John Wiley & Sons, Inc.).

What's the Matter?

Remember the periodic table? That big chart of elements that you may have had to memorize? Fortunately, you don't need to do that here (phew!); you only need to know why it has so many letters. Each box represents an *element*, the fundamental unit making up all stuff on Earth. That stuff is what we call *matter*. Elements are like LEGOs: They come in many shapes, sizes, and colors, but are all made of the same plastic and attach together in the same way. Each variety of brick, a red 4x2 rectangle for example, is an element; we'll say that's carbon. There are also little, blue 2x1 blocks, which could be hydrogen. You could click the LEGOs together easily and make, say, a couch. Bring in more bricks of different elements, and you could build a room — a whole house even. You could randomly decide to remodel and be done quickly. You could bring in other contractor LEGO builders to help you do the work faster. The analogy is starting to get a bit out of control here, but the point is this: Physiology is just like playing with LEGOs. Each brick is an *atom*, and the atoms are clicked together (making a *bond*) to create *molecules* — the innumerable structures you can build.

So that's you . . . a bunch of LEGO furniture assembled into houses (cells), which are organized into neighborhoods (tissues), which create LEGO cities (organs), which link to form states (body systems) and finally a big LEGO you! (Are you imagining what you'd look like if you were made of LEGOs?!)

REMEMBER

You often see the word *molecule* used interchangeably with the term *compound*. True, both things are made of connected atoms. But compounds by definition include atoms of different elements (different shapes and colors of LEGOs). Molecules are any elements linked together, even if it's just two Hydrogens (the simplest molecule possible).

WARNING

So all compounds are considered to be molecules, but not all molecules are compounds. Be careful with the difference, because the words don't mean exactly the same thing.

Now, of course, building molecules is far more complicated than that. Some LEGOs just won't attach to others because of their unseen structures in our analogy. (You've gotta use your imagination here.) They're all made of the same three components: *protons* (positively charged particles), *neutrons* (neutral), and *electrons* (negative). The numbers of these particles make the elements (LEGO blocks) different and influence how they behave (which other LEGOs they'll attach to).

TIP

Here's what matters most about this matter:

>> **The number of protons defines the element.** Protons will never be lost or gained, as the change would make the atom a different element. (This change is possible, but not in the human body; it's what happens in nuclear fission and fusion.)

>> **Electrons move around in a cloud that surrounds the atom's nucleus.** The *nucleus* is the center mass where the protons and neutrons (which don't move) reside. For this reason, electrons can interact with other atoms and their electrons. So when you attach two atoms, you create a bond. Often, the atoms share the electrons, creating what is called a *covalent bond*.

If you have that information down, you'll feel much more comfortable as this chapter progresses.

Keep Your Ion the Prize

All the molecules that make us up are going to interact, staying put to create our structures and moving around to perform the reactions that keep us alive. Large compounds to single atoms all have their roles to play.

Try on some ions

Some atoms never really grasped the old "sharing is caring" adage. They end up grabbing electrons wherever they're able to and keeping them to themselves; they're greedy. Because electrons are negatively charged, now the entire atom is negatively charged, or an *anion*. Other ions are martyrs, readily giving up electrons to the greedy atoms. These ions become positively charged, or *cations*.

Sodium (Na), for example, is an electron donor, becoming a cation: Na^+. Chlorine is an electron acceptor, becoming an anion: Cl^-. Just like with magnets, the opposite charges attract. This connection, called an *ionic bond*, is another way to build molecules. In this case, we get NaCl, which you know as the delicious little white crystals you like to sprinkle on your food (table salt).

TECHNICAL STUFF

The number of neutrons in an atom can and does vary as well. Since they're not charged particles, a change in number leaves the net charge as zero, or neutral. It does, however, change the mass of the atom, which will affect its behavior; these variants are called *isotopes*. Although they're important in nature as well as medical technology (especially imaging), isotopes don't play a major role in our bodies' physiology.

Being ionic isn't limited to atoms; whole molecules can carry charges as well. (Bicarbonate and phosphate molecules are examples commonly found in our bodies.) When you tally the total number of protons and electrons, they're unequal. Fewer electrons means that the molecule as a whole is positive; more makes it negative. There are numerous reasons why ions are so important for our bodies processes to function properly, and you'll find them popping up all over this book. But for now, we need them to talk about one specific thing: water.

Water, water everywhere

Water is not an ionic molecule. Let me repeat that for effect: Water *is not* an ionic molecule. It does, however, carry a charge.

I know what you're thinking: "You literally just told us that atoms and molecules that carry charges are ionic!" But here's the catch: "carrying a charge" does not mean the same thing as "being charged" (much like the difference between compounds and molecules). When you tally the number of protons from the two hydrogen (H) atoms and the single oxygen (O), you get 10. The same is true when you count the number of electrons. The charges balance each other out, so the molecule as a whole is neutral. Oxygen, however, is an electron hog. It shares, as any good kindergarten graduate does, but it's going to hold on to the electrons while letting the hydrogen play with them too. This results in an uneven distribution of the charge, though the overall charge remains neutral. Because the O is hogging the electrons, that side of the molecule carries a partial negative charge, while the H sides will be partially positive. (Partial charges are indicted as δ⁻ or δ⁺.) We describe molecules like this as being *polar*.

TIP When I see the word *polar*, the first thing I think of is bears. Then I think of the North and South Poles, which is definitely more helpful because they're located on opposite ends of the Earth. They're obviously not charged, but the opposing poles in magnets are. A magnet itself is not charged, but one end is negative and one is positive. If you line two magnets up with opposite charges, they click together. If you line them up with the same charges, they repel, and try as you might, you're not getting those ends of the two magnets to stick. This is how polarity works in molecules too.

Still not quite grabbing on to why this matters? Your body is about 55 percent water. (Well, mine is; the younger you are, the higher the percentage is, and vice versa.) Because like charges repel and opposites attract, this has a major impact on our physiology — especially since water definitely isn't the only polar molecule we'll find in large amounts in our bodies. Sugars and salts are polar; fats are not. We'll look at the implications in Chapter 3, but for now, we haven't finished our glass of water (and we really want that dessert!).

Speaking of a glass of water, remember that the water inside that glass is actually a bunch of water molecules, H_2O (more than 1.5 sextillion in a single drop!). Each one can move around, but the polarity of the individual molecules arranges them in a predictable pattern. And just as the proverbial girl next door chooses the rebellious jerk over the kind and caring sweetheart, opposites attract. As you can see in Figure 2-1, the molecules are held tightly together even though they're not actually bound. These are called *polar bonds* or, in the case of water, *hydrogen bonds*.

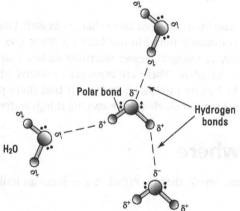

FIGURE 2-1:
The hydrogen
bonding
of water
molecules.

As a result, polar and ionic molecules interact with water (*dissolve*, in other words), while nonpolar ones, like oils and fats, do not.

Remaining neutral

Because our cells are filled with and surrounded by, essentially, water, we must also keep pH in check. With all the chemical reactions that happen, we're destined to form compounds that can affect our bodies' pH levels.

When we see the term pH, we tend to think of acidity. In actuality, though, pH is an entire scale; acids are only one side of it. What it actually tells us is the potential of Hydrogen, scaled from 0 to 14. Neutral is defined as 7, a low number indicates the level of acidity, whereas a high number indicates the level of alkalinity, or how basic a molecule is. What matters here isn't the details of how the numbers are determined or the structural differences between acids and bases. What matters is that the stronger the acid or base, the more damage is done to your molecules, which is especially bad for the ones that create structures, such as a cell membrane (the outermost layer of a cell). Your blood and body fluids have a very narrow range of tolerance, from 7.35 to 7.45. Unfortunately, many of our bodies' daily chemical reactions create acids and bases. Although the two can counteract, or neutralize, each other, it would be better to keep the pH from changing in the first place. Chemicals that can do this are called *buffers*; our most common one is bicarbonate.

PRACTICE

Let's see whether all that information got all the compounds in your brain reacting:

1. How do covalent and ionic bonds differ?

_____

_____

For questions 2–7, identify the statement as true or false. When the statement is false, identify the error.

2. All molecules are compounds.

_____

3. Hydrogen bonds connect molecules in way similar to ionic bonding.

_____

4. Cations are positively charged ions.

_____

5. Atoms of different elements bond through the neutrons surrounding the nucleus.

_____

6. Alkaline molecules, like many soaps, have pH levels less than 7.

_____

7. Buffers don't neutralize acids and bases, they prevent the pH from changing in the first place.

_____

_____

8 Ice floats in a glass of water because in a solid state, it is less dense. This is due to the water molecules' polarity. Explain how the water molecules become oriented to one another, creating a crystalline-like structure.

_____

_____

_____

Building Blocks That Build You

Though the processes of life may appear to be miraculous, biology always follows the laws of chemistry and physics. Biochemical processes are much more varied and more complex than other types of chemistry, and they happen among molecules that exist only in living cells. Molecules many thousands of times larger than water or carbon dioxide are constructed in cells and react to one another in seemingly miraculous ways. This section talks about these large molecules, called *macromolecules*, and their complex interactions.

REMEMBER

The four categories of these macromolecules, often called the *biomolecules of life,* include carbohydrates, lipids, proteins, and nucleic acids. All are made mainly of carbon (C), with vary-ing proportions of oxygen (O) and hydrogen (H). Many incorporate other elements, such as nitrogen (N), phosphorus (P), and sulfur (S).

Macromolecules, as the term suggests, are huge. So instead of LEGO tables and chairs, we're talking about entertainment centers and appliances. Each category of macromolecules is built from a specific set of LEGOs. We call these repeating subunits (molecules) *monomers.* Macromolecules, therefore, are *polymers* (from *polys,* the Greek word meaning *many* or *much*), and their characteristic monomers are arranged in specific ways.

Carbohydrates

Carbohydrate molecules consist of carbon, hydrogen, and oxygen in a ratio of roughly 1:2:1. The simplest of these are called *monosaccharides,* or simple sugars. Monosaccharides are the monomers of the larger carbohydrate molecules that create cell structures and store energy. Key monomers include glucose, fructose, and galactose. These three have the same numbers of carbon (6), hydrogen (12), and oxygen (6) atoms in each molecule — written as $C_6H_{12}O_6$ — but the bonding arrangements are different, giving them different properties. (See Figure 2-2.) You also see in the figure that these two sugars, glucose and fructose, can combine to make sucrose; thus we call that a *disaccharide.*

REMEMBER

Sucrose is what we generically call sugar. It tastes so much better than monosaccharides — likely so we'll eat more of it! We get more bang for our buck this way because we get two energy molecules instead of just one (after it's split apart in our bodies, of course).

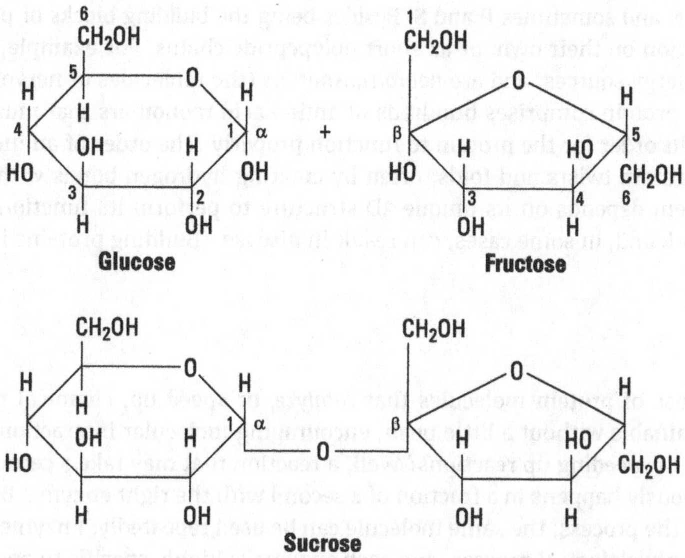

Glucose

Fructose

FIGURE 2-2:
Mono- and
disaccharides.

Sucrose

© John Wiley & Sons, Inc.

Monosaccharides can be easily linked into long chains, or *polysaccharides*. These are especially useful for providing structural support to our cells and tissues. *Glycogen* is a common polysaccharide found in the body; it breaks down into individual monomers of glucose, which cells use to generate usable energy.

Lipids

Lipids are polymers of glycerol and fatty acids, the most common and familiar of which are fats, or *triglycerides*. Fats are incredibly efficient energy sources (which is unfortunate for us but explains the body's propensity to store them!). Due to their general lack of polar bonds, they're insoluble in water. This makes them particularly useful for forming cells and the structures within them. Indeed, the entire outer membrane of a cell is formed from *phospholipid* molecules. (See Chapter 3.) *Steroids*, a special class of hormones, are also lipids, which gives them their unique function. (Hormones are discussed in Chapter 9.) *Cholesterol* and *vitamins A and D* are also common lipids found in the body.

Proteins

Proteins, which are polymers of *amino acids*, are among the largest molecules in our bodies. The amino acid monomers are arranged in a linear chain, called a *polypeptide*, and are folded and refolded into a globular 3D form. You may be familiar with *collagen* and *keratin* as structural components of our skin and hair; these are both proteins. Most hormones are proteins, and so are the *antibodies* we create to fight germs.

Amino acids

Twenty different amino acids are used by our cells to build proteins. Amino acids themselves are, by the standards of nonliving chemistry, huge and complex. In addition to C, H, and O, they contain atoms of N and sometimes P and S. Besides being the building blocks of proteins, many amnio acids function on their own, or as short polypeptide chains. For example, they work as buffers, can be energy sources, and are *neurotransmitters* (the molecules of nervous communication). A typical protein comprises hundreds of amino acid monomers that must be attached in exactly the right order for the protein to function properly. The order of amino acids determines how the protein twists and folds, often by creating hydrogen bonds with other amino acids. Every protein depends on its unique 3D structure to perform its function. A misfolded protein won't work and, in some cases, can result in disease. (Building proteins is discussed in Chapter 3.)

Enzymes

Enzymes are a class of protein molecules that *catalyze*, or speed up, chemical reactions. Life wouldn't be sustainable without a little push, encouraging molecular interactions. How effective are enzymes in speeding up reactions? Well, a reaction that may take a century or more to happen spontaneously happens in a fraction of a second with the right enzyme. Better yet, they aren't used up in the process; the same molecule can be used repeatedly. Enzymes are involved in virtually every physiological process, and each enzyme is highly specific to one or a very few individual reactions. Your body has tens of thousands of different enzymes.

Nucleic acids

The nucleic acids *deoxyribonucleic acid* (DNA) and *ribonucleic acid* (RNA) are polymers made of *nucleotides* that are arranged in a chain, one after another, and another . . . and another. DNA molecules are thousands of nucleotides long. (See Figure 2-3.) We have 46 individual strands of DNA in our cells containing our *genes*, the instructions for everything about us.

The monomer of DNA, a *nucleotide*, is made up of a sugar molecule, a phosphate group (one phosphorous atom attached to four oxygens), and a nitrogenous base. The sugar molecule is either deoxyribose (in DNA) or ribose (in RNA). The nitrogenous base is one of five:

>> Cytosine (C), guanine (G), adenine (A), and thymine (T) in DNA or uracil (U) in RNA

The bases connect in specific pairs to create the characteristic ladder. The *complementary pairs* are

>> C with G

>> A with T in DNA

>> A with U in RNA

The structural similarities between DNA and RNA allow them to work together to produce proteins within cells. That is, each one of our genes, which is a specific section of nucleotides, is really just instructions for building a protein. That's really all we are, just a big ol' pile of proteins with some sprinkles on top.

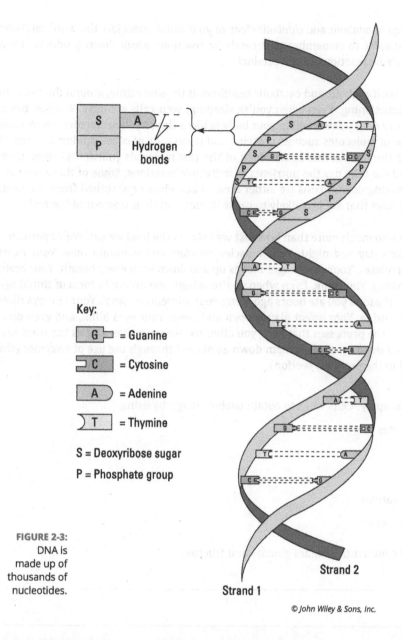

Key:

G▭ = Guanine

◖C = Cytosine

A▭ = Adenine

◗T = Thymine

S = Deoxyribose sugar

P = Phosphate group

FIGURE 2-3:
DNA is
made up of
thousands of
nucleotides.

Strand 1

Strand 2

© John Wiley & Sons, Inc.

Metabolism: It's More Than Just Food

Contrary to the most commonly cited definition of the word, the term *metabolism* refers to all the chemical reactions that happen in the body. That includes much, much more than breaking down and absorbing nutrients from the food we eat. We need to build a mind-numbingly large number of proteins; we have structural carbohydrates to make; we have fats to break down as well as byproducts from other chemical reactions. All these processes are part of our metabolism. These chemical reactions fall into two general categories: *anabolic reactions*, which make things, and *catabolic reactions*, which break things down.

TIP

To keep the meanings of *anabolic* and *catabolic* clear in your mind, associate the word catabolic with the word *catastrophic* to remember that catabolic reactions break down products. Then you'll know that anabolic reactions create products.

Your body performs both anabolic and catabolic reactions at the same time, around the clock, to keep you alive and functioning. Even when you're sleeping, your cells are busy. You just never get to rest (until you're dead). Every cell in your body is like a tiny, 24-hour factory, converting raw materials to useful molecules such as proteins and thousands of other products, many of which are discussed throughout this book. Much of the raw materials (nutrients) come from the food you eat, and the cells use the nutrients in metabolic reactions. Some of these cells are also tasked with breaking down toxins we either created ourselves or absorbed from our food, forming waste molecules that are less likely to cause damage on their way out of the body.

The *metabolic rate* is also much more than how fast we process the food we eat. We're performing chemical reactions day and night. Your muscles contract and maintain tone. Your heart beats. Your blood circulates. Your diaphragm moves up and down with every breath. Your brain keeps tabs on everything. You think. Even when you're asleep, you dream (a form of thinking). Your intestines push the food you ate hours ago along your alimentary canal. Your kidneys filter your blood and make urine. Your sweat glands open and close. Your eyes blink, and even during sleep, they move. The processes that keep you alive are always active. And as the word *rate* implies, we can speed them up or slow them down as needed through the use of enzymes (the proteins mentioned in the previous section).

PRACTICE

9 Choose all that apply. Cells work to obtain usable energy by using

(a) carbohydrates

(b) proteins

(c) lipids

(d) monosaccharides

(e) polysaccharides

10 Compare and contrast the sugars glucose and fructose.

_____

_____

_____

11 Give three examples of reactions that would be considered to be part of our metabolism.

_____

_____

_____

Charging Your Batteries

Perhaps the most important chemical reactions we perform are those that give our cells energy. You likely know that the main reason we need to eat food is to get energy, which isn't wrong. But a seasoned student of A&P knows that the word *energy* refers to something quite specific. We need the energy from food to make energy for cells. Simply put, we need *glucose* to create a source of energy that cells can actually use.

The molecule that cells are able to use for energy is a nucleic acid called *adenosine triphosphate* (ATP). As shown in Figure 2-4, ATP contains three phosphate groups aligned in a row. They are attached to an adenosine molecule, which is the nucleotide *adenine* linked to a *ribose* sugar. Breaking the bond between the second and third phosphate groups lets a cell access the energy. This leaves behind *adenosine diphosphate (ADP)* and a *phosphate* (P_i) by itself.

When a cell needs energy, it removes one of the phosphate groups to release it. When the bond to the last P breaks, the phosphate group pops off, giving the cell energy to perform a specific action, such as bringing a molecule in. What remains is now ADP, which is a nearly dead battery. Fortunately, we're able to recharge it easily so we can use it all again. This cycle is illustrated in Figure 2-5.

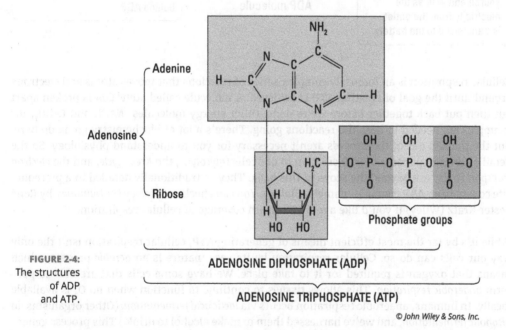

FIGURE 2-4:
The structures
of ADP
and ATP.

© John Wiley & Sons, Inc.

Most ATP is created via *cellular respiration*, an elaborate process that occurs in and around the mitochondria of a cell. This series of reactions cycles numerous molecules and ends with the production of 38 ATP molecules. This process, though, requires energy sources of its own, which we gain from the food we eat. Our go-to energy source for this is *glucose*, a simple sugar readily available in the food we eat.

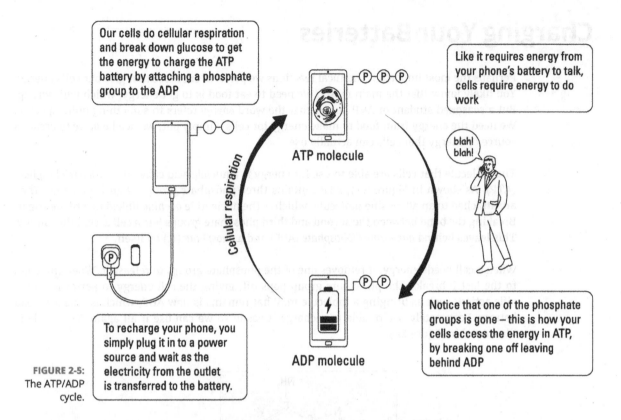

Our cells do cellular respiration and break down glucose to get the energy to charge the ATP battery by attaching a phosphate group to the ADP

Like it requires energy from your phone's battery to talk, cells require energy to do work

blah! blah!

ATP molecule

Cellular respiration

Notice that one of the phosphate groups is gone – this is how your cells access the energy in ATP, by breaking one off leaving behind ADP

To recharge your phone, you simply plug it in to a power source and wait as the electricity from the outlet is transferred to the battery.

ADP molecule

FIGURE 2-5:
The ATP/ADP cycle.

TECHNICAL STUFF

Cellular respiration is an intensely complex series of reactions that moves atoms and electrons around until the goal of creating ATP is achieved. A molecule called *acetyl CoA* is broken apart but then put back together before we're done. Other energy molecules (*NADH* and *FADH$_2$*, for example) are needed to keep the reactions going. There's a lot of biochemistry going on here. But the ins and out of the process aren't necessary for you to understand physiology. So the detail of the components of ATP creation in our cells (*glycolysis*, the *Krebs cycle*, and the *electron transport chain*) are beyond the scope of this book. They're traditionally detailed in a prerequisite course to an A&P course — namely, Biology. You can check out *Biology For Dummies*, by Rene Fester Kratz (Wiley) if you'd like a more in-depth coverage of cellular respiration.

While it's by far the most efficient means of generating ATP, cellular respiration isn't the only way our cells can do so. Cellular respiration by its very mature is an *aerobic* process, which means that oxygen is required for it to take place. We have some cells that are able to perform *anaerobic respiration*. This allows tissues to continue to function when no O_2 is available locally. In humans, anaerobic respiration occurs via *lactic acid fermentation*. (Other organisms do *alcoholic fermentation*, and we've harnessed them to make alcohol to drink.) This process generates two ATP molecules for each glucose molecule used. Besides being less efficient, this series of chemical reactions creates lactic acid, which results in a drop in pH within that area. Thus, we can't use this process for long; if we did, the lactic acid buildup would start interfering with all the other metabolic reactions occurring in that tissue.

It sure would be nice if cells could just use the darn glucose, wouldn't it? But think of this: suppose that your cellphone used the same type of batteries as your TV remote control. The number of dead batteries you'd generate would be obscene. Now multiply that number by all the people in your city who have cellphones. See where this analogy is going? Mountains of batteries would be ruining the landscape and polluting the environment because they contain toxic chemicals. (Which reminds me — stop throwing your dead batteries in the trash! Take them to your city's recycling or hazardous-waste center.) Though it's necessary in certain scenarios, we can't use glucose for the same reason. (For more on the complications of having too much or too little glucose, see Chapter 9.) ATP molecules are rechargeable batteries. Aerobic and anaerobic respiration aren't creating ATP from scratch; they're recharging it.

Although anaerobic respiration is useful for energy creation in muscles when O_2 has been used up, the process hardly seems to be worthwhile; anaerobic respiration provides only 2 ATP molecules per glucose compared with the 38 of aerobic respiration. Tack on the creation of lactic acid, which causes *muscle fatigue* (inability of fibers to contract) as a byproduct, and you may wonder why this process even exists! Well, we need to exercise, and some of the muscle fibers (but not all) will inevitably run out of oxygen. Luckily for us, the liver has a built-in mechanism to rid the body of lactic acid, though it may not happen as quickly as we'd like.

PRACTICE

Hopefully you're not too fatigued for a couple of review questions. Give these questions a shot, check your answers and process them, and then take a well-deserved break! Come back later to try your hand at the chapter quiz.

12   How do cells store and access energy to power their functions?

_____

_____

_____

13   Although we would love to say that the purpose of oxygen is "to breathe," that would be wrong. We breathe because we need the oxygen. What do we need it for?

_____

_____

_____

Practice Questions Answers and Explanations

1. **Covalent bonds occur when two atoms actually share electrons while ionic ones transfer their electrons and then stay together due to their charge attraction.** Remember that electrons are located in the cloud surrounding the atom's nucleus, which contains the protons and neutrons. An electron can exist in both clouds in an area where they overlap; thus, they stay attached. Other times, one atom gives an electron (or electrons) to another. This process has to do with atoms trending toward stability, which is caused by the total number of electrons and how they're distributed into the cloud (the *orbitals*, if that connects to any old memories from chemistry class). Now one atom (the one that gained an electron) is negative, and the other (the one that gave it away) is positive. The atoms stay connected because of the attraction between those opposite charges.

2. **False.** All compounds are molecules. Compounds have the requirement of having at least two atoms of different elements.

3. **True.** They are both charge attractions, but hydrogen bonds occur because in some molecules (polar ones), the hydrogens carry a partial positive charge.

4. **True.**

5. **False.** Atoms bond using electrons surrounding the nucleus. Students who are uncomfortable with chemistry commonly make this mistake. It seems to be logical that the two charged particles (protons and electrons) would pile in the middle, but that's not how it works. The positively charged protons are bundled with the neutrons that aren't charged. The electrons orbit around them.

6. **False.** Alkaline molecules have a pH greater than 7. Alkalis are more familiarly called *bases*, and soap is indeed basic. But acids have the low numbers on the pH scale.

7. **True.**

8. *Polarity* **means that the water molecule as a whole is neutral, but one side is a little positive, and the other is a little negative. The partial charges can align with the opposite partial charges in a different molecule. One oxygen can interact with four total hydrogens, the two it's bound to, the H of one molecule and the H of another molecule (as opposed to both Hs from the same water molecule). This interaction creates a crystalline structure (like lattice fencing) that holds the individual water molecules farther apart than in a liquid. (Usually, molecules are held together more closely in a solid state.) So ice is less dense and floats.**

9. **All of these.** I tried to trick you; did my trick work? First of all, monosaccharides and polysaccharides are categories of carbohydrates. Monosaccharides, like glucose, are the most familiar energy sources. Glucose can be linked into a huge molecule called *glycogen*, which is a polysaccharide. Lipids and proteins are also made of molecules that are used for parts of a cell's energy-making process.

10. **Glucose and fructose are both monosaccharides with six carbons; their chemical formula is $C_6H_{12}O_6$. The difference is in how the carbons are connected. (Glucose has five Cs in its ring; fructose has only four.)** The different structures cause different behaviors, but more notably, they react with different sensory cells, which is why glucose tastes bitter and gross, whereas fructose is super-sweet and delicious.

11) There isn't a concrete answer to this question. The purpose of my asking it was to prompt you to try and make the connection between cells (and their work) and the energy molecule ATP. Building proteins and other molecules in the cell requires energy. Stockpiling nutrients and ions takes energy (because storing them indicates that the molecules are being moved against their concentration gradient). Muscle contraction and nervous communication require energy. Almost anything you can think of a cell doing in some way requires energy.

12) **The ATP/ADP cycle.** ADP is often found floating around aimlessly inside a cell, waiting to be contacted (so to speak); phosphates are too. When cells bring in glucose, the glucose makes its way to the mitochondria, where cell respiration occurs. The glucose molecule is broken apart and is moved through a long chain of reactions that eventually release electrons, which are then used to reattach a phosphate to the ADP, creating ATP. The ATP also floats around in the cell until it's needed. To access the energy stored within, the phosphate group (P_i) is broken off. The cell can now power things like pumps to move things in and out of itself (and much, much more, as mentioned in question 11). ATP is broken down into ADP and P_i, and the process starts over.

13) **Oxygen molecules are required for our most efficient way to create ATP: aerobic cellular respiration. The byproducts are water, which we want, and CO_2, which we get rid of by exhaling. Cells need massive amounts of ATP, so they need massive amounts of O_2.** We can do anaerobic respiration to generate ATP, but the process is far less productive and it generates lactic acid, which is a waste we can't exhale. Lactic acid must be processed by the liver. That processing takes time (and, incidentally, oxygen), so it tends to build up and disrupt everything in the area — because, you know, it's an acid.

If you're ready to test your skills a bit more, take the following chapter quiz that incorporates all the chapter topics.

Whaddya Know? Chapter 2 Quiz

Quiz time! Complete each problem to test your knowledge of the various topics covered in this chapter. You can find the solutions and explanations in the next section.

1 Which of the following statements is *not* true of DNA?

(a) DNA is found in the nucleus of the cell.

(b) DNA can be replicated.

(c) DNA contains the nitrogenous bases adenine, thymine, guanine, cytosine, and uracil.

(d) DNA forms a double-helix molecule.

(e) DNA is made of monomers called nucleotides.

2 Fill in the blanks: By the end of aerobic cellular respiration, a single molecule of glucose can be converted into _____ ATP molecules whereas anaerobic respiration gets you _____.

3 The formation of chemical bonds is based on the tendency of an atom to

(a) move protons into vacant electron orbit spaces.

(b) stabilize its electron cloud.

(c) radiate excess neutrons.

(d) pick up free protons.

4 What are buffers, and what is the benefit of having them?

_____

_____

_____

5 Which of the following is NOT a bulk element found in most living matter?

(a) Carbon

(b) Oxygen

(c) Nitrogen

(d) Potassium

(e) Sulfur

6 What category of biomolecule does the energy molecule ATP fall into?

   (a) Carbohydrate

   (b) Lipid

   (c) Protein

   (d) Nucleic acid

   (e) None of these

7 What is the most likely pH for your stomach (gastric) acid?

   (a) 2

   (b) 6

   (c) 7

   (d) 8

   (e) 13

For questions 8–13, fill in the blanks to complete the sentences.

Different isotopes of the same element have the same number of 8 _____
and 9 _____ but different numbers of 10 _____.
Isotopes also have different masses. An atom that gains or loses an electron is called a(n)
11 _____. If an atom loses an electron, it carries a(n)
12 _____ charge. The number of 13 _____ in an
atom always stays the same because it defines the element.

14 Choose all that apply. Amino acids

   (a) help reduce carbohydrates.

   (b) are the building blocks of proteins.

   (c) modulate the production of lipids.

   (d) catalyze chemical reactions.

   (e) control nucleic acids.

15 Covalent bonds are a result of

   (a) neutralizing the nucleus charge.

   (b) sharing one or more electrons between atoms.

   (c) removing neutrons from both atoms.

   (d) reactions between protons.

   (e) transferring electrons to another atom.

16 Which statement best describes the result of prolonged use of anaerobic respiration by muscle cells?

(a) The CO_2 that results is removed by heavy breathing.

(b) Lactic acid buildup causes muscle cramping.

(c) The byproducts are removed by the kidneys.

(d) Fibers temporarily lose their ability to contract.

(e) Muscle cells remain clinched until the wastes are cleared.

17 Polysaccharides

(a) can be reduced to fatty acids.

(b) contain nitrogen and phosphorus.

(c) are complex carbohydrates.

(d) are monomers of glucose.

(e) contain adenine and uracil.

For questions 18–23, fill in the blanks to complete the sentences.

Molecules like water are 18 _____ because they share their electrons unequally. This is due to oxygen having a higher 19 _____ than hydrogen. As a result, the oxygen side of the molecule has a 20 _____ charge, while the hydrogen sides have a 21 _____ charge. When exposed to other polar molecules, 22 _____ charges attract, and they form 23 _____ bonds.

24 What category of biomolecule are enzymes? What do they do? Why are they unique?

_____

_____

_____

Answers to Chapter 2 Quiz

1. **(c) DNA contains the nitrogenous bases adenine, thymine, guanine, cytosine, and uracil.** It does contain A, T, G, and C, but uracil is found only in RNA.

2. **38, 2.**

3. **(b) stabilize its electron cloud.** It's, of course, more complicated than that because the "cloud" is actually different levels called orbitals, and each one has a different maximum number of electrons that get filled in a specific order. Stable would be when their outermost orbital is full or contains the maximum number of electrons. We don't need to worry about the details of this, but it explains why bonding *just happens*. It's also why you see trends in compounds. Oxygen usually attaches to two other atoms, for example, but carbon will attach to four.

4. **Buffers are molecules (usually weak acids or bases) that resist pH changes in a solution.** The solution in this case is our body fluids. The benefit is that it removes the need to constantly balance the pH by releasing bases when pH drops or acids when it raises.

5. **(d) potassium.** I tried to trick you! The P in the list is phosphorus; the symbol for potassium is K.

6. **(d) Nucleic acid.** Adenosine should connect in your head to adenine, which is one of the four bases in DNA. In fact, adenosine in made from adenine.

7. **(a) 2.** It's not that you should know that the pH is 2 (yet); rather that's the only choice that's an actual acid. A pH of 6 is the acidic side of neutral, which is 7. Higher numbers are alkaline/basic.

8. **protons or electrons.**

9. **electrons or protons.**

10. **neutrons.**

11. **ion.**

12. **positive.**

13. **protons.**

14. **(b) the building blocks of proteins.** This is the only reasonable function of amino acids; they don't catalyze chemical reactions because enzymes are full, 3D proteins.

15. **(b) sharing one or more electrons between atoms.**

16. **(d) Fibers temporarily lose their ability to contract.** When people say you get cramps from lactic acid buildup, that's an oversimplification. Lactic acid prevents a muscle fiber from contracting. A cramp is the exact opposite, an involuntary contraction. Muscle fatigue can also be very painful.

17. **(c) are complex carbohydrates.** Choice (d) didn't trip you up, did it? It's backward; glucose can be the monomer for a polysaccharide.

18. **polar.**

19. **electronegativity.**

(20) **partial negative.** Make sure to include *partial* here; it doesn't seem to be important, but it matters.

(21) **partial positive.** Make sure to include *partial* here; it doesn't seem to be important, but it matters.

(22) **opposite.**

(23) **polar.** Hydrogen bonds isn't entirely correct. Although that happens with water, the attraction between partial charges doesn't have to involve hydrogen. *Polar bond* is a more generic term.

(24) **Enzymes are proteins, and their ability to function depends on their 3D structure.** They function to speed up, or catalyze, metabolic reactions. What makes them unique is that they aren't used up in the process. They can grab on to a molecule and split it in two; let go of the pieces; and repeat the process indefinitely.

Chapter 3

The Secret Life of Cells

Chemical reactions are not random events. Any reaction takes place only when all the conditions are right for it. In other words, nothing happens unless the required molecules are close together in the right quantities; the fuel for the reaction is present in sufficient amount and in the right form; and all the environmental variables are within the right range, including temperature, salinity, and pH. *Homeostasis* is the term physiologists use to reference the balance of all the variables. And it is the job of our cells to keep tabs on all this, making adjustments when necessary, coordinating with other cells, and using numerous homeostatic mechanisms to keep everything in check; otherwise, all the reactions that comprise our metabolism can't occur. The cells too must be functioning in tip-top shape; otherwise, we'd have to replace them with newer, shinier models that we'd have to break in and train.

Keeping Same Things the Same: Homeostasis

The basic model for homeostasis is

Input ⇨ Integration ⇨ Output

Let's look at it through an example. A change in our biochemistry occurs: Your blood sugar raises after you eat a meal, for example. This is a deviation from the usual *set point* — the typical amount of glucose floating around in your blood. There are cells in your body whose job it is to monitor these levels. (In this case, the cells are found in the pancreas.) Not thrilled about this particular change, cells in the pancreas release insulin, which works to lower the blood sugar. (For details on blood glucose homeostasis, check out Chapter 9.)

The *input* was the change in blood sugar concentration. The *integration* step occurs when the pancreas notices the increase. (Our body's integrators are always either glands or the brain.) Perhaps the increase was relatively slight, close enough to the set point to wait it out. Then nothing happens; that's the integration part. We may not need to invest energy and enlist other cells and organs to solve the problem because it really isn't a problem. If the increase warrants action, however, a response occurs that leads back to normal. In this case, the release of insulin leads to blood sugar levels dropping back to the set point.

Because the body's response is in the opposite direction of the change, we call this process *negative feedback*. Our body also employs *positive feedback* mechanisms. Here, by definition, the response that occurs is in the same direction as the change. If our body's response to increased blood sugar was to release more into it, further increasing the amount of glucose in the blood, that would be positive feedback. Because we're trying to keep everything in balance, this sounds like a terrible idea. Indeed, there aren't many examples of this in the human body, as pushing a value away from the set point defeats the purpose of its existence. The classic examples are blood clotting and childbirth.

Oh, no! You're bleeding! (Play along for me.) That definitely deviates from your set point of not bleeding. This happens because you've broken a blood vessel (hopefully just a little one), but the usually smooth, continuous tube now has rough spots where it has split. Elements in your blood called *platelets* get stuck on the jagged little edges. This in turn leads to the production of more platelets, so even more get stuck, forming a clot. (This process is discussed in greater detail in Chapter 10.) Blood clotting leads to more blood clotting, which uses positive feedback to return us to the normal state of the blood staying in its vessels. Eventually, though, we have to turn off this positive feedback loop; otherwise, we'd clot our blood vessels closed. (Which is certainly not good!)

WARNING Be careful that you don't associate *positive* with *increase* and *negative* with *decrease*. That's not how the language works here. *Positive* means that the change and response are moving in the same direction; *negative* means that the response opposes the change. When our pH drops, becoming more acidic, our body works quickly to bring it back up. So while pH is increasing in value, this is a negative feedback mechanism.

Maintaining a constant temperature: Thermoregulation

Our metabolic reactions require that the temperature of the body be within a certain range; therefore, we must work to maintain a relatively constant internal body temperature. As our cells generate ATP to do work, they create heat as a byproduct. At the same time, we lose heat from our body's surfaces. The heat loss and creation must balance each other out to keep a cozy 98.6°F (37°C) within our body's tissues.

We control our body temperature by employing adaptations that conserve the heat generated by metabolism within the body in cold conditions or dissipate that heat out of the body in overly warm conditions. Keep in mind, though, that the stimulus comes from inside the body. You don't start shivering as soon as cold air hits you, for example; that reaction happens only when your body temperature starts to drop. Specific adaptations for *thermoregulation* include the following:

>> **Sweating:** Sweat glands open their pores to release sweat, which is mostly water, onto the skin. The body's heat causes the water to evaporate, cooling you down.

>> **Blood circulation:** Blood effectively holds a great deal of the body's heat. Blood vessels near the skin dilate (enlarge) to spread the heat to the surface, carrying it away from your core. That's why your skin flushes (reddens) when you're hot: red is the color of your blood visible at the surface of your skin. To conserve heat, blood vessels constrict (narrow). This reduces the amount of blood flow to your body surfaces, lessening the amount of heat loss that occurs there.

>> **Muscle contraction:** When closing pores and constricting blood vessels aren't enough to conserve heat in cold conditions, your muscles perform quick, repeated contractions — the familiar shivering. When ATP is used for energy, heat is a byproduct of the chemical reaction. Every little contraction for each individual muscle cell requires ATP, so all that shivering ends up generating a great deal of heat.

>> **Insulation:** Regions of fatty tissue under the skin provide insulation, holding the warmth of the body in. Body hair aids in this process too (though not enough to keep us from needing nice warm winter coats).

Swimming in H_2O: Fluid balance

A watery environment is a requirement for most of our metabolic reactions. (The rest need a lipid, or fatty, environment.) Thus, the body contains a lot of water — in your blood, in your cells, in the spaces between your cells, in your digestive organs, here, there, and everywhere. This isn't pure water, though. The water in your body is a *solvent* for thousands of different ions and molecules (*solutes*). The quantity and quality of the solutes change the character of the solution. Because solutes are constantly entering and leaving the solution as they participate in (or are generated by) chemical reactions, the characteristics of the watery solution must remain within certain bounds for the reactions to continue happening:

>> **Changes in the composition of urine:** The kidney is a complex organ that has the ability to measure the concentration of many solutes in the blood, including sodium, potassium, and calcium. Very importantly, the kidney can measure the volume of water in the body by sensing the pressure of the blood as it flows through. (The greater the volume of water, the higher the blood pressure.) If changes must be made to bring the volume and composition of the blood back into the ideal range, the various structures of the kidney incorporate more (or less) water, sodium, potassium, and so on into the urine. That's why your urine is paler or darker at different times. (I discuss functions of the urinary system in greater detail in Chapter 15.)

>> **The thirst reflex:** Water passes through your body constantly, mainly in through your mouth and then out through various organ systems, including the skin, the digestive system, and the urinary system. If the volume of water falls below the optimum level (dehydration), and the kidneys alone can't regain the balance, the mechanisms of homeostasis intrude on your conscious brain to make you uncomfortable. You feel thirsty. You ingest something watery. Your fluid balance is restored, and your thirst reflex leaves you alone.

Division of Labor

All measurements that we take as inputs in our model of homeostasis are made by cells. Collectively, we call such cells *receptors* because they're able to notice (or receive) the change and communicate it. A decision about action (or inaction) occurs in cells assembled into a *control center* — often the brain but sometimes a gland. Regardless, the control center is where the input is connected with the output. The final response depends on cells that we call *effectors*, which either release a molecule or perform a specific chemical reaction. What this means is that all of our chemical reactions, our metabolism, happens within our cells. We have to peer into them and take a look at their *organelles* (internal structures) to see how they work before we can move on to tissues and organs.

Mission Control: The nucleus

The cell nucleus is the largest organelle, on average accounting for about 10 percent of the total volume of the cell. It holds 46 chromosomes, two of each of the 23 different ones that house your *genome* (the varieties of all your genes).

TECHNICAL STUFF

All cells have one nucleus, at least at the beginning of their life cycle. As a cell develops, it may lose its nucleus, as do red blood cells. Other cells, like those of skeletal muscle, merge with other cells, retaining all the nuclei. In these cases, the cells are referred to as *multinucleated*.

The nucleus is a space created by the *nuclear envelope*, which is a selectively permeable membrane made from lipids. Its large pores allow for relatively free movement of molecules and ions, including large proteins, but the envelope still manages to keep the chromosomes contained inside.

Also within the nucleus is the *nucleolus*, a small spherical body that stores RNA molecules and produces *ribosomes*, which are needed to begin protein synthesis.

The literal function of DNA is to be a copy of the instructions for building specific proteins.

REMEMBER

Outside the nucleus is the *cytoplasm*, a mixture of macromolecules (such as proteins and RNA), small organic molecules (such as glucose), ions, and water. The fluid part of the cytoplasm, called the *cytosol*, differs in consistency based on changes in temperature, molecular concentrations, pH, pressure, and agitation. Within the cytoplasm lies a network of fibrous proteins collectively referred to as the *cytoskeleton*. Rather than being rigid or permanent, it changes and shifts according to the activity of the cell. The cytoskeleton maintains the cell's shape; enables it to move; and anchors its organelles, including the nucleus.

The workhorses: Organelles and their functions

Organelles (literally translated as *little organs*) are nestled inside the cytoplasm except for the two organelles that move: *cilia* and *flagellum*, which are on the cell's exterior. Each organelle has different responsibilities for producing materials used elsewhere in the cell or body. Here are the key organelles and what they do:

>> **Centrioles:** These bundles of microtubules separate genetic material during cell division.

>> **Cilia:** These short, hairlike cytoplasmic projections on the external surface of the cell move materials across the surface of the cell, often sweeping substances through a tube. Although cilia can provide momentum for a cell to move, like the movement of single-celled organisms, that is not the case in our bodies.

>> **Endoplasmic reticulum (ER):** The ER is the site of lipid and protein synthesis. This organelle, which makes direct contact with the cell nucleus, is composed of membrane-bound canals and cavities that aid in the transport of materials throughout the cell. The two types of ER are *rough* (dotted with ribosomes on the outer surface, in other words) and *smooth* (with no ribosomes on the surface). The rough ER manufactures membranes and secretory proteins in collaboration with the ribosomes. The smooth ER has a wide range of functions, including synthesizing carbohydrates and lipids, and breaking down drugs.

>> **Flagellum:** This whiplike cytoplasmic projection lies on the cell's exterior surface. In humans, it's primarily found on sperm cells, which use it for locomotion.

>> **Golgi apparatus (or body):** This organelle, which consists of a stack of flattened sacs, is found near the rough ER. It receives assembled proteins and modifies them into the final, functional form. It's also responsible for the storage, modification, and packaging of proteins for secretion to various destinations within the cell or for release.

>> **Lysosome:** A tiny, membranous sac containing acids and digestive enzymes, the lysosome breaks down large molecules such as proteins, carbohydrates, and nucleic acids into materials that the cell can use. The enzymes can be used to open vesicles (break down their lipid bilayer) to access the materials stored within them. It also destroys foreign particles, including bacteria and viruses, and helps remove nonfunctioning structures from the cell.

>> **Mitochondria:** Affectionately called the "powerhouse of the cell," this (usually rod-shaped) organelle is the site of cellular respiration. It uses glucose to generate ATP to power the cell processes that require energy.

>> **Ribosomes:** These tiny structures are found either attached to an endoplasmic reticulum or just floating in the cytoplasm. Composed of 60 percent RNA and 40 percent protein, they translate the genetic information in messenger RNA molecules to begin the process of building proteins.

>> **Vesicles:** These small spheres of membrane carry products around the cell. *Transport vesicles* carry newly made proteins from the ER to the Golgi apparatus, and *secretory vesicles* carry the Golgi products to the cell membrane.

Let's see how well you've balanced and organized this chapter's information so far.

PRACTICE

1 Compare and contrast the following terms: cytosol, cytoskeleton, and cytoplasm.

_____

_____

_____

② What is the role of the nuclear envelope? Choose all that apply.

(a) Copy chromosomes to prepare for division.

(b) Assemble and store ribosomes.

(c) Allow proteins and nucleic acids to go in and out of the nucleus.

(d) Create a protein construction area within the cell.

(e) House genetic material.

For questions 3–15, use the terms that follow to identify the cell structures and organelles shown in Figure 3-1.

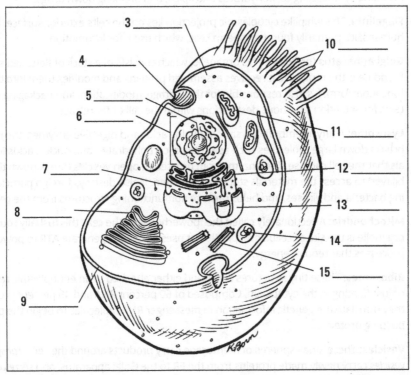

FIGURE 3-1: A cutaway view of an animal cell and its organelles.

Illustration by Kathryn Born, MA

(a) Centriole

(b) Cilia

(c) Cytoplasm

(d) Golgi apparatus

(e) Lysosome

(f) Mitochondrion

(g) Nucleolus

(h) Nucleus

(i) Cell (plasma) membrane

(j) Ribosomes

(k) Rough endoplasmic reticulum

(l) Smooth endoplasmic reticulum

(m) Vesicle (formation)

For questions 16–21, identify the organelle described.

16 _____ Long, whiplike organelle for locomotion

17 _____ Organelle where protein modification occurs

18 _____ Membranous sacs containing digestive enzymes

19 _____ Site of cellular respiration

20 _____ Site of ribosome production

21 _____ Structure used to transport molecules in, out, or around the cell

Controlling the Ins and Outs

The *cell membrane* is much more than just a line that creates the shape of a cell. Think of it as a gatekeeper, guardian, or border guard keeping the cell's cytoplasm in place and letting only select materials enter and depart the cell. This *semipermeability*, or *selective permeability*, is a result of a double layer *(bilayer)* of *phospholipid* molecules mingling with a variety of protein molecules. (See Figure 3-2.) The outer surface of each layer is made up of tightly packed polar (unevenly charged), *hydrophilic heads.* Inside, between the two layers, you find nonpolar, *hydrophobic tails* consisting of fatty-acid chains.

TIP

It certainly helps to remember that *phobia* means *fear of*; thus, *hydrophobic* becomes *fear of water.* Conversely, *-phile* means *lover of*; for instance, a bibliophile is a lover of books. Realize, though, that fearing versus liking water refers to the molecules' capability to react with water. Because our cells are surrounded by and filled with water, it makes sense that we'd want the lipid heads to interact with it willingly. However, if the entire phospholipid molecule were hydrophilic, the membrane would dissolve, and we'd have no cell.

TECHNICAL STUFF

The membrane's interior is made up of nonpolar fatty-acid molecules (the phospholipid tails) creating a dry middle layer. Lipid-soluble molecules can pass through this layer, but water-soluble molecules such as amino acids, sugars, and many proteins cannot. These molecules must enter the cell through special channels or through energy-requiring processes.

Since the phospholipids give the membrane a rather fluid structure, cholesterol molecules are embedded to provide support. The numerous embedded proteins aid in membrane stability as well. Some proteins span the entire membrane, creating channels and pumps (the latter requiring ATP to function). Other proteins stick up off the surface for recognition, often with a carbohydrate chain attached, creating a *glycoprotein.* These recognizable chains (refer to Figure 3-2) can attach directly to the lipids as well, forming *glycolipids.* Proteins also create the receptors that bind anything from small molecules to entire cells (though each receptor is incredibly specific, binding only what it is designed to bind).

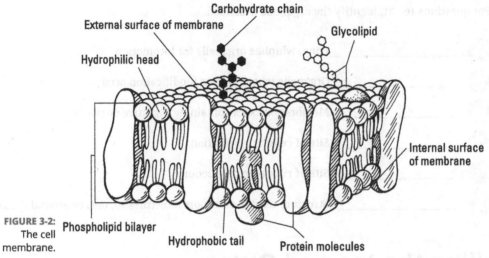

Carbohydrate chain
External surface of membrane
Glycolipid
Hydrophilic head
Internal surface of membrane

FIGURE 3-2: The cell membrane.

Phospholipid bilayer
Hydrophobic tail
Protein molecules

Illustration by Kathryn Born, MA

REMEMBER

The cell membrane is designed to hold the cell together and to isolate it as a distinct functional unit. Although the membrane can spontaneously repair minor tears, severe damage to the membrane causes the cell to disintegrate. The membrane is picky about which molecules it lets in or out. It allows movement across its barrier by *diffusion, osmosis,* or *active transport.* (For more on these concepts, see the next few sections.)

Diffusion

Atoms, ions, and molecules are in constant motion, taking a straight-line path until they inevitably crash into another particle and bounce off in a new trajectory. As a result, particles naturally flow to areas of lower concentration. (More collisions means more movement.) This process is *diffusion* — the movement of molecules from high to low concentration. The kinetic energy of the particles is the driving force moving them down their *concentration gradient.* Even after *equilibrium* has been reached (when molecules are equally distributed throughout), diffusion continues. There just isn't a net change; when one molecule randomly moves out, another one will move in to keep the balance.

So what does this have to do with cells? As long as the molecules can move through any barricade that may be in their way (such as the cell membrane), the body doesn't need to invest any energy in their movement. Diffusion is an easy way to move things into and out of a cell, passing through the cell membrane. Because the process requires no energy, it is a form of *passive transport.* Small, nonpolar molecules such as oxygen (O_2) and carbon dioxide (CO_2) simply squeeze between phospholipid heads, entering or exiting the cell, depending on where the concentration of the molecule is lower. Because the inside of the membrane is hydrophobic, polar molecules and ions are unable to simply diffuse (in or out). They can, however, still move passively down their concentration gradient; they just require a protein channel to move through. This process is termed *facilitated diffusion,* as these molecules, such as glucose, need a protein to *facilitate* (allow for) their movement. Unlike the lipid-soluble molecules that use simple diffusion, we can control these water-soluble ones by opening and closing the protein channels.

TECHNICAL STUFF

Our cells can open and close protein channels, or *gates*, in many ways without investing energy. The three main methods of this are

>> **Ligand-gated channels:** These channels open and close in response to the binding of a specific molecule. Ion channels are often ligand-gated.

>> **Voltage-gated channels:** These channels open and close in response to an electrical charge. Voltage-gated channels are responsible for the electrical impulse of nervous communication.

>> **Mechanically gated channels:** These channels open and close in response to a physical change, such as pressure. This is how touch receptors work.

Osmosis

Osmosis, simply put, is the diffusion of water. Because our bodies are 50–65 percent water, the movement of water is of the utmost importance to us. More specifically, osmosis is the passive diffusion of solvents through a selectively permeable membrane from an area of higher concentration to an area of lower concentration. Solutions are composed of two parts:

>> **Solvent:** The *solvent* is the liquid in which a substance is dissolved. In our bodies, the solvent is always water.

>> **Solute:** A *solute* is the substance dissolved in the solvent.

Typically, a cell contains a roughly 1 percent saline solution — in other words, 1 percent salts (solutes) and 99 percent water (solvent). Water is a polar molecule that can't pass through the lipid bilayer. Embedded in many cell membranes, however, are numerous tiny proteins called *aquaporins.* Water moves readily through these little holes via osmosis, but many solutes don't because they're too big. Because these pores exist and are always open, we can't control the flow of water directly.

Transport by osmosis is influenced by the solute concentration of the water — the solution's *osmolarity,* in other words. In our bodies, water easily moves to areas of higher osmolarity; water always goes where there is less water, in other words. As water diffuses into a cell, *hydrostatic pressure* builds within the cell. Eventually, the pressure within the cell becomes equal to, and is balanced by, the *osmotic pressure* outside. So if we want to move water, we have to change the amount of solutes first.

Changing the solute concentration outside of the cell, changes the whole solution relative to the solution inside of the cell. Thus we have three scenarios:

>> An *isotonic* solution has the same concentration of solute and solvent as what's inside a cell, so a cell placed in isotonic solution experiences equal flow of water into and out of the cell, maintaining equilibrium.

>> A *hypotonic* solution has a lower solute concentration than that of the cell, which means that the area with the highest osmolarity (less water) is inside the cell. If a human cell is placed in a hypotonic solution, such as distilled water, water molecules move down the concentration gradient into the cell. This process causes the cell to swell and can continue until the cell membrane bursts.

>> A *hypertonic* solution has a higher osmolarity outside the cell. So the membrane of a human cell placed in 10 percent saline solution (10 percent salt and 90 percent water) would let water flow out of the cell (to where osmotic pressure is lower because water concentration is lower), causing it to shrivel.

TIP

The suffix *-tonic* refers to the solutes in the solution. Pair it with the prefix (either *hyper-* for *high* or *hypo-* for *low)*, and you'll always be able to figure out what's going to happen. Concentrations are just percentages, after all, so *high solute* has to mean *low water*. And where does water always go? To where there's less of it. Also, if you remember "*Hypo-* gets big like a hippo," you'll know whether you have your terms backward.

Active transport

Often, our bodies want to maintain an imbalance of concentration inside versus outside a cell. This is particularly true for ions, whose imbalance is the foundation of nervous physiology. Our ability to move ions allows us to influence osmosis. Because the ions in this case are being forced against their concentration gradient, this movement requires energy. *Active transport* occurs across a semipermeable membrane, pumping molecules from the area of lower concentration to the area of higher concentration. Active transport also lets cells obtain nutrients that can't pass through the membrane by other means. These *carrier* or *transport* proteins use ATP to pump molecules against their concentration gradients.

Some molecules (many proteins, for example) are far too large to be transported through the cell membrane, even through another protein; the hole it would need to move through would cause the cytoplasm to spill out. Bringing these molecules in, *endocytosis*, or out, *exocytosis*, requires energy, but a concentration gradient is irrelevant for this type of active transport. Large molecules such as hormones are manufactured in a cell but must be released to carry out their function. The product (hormone) is packaged in a vesicle. The outer wall of the vesicle is a phospholipid bilayer, just like the cell membrane. When the vesicle reaches the cell membrane, its phospholipids align with those of the membrane. Thus, the contents of the vesicle are released into the extracellular fluid without the cell's ever being "opened."

Endocytosis occurs in much the same way, but in reverse. During *pinocytosis,* a fluid droplet is pulled into the cell as the membrane pinches in. (Again, phospholipids realign, never exposing the inside of the cell to the outside environment.) *Phagocytosis* is similar to pinocytosis but occurs with solid particles. When the particle makes contact with the cell, the neighboring phospholipids project outward, engulfing the particle. To target specific molecules, such as cholesterol, cells use *receptor-mediated endocytosis*. Here, the targeted molecule binds to the receptor, and the membrane extends around them both, bringing them into the cell.

PRACTICE

For questions 22–25, fill in the blanks to complete the following sentences.

The lipid bilayer structure of the cell membrane is made possible because phospholipid molecules contain two distinct regions. The (22) _____ region is attracted to water. The (23) _____ of the phospholipids thus create an internal and external membrane surface. The (24) _____ region is repelled by water, so the (25) _____ of the phospholipids create an internal layer that won't interact with water.

(26) Explain what would happen to cells in a hypertonic solution. What could cause this situation in the human body?

_____

_____

_____

(27) Which of the following processes require energy from ATP? Choose all that apply.

(a) Active transport

(b) Diffusion

(c) Exocytosis

(d) Facilitated diffusion

(e) Osmosis

(f) Phagocytosis

(28) Transcytosis occurs in layers of cells that form barriers. One end of a cell has receptors to bind a specific molecule. The cell brings that molecule in but then immediately sends it out on the other side. Explain the two processes that make up transcytosis.

_____

_____

_____

Divide and Conquer

Have you ever had so many places to be that you wished you could just divide yourself in two? If only we could be more like the cells that make us up! Cell division is how one "mother" cell becomes two identical twin "daughter" cells. Cell division takes place for several reasons:

>> **Growth:** We all started out as a single cell: the fertilized egg. That one cell divides (and divides and divides), eventually becoming an entire complex being.

>> **Injury repair:** Uninjured cells in the areas surrounding damaged tissue divide to replace those that have been destroyed.

>> **Replacement:** Cells eventually wear out and cease to function. Their younger, more functional neighbors divide to take up the slack.

REMEMBER

A word to the wise: Cells are living things, so they mature, reproduce, and die. They are not always actively engaged in reproduction. They spend most of their lives doing their own little cell thing until they're signaled to divide. This phase is called *quiescence*. (Be careful not to confuse the *cell cycle* with cell division.)

Walking through the cell cycle

The cell cycle has two parts: *interphase*, where it undergoes changes to prepare for division and *M phase*, where the cell actually divides. A cell spends most of its life in interphase, the part of the cell cycle that precedes division. This begins with quiescence, termed G_0, followed by three phases of preparation: G_1, S, then G_2. (I'll detail the events of these phases in the next section.) The length of time spent in interphase is highly variable, as our cells have drastically different programmed life spans. Certain types of white blood cells and the cells that line parts of our digestive tract often don't live to see 24 hours. Others, such as neurons and cardiac muscle cells, have the same life span as you. Regardless, each cell starts its life in G_0.

WARNING

The G_0 phase doesn't have an agreed-upon home. Some people view this phase as the first of four that comprise interphase (as I just presented it). Other people see it as an offshoot, a sort of limbo phase, because many cells never continue the cell cycle, leaving interphase with three parts. Either way, a cell doesn't immediately begin to divide again as soon as it was created.

The division part of the cell cycle, M phase, includes *mitosis* (division of the nucleus) and *cytokinesis* (division of the cell). On completion, the two new daughter cells begin their quiescent phase. From here, there are three options — though the choice is genetically predetermined:

>> **Undergo apoptosis:** Apoptosis, or programmed cell death, is the fate of cells that have reached the end of their life span. Enzymes trigger this process, and the cells are dismantled from the inside out (though it looks more like they're exploding). This also occurs when too many new cells have been created.

>> **Differentiate:** Cells undergo *differentiation*, the process where they change and mature to take on their specialized functions. There's usually no turning back from here; the cells live but never divide.

>> **Prepare for division:** A chemical signal triggers the cell to begin G1, the start of interphase (or phase two, depending on how you look at it). There's still much work to be done before cell division can officially begin.

Waiting for action: Interphase

Interphase is the period when the cell isn't dividing but is continuing in the cell cycle. Interphase begins when the new cells are done forming and ends when the cell is prepared to divide. Although this phase is called the "resting stage" of the cell cycle, there's constant activity in the cell during interphase, which is broken into three subphases:

>> **G_1:** Stands for *growth* or *gap*. During G_1, the cell continues carrying out its functions while growing in size and creating the structures needed for mitosis (the centrioles in particular).

>> **S:** Stands for synthesis. *DNA synthesis* or *replication* occurs during this subphase. Each DNA molecule inside the cell's nucleus is copied, creating two *sister chromatids.* At the end of the S phase, the cell has two copies of every piece of DNA (chromosome) so that each daughter cell can receive a full set.

>> **G_2:** Stands for *gap.* Enzymes and proteins needed for cell division are produced during this subphase. The cell continues to grow, more rapidly this time, and additional organelles are made to dole out to the daughter cells. Then the cell officially begins the division process, or *M phase*.

Sorting out the parts: Prophase

Chromosomes cannot be evenly distributed if they're disorganized and packed away in the nucleus. As the first phase of mitosis, one of the key events of *prophase* is the dismantling of the nucleus. Each pair of centrioles moves to opposite ends of the cell, forming poles. They begin to produce protein filaments that form *spindle fibers* radiating out from the poles.

At the same time, the *chromatin* threads (pieces of DNA) shorten and coil, wrapping around proteins called *histones* to condense their structure into *chromatids.* The two identical chromatids are linked in the middle by a *centromere* creating the X-shaped *chromosome* you're familiar with. Now that the DNA is more organized, it can be easily moved to distribute the two identical sets between the two cells.

Dividing at the equator: Metaphase

Now that the nucleus is gone, the chromosomes migrate to the middle of the cell and line up along the *equator*, with each centromere on the invisible center line. The spindle fibers continue to grow and by the end of metaphase, they attach each centromere to the centrioles at each pole.

Packing up to move out: Anaphase

In *anaphase,* the centromeres are split apart by the centrioles reeling in the spindle fibers. The sister chromatids (identical copies of DNA) are pulled to the opposite poles. The cell begins to elongate, and a slight furrow develops in the cytoplasm, showing where cytokinesis will eventually take place.

Pinching off: Telophase

Telophase occurs as the chromosomes reach the poles and the cell nears the end of division. Having served their purpose, the spindle fibers disappear. Each newly forming cell begins to synthesize its own nucleus to enclose the separated chromosomes. The coiled chromosomes unwind, becoming chromatin once again. There's a more pronounced pinching, or furrowing, of the cytoplasm into two separate bodies, but there continues to be only one cell.

Splitting up: Cytokinesis

Cytokinesis means that it's time for the big breakup. The furrow lengthens from each side until it splits the elongated single cell into two new cells. Each new cell is smaller and contains less cytoplasm than the mother cell, but the daughter cells are genetically identical to each other and to the mother cell. Cytokinesis begins during anaphase, and the two cells are separated by new cell membrane around the same time that telophase ends.

TIP

Try remembering the phases this way:

Interphase = intermission, Prophase = prepare, Metaphase = middle, Anaphase = apart, Telophase = twins, Cytokinesis = complete

Differentiation

These news cells may stay linked until they divide or die, or they may physically separate from each other. And though they've graduated from Cell Cycle University, they aren't necessarily ready to join the workforce. The cells destined for more specialized jobs still need to undergo *differentiation*, which I suppose would be like an internship.

To differentiate, or become specialized, cells are exposed to chemical triggers that lead to differing *gene expression*; some genes are turned on while others are turned off. This is why cardiac muscle cells and nervous cells are dramatically different despite having the same DNA. The field of *epigenetics* strives to understand how gene expression is controlled, and there is still much to be learned!

Cell differentiation is often a process with numerous intermediaries between the stem cell and the final, specialized one. A *stem cell* is an undifferentiated cell that can give rise to cells that can then specialize (or they can just make more of themselves). Only very early in embryonic development do you find truly *totipotent* stem cells — those that can develop into any cell type in our bodies, including that of the placenta. As the embryo begins to organize, the stem cells are *pluripotent*, which means that they can develop into any cell type except the placenta. When the embryo starts to form its layers (the endoderm, mesoderm, and ectoderm), the stem cells are considered to be *multipotent*. (For more on embryonic development, check out Chapter 17.) They have started to follow one of three pathways. Cells in the mesoderm can develop into muscles and bones (among other things), but they can't become skin; cells in the ectoderm are on that path.

And now we're back at the beginning: G_0. Which means that it's time for you to "interphase" with some review questions.

PRACTICE

29   When the cell cycle is complete, the two newly formed daughter cells are

(a) the same size as the mother cell.

(b) genetically unique.

(c) unequal in size.

(d) genetically identical to the mother cell.

For questions 30–34, fill in the blanks to complete the following sentences.

When a cell enters S phase, its DNA is in long strands called 30 _____.

Once replicated, the identical copies are linked together with proteins, creating a

31 _____. During the first phase of mitosis, 32 _____

the DNA condenses, taking on a more manageable form: a 33 _____.

In the process of dividing the nucleus, these DNA bundles are pulled apart, distributing

identical 34 _____ to each side of the cell.

35   What is the difference between telophase and cytokinesis?

_____

_____

_____

36   Which of the following is true of metaphase? Choose all that apply.

(a) The nuclear membrane reappears.

(b) The chromosomes move to the poles.

(c) The chromosomes align on the equatorial plane.

(d) It's composed of subphases G_1, S, and G_2.

(e) Chromatin coils up, making it easier to move.

37   Identify the event that *does not* happen during anaphase.

(a) Early cytokinesis occurs, with slight furrowing.

(b) The cell goes through the growth subphase.

(c) Centrioles shorten the spindle fibers.

(d) Chromatids move to the outside of the cell.

(e) The centromeres split.

For questions 38–42, identify the parts of M phase illustrated in Figure 3-3.

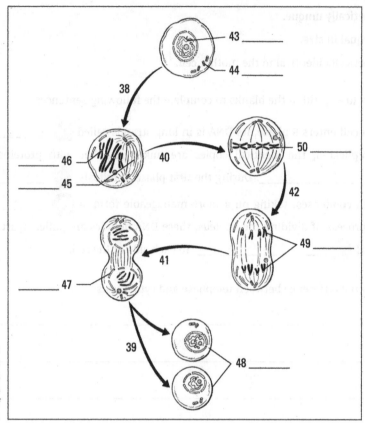

FIGURE 3-3:
Cell structures
and changes
that make up
the stages of
the cell cycle.

Illustration by Kathryn Born, MA

38 _____

39 _____

40 _____

41 _____

42 _____

For questions 43–50, use the choices to identify the illustrated structures in Figure 3-3.

(a) Centriole   (e) Equator

(b) Centromere   (f) Nuclear envelope

(c) Chromatin   (g) Sister chromatids

(d) Daughter cells   (h) Spindle fiber

Practice Questions Answers and Explanations

1. *Cytoplasm* is the term that refers to the filling of a cell. It's the shape-shifting material that the organelles all "float" around in. The fluid part of the cytoplasm is called *cytosol*. The *cytoskeleton* consists of the protein pieces embedded in the cytoplasm that help the cell keep its shape.

2. **(c) Allow proteins and nucleic acids to go in and out of the nucleus** and **(e) House genetic material.** The nuclear envelope is the membrane that creates the nucleus. It controls what can enter and leave, which includes nucleic acids; but the chromosomes as a whole cannot.

3. **(c) Cytoplasm**

4. **(m) Vesicle (formation).** You wouldn't be able to tell the difference between a vesicle and a lysosome just by looking, but vesicles are always the ones used to bring things into or out of a cell.

5. **(g) Nucleolus**

6. **(h) Nucleus**

7. **(j) Ribosomes.** They can be free or attached to the ER.

8. **(d) Golgi apparatus**

9. **(i) Cell (plasma) membrane**

10. **(b) Cilia**

11. **(f) Mitochondrion.** (*Mitochondria* is plural.)

12. **(k) Rough endoplasmic reticulum**

13. **(l) Smooth endoplasmic reticulum**

14. **(e) Lysosome**

15. **(a) Centriole**

16. **Flagellum**

17. **Golgi apparatus**

18. **Lysosomes**

19. **Mitochondria**

20. **Nucleolus**

21. **Vesicle**

22. **hydrophilic**

23. **heads**

24. **hydrophobic**

25. tails

26. **A hypertonic solution by definition has a higher concentration of solutes outside the cell than inside. This means that the concentration of water is lower outside. Because water moves to where there is less water (osmosis) through the aquaporins, the cell would begin to shrivel due to water loss.** This is the major issue with dehydration; it "concentrates" the blood, making it a hypertonic solution.

27. **(a) Active transport, (c) Exocytosis, and (f) Phagocytosis.** Diffusion and osmosis both move down concentration gradients, so no energy is needed. The only difference with facilitated diffusion is that the molecules must move through a channel, but they're still diffusing.

28. First, receptor-mediated endocytosis occurs to bring the molecule into the cell. Because the cell membrane spreads around the molecule, it enters already packaged in a vesicle. Rather than unpack its contents, the cell simply pushes the vesicle to the opposite end, where it fuses with the cell membrane, releasing the molecule via exocytosis.

29. **(d) genetically identical to the mother cell**

30. chromatin

31. centromere

32. prophase

33. chromosome

34. sister chromatids

35. **Telophase is the final step of mitosis, which is division of the nucleus. It's basically the reverse of prophase. Cytokinesis, which is division of the cell, is the final step of the cell cycle. Here, the cell membrane is extended down the middle of the elongated cell (with two new nuclei), creating the two new, identical daughter cells.** Cytokinesis usually starts during anaphase; the confusion exists because it occurs alongside telophase and both processes end at the same time.

36. **(c) The chromosomes align on the equatorial plane.**

37. **(b) The cell goes through the growth subphase.** This event is part of interphase.

38. Prophase

39. Cytokinesis

40. Metaphase

41. Telophase

42. Anaphase

43. (c) Chromatin

44. (a) Centriole

45. (h) Spindle fiber

46. (b) Centromere

(47) **(f) Nuclear envelope**

(48) **(d) Daughter cells**

(49) **(g) Sister chromatids**

(50) **(e) Equator**

If you're ready to test your skills a bit more, take the following chapter quiz, which incorporates all the chapter topics.

Whaddya Know? Chapter 3 Quiz

Think you can keep your balance while dividing and conquering this chapter quiz? See how you do, but don't check as you go. (It's better for your learning that way!) You'll find the answers in the next section.

1   Which of the following is an example of positive feedback?

(a) When blood glucose drops and glycogen is released to bring it back up

(b) When uterine contractions lead to stronger contractions for childbirth

(c) When the set point for body temperature increases to cause fever

(d) When heart rate increases to provide more oxygen to cells everywhere

2   The nucleolus

(a) packages and houses DNA.

(b) enables large molecule transport.

(c) forms a membrane around the nucleus.

(d) assembles ribosomes.

(e) supports the contents of the nucleus.

3   What is a hypotonic solution?

(a) A solution that has a lower concentration of water than exists in the cell

(b) A solution that has a greater concentration of water than exists in the cell

(c) A solution that has a lower concentration of solute inside the cell

(d) A solution that has a greater concentration of solute outside the cell

(e) A solution that constantly varies in concentration compared to what exists in the cell

4   The G_1 subphase of interphase is

(a) the period of DNA synthesis.

(b) the most active phase.

(c) the phase between S and G_2.

(d) the phase immediately before mitosis begins.

(e) part of cell division.

For questions 5-9, match the terms with their descriptions.

(a) Apoptosis

(b) Differentiated

(c) Multipotent

(d) Pluripotent

(e) Totipotent

5 _____ Cell that can become any type of other cell

6 _____ Partially differentiated stem cell; can still become numerous types but is limited

7 _____ Sequence of events that leads to the death of a cell

8 _____ Cell that carries out a specific specialized function

9 _____ Cell that can become most other cells, except placenta

10 Which event *does not* occur during telophase?

(a) The chromosomes uncoil.

(b) The chromosomes reach the poles.

(c) The spindle fibers degenerate.

(d) The nuclear membrane reforms.

(e) The cell splits in two.

11 What's the difference between facilitated diffusion and simple diffusion?

_____

_____

12 Which is the correct order of phases of the cell cycle and mitosis?

(a) Prophase, metaphase, anaphase, telophase, interphase, cytokinesis

(b) Interphase, cytokinesis, telophase, prophase, anaphase, metaphase

(c) Cytokinesis, interphase, telophase, anaphase, metaphase, prophase

(d) Interphase, prophase, metaphase, anaphase, telophase, cytokinesis

(e) Prophase, metaphase, cytokinesis, anaphase, telophase, metaphase

13 Which organelle is the site of protein synthesis?

(a) Smooth endoplasmic reticulum

(b) Golgi apparatus

(c) Nucleus

(d) Rough endoplasmic reticulum

(e) Mitochondria

14 Which of the following can change the consistency of cytoplasm?

(a) Changes in pH

(b) Temperature

(c) Pressure

(d) Molecule concentration

(e) All of the above

For questions 15–19, identify the statement as true or false. If it's false, identify the error.

15 During metaphase, each chromosome consists of two duplicate chromatids.

_____

16 The nucleus dissolves during S phase so that each chromosome can be copied.

_____

17 Cholesterol molecules can be found sticking up out of a cell membrane for recognition of self versus nonself cells.

_____

18 The tails of the phospholipids that create the cell membrane are hydrophobic, which helps maintain its structure since it's surrounded by water.

_____

19 Cilia are external organelles that allow cells to move through body fluids.

_____

20 In the nucleus of a cell that isn't dividing, DNA

(a) moves about freely.

(b) is packaged within chromatin threads.

(c) regulates the activity of mitochondria.

(d) is organized into chromosomes.

21 Injecting a large quantity of distilled water into a human's veins would cause many red blood cells to

(a) swell and burst.

(b) shrink or shrivel.

(c) carry more oxygen.

(d) aggregate.

(e) remain normal.

22 What is the function of a mitochondrion? Choose all that apply.

(a) To modify proteins, lipids, and other macromolecules

(b) To produce energy through aerobic respiration

(c) To keep the cell clean and recycle materials within it

(d) To produce and export hormones

(e) To assemble amino acids into proteins

23 This type of protein channel opens in response to a change in charge to let ions diffuse.

(a) Voltage–gated

(b) Ligand–gated

(c) Mechanically gated

(d) Ion pumps

(e) Protein pumps

24 What structures disappear during telophase?

(a) Spindle fibers

(b) Nuclear membranes

(c) Nucleolei

(d) Cytoplasm

(e) Chromatin

25 Which organelle translates genetic information to direct protein synthesis?

(a) Ribosome

(b) Lysosome

(c) Centriole

(d) Nucleus

(e) Vesicle

26 What options does a cell have once it enters G$_0$?

_____

_____

_____

27 Cytokinesis can be described as

(a) the period of preparation for cell division.

(b) the division of the cytoplasm to surround the two newly formed nuclei.

(c) the stage of alignment of the chromatids on the equatorial plane.

(d) the initiation of cell division.

(e) the dissolution of the cell's nucleus.

For questions 28–31, match the transport term with the description.

(a) Active transport

(b) Exocytosis

(c) Pinocytosis

(d) Phagocytosis

28 _____ Droplets of liquid are brought into the cell.

29 _____ The cell membrane extends around a molecule, engulfing it into the cell.

30 _____ Molecules are pumped from areas of low concentration to high concentration.

31 _____ Molecules are released from the cell by a vesicle fusing with the cell membrane.

32 What happens to DNA during interphase? Choose all that apply.

(a) It's duplicated during the S subphase.

(b) It's pulled apart during the G$_1$ subphase.

(c) It unwinds rapidly during the G$_2$ subphase.

(d) It gathers in the nucleus.

(e) It's checked for errors.

33 Which of the following is least likely to be an effector in a homeostatic mechanism?

(a) Pancreatic cell

(b) Pain receptor

(c) Back muscle

(d) Blood vessel

(e) Cardiac muscle cell

Answers to Chapter 3 Quiz

1. **(b) When uterine contractions lead to stronger contractions for childbirth.** Positive versus negative feedback isn't about an increase or a decrease; all that changing the set point does is shift the target for that value. Answer (d) is incorrect because increasing heart rate is an effect — a solution in the balancing of our oxygen levels.

2. **(d) assembles ribosomes.** All the choices relate to the nucleus, but the nucleolus is the structure within that builds ribosomes.

3. **(b) A solution that has a greater concentration of water than exists in the cell.**

TIP

The prefix *hypo-* means *under* or *below normal*. The prefix *hyper-* means *excess* or *above normal*. Someone who has been out in the cold too long suffers hypothermia — literally, insufficient heat. So a solution, or tonic, with very few particles would be hypotonic.

4. **(e) part of cell division.** None of the other statements is accurate.

5. **(e) Totipotent**

6. **(c) Multipotent**

7. **(a) Apoptosis**

8. **(b) Differentiated**

9. **(d) Pluripotent**

10. **(e) The cell splits in two.**

11. **Both are passive methods of transport; they don't require energy. Molecules are moving from the area of high to low concentration. The difference is that some molecules are not able to pass through the phospholipids of the cell membrane directly (because they're polar or charged, for example).** In this case, a protein is used to create a pathway for the molecule to get in or out of the cell. It *facilitates* the process, hence the term *facilitated diffusion*.

12. **(d) Interphase, prophase, metaphase, anaphase, telophase, cytokinesis**

13. **(d) Rough endoplasmic reticulum.** Ribosomes isn't a choice, and it's the attachment of ribosomes that makes the RER rough.

14. **(e) All of the above.** In addition, agitation, or just being jostled, can change cytoplasmic consistency.

15. **True**

16. **False.** Chromosomes are duplicated within the nucleus; it dissolves during prophase.

17. **False.** Cholesterol is embedded between phospholipid layers; carbohydrates stick out.

18. **True**

19. **False.** Our ciliated cells are not mobile; the function of cilia is to sweep things past the cell.

20. **(b) is packaged within chromatin threads.**

(21) **(a) swell and burst.** With a hypotonic solution outside the cell, the membrane would allow osmosis to continue past the breaking point. Remember the phrase "*Hypo-* makes cells big, like a hippo."

(22) **(b) To produce energy through aerobic respiration.** That's it. Knowing that the mitochondria is the powerhouse of the cell is great, but make sure you know what that entails.

(23) **(a) Voltage-gated**

(24) **(a) Spindle fibers.** Telophase is focused on rebuilding, but the spindle fibers have served their purpose and are now in the way.

(25) **(a) Ribosome**

(26) **It can undergo apoptosis because it isn't needed. (A cell might divide because its neighbor died, for example, but there's only enough room for one cell, not two.) It can undergo differentiation so it can specialize to perform its unique function. Or it can simply exist until it is triggered to divide.**

(27) **(b) the division of the cytoplasm to surround the two newly formed nuclei**

(28) **(c) Pinocytosis**

(29) **(d) Phagocytosis**

(30) **(a) Active transport**

(31) **(b) Exocytosis**

(32) **(a) It's duplicated during the S subphase** and **(e) It's checked for errors.**

(33) **(b) Pain receptor.** Effectors are the cells that carry out the effect – the response of the negative feedback mechanism. Receptors are the receiving cells; they start the whole process.

Chapter **4**

Joining Forces: Cells Organize into Tissues

O h, what tangled webs we weave! The study of tissues is called *histology*, but you may be surprised to find out that the Greek word *histo* doesn't translate as *tissue*; instead, it means *web*. After reviewing the cell and cellular division (see Chapter 3), the logical next step is to look at what happens when groups of similar cells "web" together to form tissues. The four types of tissue in the body are as follows:

» Epithelial tissues, which form coverings and linings

» Connective tissues, which link differing tissue types

» Muscle tissues, which cause movement

» Nervous tissues, which carry out communication

In some tissues, like many of the connective tissues, the cells do not make physical contact with each other. The space between them can be filled with various fibers or simply left empty, though that "empty" space would be filled with *interstitial fluid,* which is basically water and all the molecules that are pushed out with it from the blood. (See Chapter 10 for an explanation of how this process happens.) In other tissues, such as epithelial tissues, the function is dependent on the cells being tightly packed together. This is done by connecting the cells through *cell junctions,* which come in three general varieties:

» **Tight junctions:** Cells are effectively sewn together with proteins leaving little to no space between them. These junctions can be found in most epithelial tissues.

>> **Desmosomes (also known as adhesive junctions):** The proteins of a desmosome are anchored to the cytoskeleton inside of a cell and extend outward through the cell membrane. They link up with the proteins from a neighboring cell's desmosome, forming a connection that is well-suited for resisting mechanical forces. These junctions can be found in the epidermis (the outer layer of skin) and the lining of the uterus.

>> **Gap junctions:** Cells are connected by proteins that form channels, allowing the two neighboring cells to share cytoplasm. Gap junctions enable the cells to pass on small molecules without investing energy. These junctions can be found in most tissues of the body but are especially important in cardiac muscle.

I discuss many of the tissue types in greater detail in the chapters dedicated to the relevant body systems. This chapter is designed to give you a quick review of the four tissue types and a practice quiz to test your knowledge of them.

Covering Up: Epithelial Tissues

Perhaps because of its unique job of lining all our body surfaces (both external and internal), epithelial tissue has many characteristics that distinguish it from other tissue types.

Epithelial tissues, arranged as they are in sheets of tightly packed cells, always have a free, or *apical*, surface that is exposed to the air or to fluid. The cells have a defined top and bottom, and can be found in a single layer or sandwiched between other cells of the same type. Cells on the apical side sometimes have cytoplasmic projections such as *cilia* (hairlike growths that can move material over the cell's surface) or *microvilli* (fingerlike projections that increase the cell's surface area for increased adsorption and more efficient absorption and secretion).

REMEMBER

*Ad*sorption involves the *ad*hesion of something to a surface, whereas *ab*sorption involves a fluid permeating or dissolving into a liquid or solid.

Opposite the apical side is the *basal* side, the bottom layer of cells. These cells are attached to the tissue's *basement membrane*, which is a layer of nonliving material secreted by the epithelial cells. It adheres to the underlying connective tissue and acts as a filter — controlling what resources get into the tissue from the layer below as well as what is allowed through this covering and into the deeper tissues.

Epithelial tissue serves several key functions, including the following:

>> **Protection:** Skin protects vulnerable structures or tissues deeper in the body.

>> **Barrier:** Epithelial tissues prevent foreign materials from entering the body and work to retain interstitial fluids.

>> **Sensation:** Sensory nerve endings embedded in epithelial tissue connect the body with outside stimuli.

>> **Secretion:** Epithelial tissue in glands can be specialized to secrete enzymes, hormones, and fluids.

>> **Absorption:** Linings of the internal organs are specialized to facilitate absorption of materials into the body.

>> **Filtration:** Epithelial tissue can filter things as they move past it. The respiratory tract, for example, filters out dirt and particles to clean the air that's inhaled. This tissue can also filter the fluids that move through them, like how blood is filtered by the kidneys.

Epithelial tissues are identified first by their layers and then by their shape. Single-layer epithelial tissue is classified as *simple*. Tissue with more than one layer is called *stratified*. The shapes come in three varieties:

>> **Cuboidal:** As the name implies, *cuboidal* cells are equal in height and width, shaped like cubes.

>> **Columnar:** *Columnar* cells are taller than they are wide.

>> **Squamous:** *Squamous* cells are thin, flat, and odd-shaped.

REMEMBER

Following are the primary types of epithelial tissues:

>> **Simple squamous epithelium:** A single layer of flat cells, this tissue facilitates rapid diffusion and filtration. The walls of our blood capillaries are typical tissue of this type.

>> **Simple cuboidal epithelium:** A single layer of cube-shaped cells, this tissue facilitates absorption and secretion. It is typically found in glands, producing secretions and creating tubes for their transport.

>> **Simple columnar epithelium:** A single layer of cells, this tissue is elongated in one dimension (like a column). Like simple cuboidal epithelium, this tissue facilitates secretion and absorption. This type of tissue is primarily found lining portions of the digestive tract.

The cells may also be *ciliated*, possessing a type of organelle called *cilia* — hairlike structures that move substances along in waves. Ciliated simple columnar epithelium can be found lining the uterine tube.

>> **Pseudostratified columnar epithelium:** This tissue is a single layer of columnar cells. Note that the prefix *pseudo* means *false*. The cells of pseudostratified epithelium do not have a uniform width; some are wider at the top than the bottom, others wider at the bottom than the top. Because of this, the tissue appears to be *stratified* (layered) because the cells' nuclei don't line up in a row as they do in simple columnar epithelium. They are a single layer, though, and also facilitate absorption and secretion. Pseudostratified tissue lines ducts in testicular structures.

More commonly, this tissue type is ciliated. It is present in the linings of the respiratory tract, functioning in a more or less identical way to the simple columnar ciliated epithelium.

>> **Stratified squamous epithelium:** This tissue is the stuff you see every day — your outer skin, or epidermis. Found in areas where the outer cell layer is constantly worn away, this type of epithelium regenerates its surface layer with cells from lower layers.

WARNING

Though the newly produced cells along the basement membrane look more cuboidal, the cells in the other layers have the characteristic squamous shape. Don't fall into the trap of calling this tissue stratified cuboidal!

>> **Stratified cuboidal epithelium:** This rare epithelium consists of two or three layers of cube-shaped cells. It can be found in sweat and mammary glands. Its primary function is protection of ducts in their neighboring tissue.

>> **Stratified columnar epithelium:** Several layers of elongated cells act mainly as protection in such structures as the conjunctiva of the eye and the pharynx (throat).

>> **Transitional epithelium:** This multilayered epithelium is referred to as *transitional* because its cells can shape-shift from cuboidal to squamous and back again. Found lining the urinary bladder, the cells stretch and flatten (appearing to be squamous) to make room for urine. When urination occurs, the cells relax and assume their original form (cuboidal).

Making a Connection: Connective Tissues

Connective tissues connect, support, and bind body structures together and are the most abundant tissues by weight. Generally, connective tissue is made up of cells that are spaced far apart within a gel-like, semisolid, solid, or fluid matrix. (A *matrix* is a material that surrounds and supports cells. In a chocolate–chip cookie, the dough is the matrix for the chocolate chips.)

Also in the matrix of connective tissues are varying proportions of fibrous proteins of two types: collagenous and elastic. *Collagenous fibers* are made of collagen, a bulky protein, and serve to provide structure. *Elastic fibers* are made of elastin, a thin protein, and serve to provide stretch.

Connective tissue has many functions and, thus, many forms; it's the most varied type of the tissue groupings. In some parts of the body, such as the bones, connective tissue supports the weight of other structures, which may not be connected to it directly. Other connective tissue, such as adipose tissue (fat), cushions other structures from impact. You encounter lots of connective tissue in the chapters to come because every organ system has some kind of connective tissue.

Based on the cell types and matrix components, connective tissues fall into two categories: proper and specialized. Over the next two sections, we'll take a closer look at each of these.

Proper connective tissue

Just as we have different kinds of tapes and glues, we have different varieties of proper connective tissues to hold things together. These tissues can be split into two categories: *loose* and *dense*. (See Figure 4-1.)

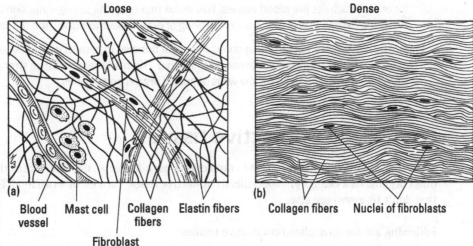

Loose

Dense

FIGURE 4-1:
Areolar tissue
(a) and dense
regular
connective
tissue (b).

(a)

Blood vessel   Mast cell   Collagen fibers   Elastin fibers

Fibroblast

(b)

Collagen fibers   Nuclei of fibroblasts

Illustration by Kathryn Born, MA

Following are the primary types of proper connective tissue:

LOOSE

>> **Areolar tissue:** This tissue exists between and around almost everything in the body to bind structures together and fill space. The thick, wavy ribbons of collagenous fibers in combination with numerous thin elastic fibers provide support. Various cells — including lymphocytes, fibroblasts, macrophages, and mast cells — are scattered throughout the matrix, leaving ample room for blood and lymphatic vessels as well as nerves. (Refer to the image on the left side of Figure 4-1.)

>> **Adipose tissue:** Composed of *adipocytes* (fat cells), this tissue forms padding around internal organs, provides insulation, and stores energy in fat molecules called *triglycerides*. These lipids fill the cells, forcing the nuclei against the cell membranes and giving them a bubble-like shape. Adipose has an intracellular matrix rather than an extracellular matrix.

>> **Reticular tissue:** In Latin, *reticulum* means "little net," hence the name of this net-like tissue. *Reticular tissue* is made up of slender, branching reticular fibers with specialized fibroblasts overlaying them. Reticular fibers are made from collagen but are thinner and have a carbohydrate attached. Its intricate structure makes it a particularly good filter, which explains why it's found inside the spleen, lymph nodes, and bone marrow.

DENSE

>> **Dense regular connective tissue:** Made up of parallel, densely packed bands or sheets of collagen (see the image on the right in Figure 4-1), this type of tissue is found in tendons, ligaments, surrounding joints, and anchoring organs. Dense regular connective tissue has great tensile strength that resists lengthwise pulling forces. However, there is little space for blood vessels and little interstitial fluid, which is why an injury to a tendon or ligament takes so long to heal.

>> **Dense irregular connective tissue:** This tissue has the same components as dense regular tissue but appears much less organized. Its fibers twist and weave around each other, forming a thick tissue that can withstand stresses applied from any direction while leaving space

for other structures like blood vessels. This tissue makes up the strong inner skin layer called the dermis as well as the outer covering of organs like the kidney and the spleen.

>> **Elastic connective tissue:** Collagenous fibers are bundled in parallel with thick bands of elastic fibers sandwiched between (like lasagna). This makes elastic tissue strong but stretchy — perfectly suited for the walls of hollow organs and arteries.

Specialized connective tissue

The other connective tissues create connections on a larger scale, like connecting organs to other organs. As a category, "specialized connective tissue" is a bit of a catch-all for the tissues that don't fit anywhere else.

Following are the specialized connective tissues:

>> **Cartilage:** These firm but flexible tissues, made up of collagen and elastic fibers, have no nerves or blood vessels. Cartilage contains openings called *lacunae* (from the Latin word *lacus,* which means *lake* or *pit)* that enclose mature cells called *chondrocytes.* The ground substance is full of proteoglycans, making it thicker than other connective tissues. There are three types of cartilage:

- *Hyaline cartilage:* The most abundant cartilage in the body, it's collagenous and made up of a uniform matrix pocked with chondrocytes. It looks glassy because the fibers within the matrix are spread so fine that you can't see them. Hyaline cartilage lays the foundation for the embryonic skeleton, forms the costal (rib) cartilages, makes up nose cartilage, and covers the adjoining surfaces of bones that form movable joints.

- *Fibrocartilage:* As the name implies, fibrocartilage contains thick, compact collagen fibers. The spongelike structure, with the lacunae and chondrocytes lined up within the fibers, makes it a good shock absorber. This type of cartilage is found in the intervertebral discs of the vertebral column.

- *Elastic cartilage:* Similar to hyaline cartilage but with a much greater abundance of elastic fibers, elastic cartilage has more tightly packed lacunae and chondrocytes between parallel elastic fibers. This cartilage — which makes up structures in which a specific form is important, such as the outer ear and epiglottis — tends to bounce back to its original shape after being bent.

>> **Osseous, or bone, tissue:** Essentially, bone is mineralized connective tissue, making it the most rigid of the group. The characteristic collagen fibers are coated in mineral deposits, giving it the hardness we associate with our bones. Cells called *osteoblasts* mineralize the matrix into repeating patterns called *osteons.* In the center of each osteon is a large opening that contains blood vessels, lymph vessels, and nerves. (Osseous tissue always makes me think of tree stumps because the layers of matrix encircle each other.) It functions as our body's framework, as protection for our vital organs, and as storage for calcium.

>> **Blood:** What? Blood is a tissue? Like other connective tissues, blood has cells surrounded by a matrix that contains proteins (though different proteins). The big difference is that the cells — *erythrocytes* (red blood cells), *leukocytes* (white blood cells), and *thrombocytes* (platelets) — are suspended in a liquid matrix, the *plasma.* This unique feature makes blood the ideal tissue for transporting materials around the body.

Flexing It: Muscle Tissues

Like the epithelial tissues, muscle tissues are made solely of cells. Called *myocytes*, the cells work to move things in and around our bodies by shortening, or *contracting*. Though each tissue type has cells of differing shapes (see Figure 4-2), they all work the same way. Following are the three types of muscle tissue:

>> **Skeletal:** Biceps, triceps, pecs — these are the muscles that bodybuilders focus on. As the name implies, skeletal muscles attach to the skeleton and are controlled throughout by the conscious (voluntary) control function of the central nervous system for movement. Muscle fibers are cylindrical, with several nuclei in each cell (which makes them *multinucleated*) and obvious *striations* (stripes that span the width of the cell) throughout. They are bundled in parallel and are roughly the length of the muscle they make up.

>> **Cardiac:** This tissue is composed of cylindrical fibers, *cardiomyocytes*, each with a central nucleus and striations. The tissue forms a branching pathway, as one cell can join with two or more fibers at one end, using gap junctions called *intercalated discs*. Cardiac muscle tissue contractions occur through the autonomic nervous system (involuntary control).

>> **Smooth:** Made up of spindle-shaped cells with large, centrally located nuclei, this muscle tissue is found in the walls of our hollow organs. Like cardiac muscle, this type of tissue contracts without conscious control. Smooth muscle gets its name from the fact that unlike other muscle tissue types, it isn't striated.

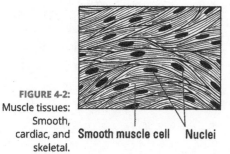
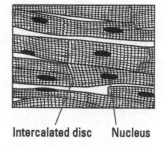
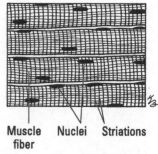

FIGURE 4-2: Muscle tissues: Smooth, cardiac, and skeletal.

Smooth muscle cell   Nuclei   Intercalated disc   Nucleus   Muscle fiber   Nuclei   Striations

Illustration by Kathryn Born, MA

Getting the Signal Across: Nervous Tissue

There's only one type of nervous tissue and only one primary type of cell in it: the *neuron*. There are numerous other cells, called *glial cells*, but they don't perform the communication function we associate with nervous tissue. The variety you see in the different varieties of nervous tissue come from the concentration of neurons in the area. Nervous structures – like a nerve versus the spinal cord versus the brain, appear different due in large part to the proper connective tissues associated with them.

A neuron is unique in that it can both generate and conduct electrical signals in the body. That process starts when sensory receptors receive a stimulus that causes an electrical signal to be sent through cytoplasmic projections called *dendrites*. From there, the signal moves through the *cell body* to the *axon*, another cytoplasmic extension. Here, an impulse is generated and spreads to the end of the axon, where it hands the signal off chemically to the next cell down the line. (I describe how all that happens in greater detail in Chapter 8.)

This chapter was short and sweet, but it had a lot of subtle details. See how well you picked up on them by taking the chapter quiz.

Whaddya Know? Chapter 4 Quiz

Quiz time! Complete each problem to test your knowledge on the various topics covered in this chapter. You can then find the solutions and explanations in the next section.

1 Why does dense irregular connective tissue twist and weave around?

(a) To fill gaps between other tissues

(b) To withstand stresses applied from any direction

(c) To prevent dead or dying cells from weakening its structure

(d) To provide stretch and elasticity

(e) To provide firm connections between muscles and bones

2 The key functions of epithelial tissue do *NOT* include

(a) protection.

(b) secretion.

(c) contraction.

(d) filtration

(e) absorption.

3 Which statement best describes how neurons communicate?

(a) Axons generate an impulse, which is passed to dendrites through the cell body.

(b) Axons receive a stimulus, which is passed to the cell body, triggering the dendrites to release a chemical.

(c) Cell bodies respond to a stimulus by generating an electrical signal to pass on to the dendrites and axons. The former passes the signal on electrically, and the latter passes it on chemically.

(d) In response to stimulus, the dendrite generates an electrical signal that is passed to the cell body. The axon spreads an impulse down its length, releasing a chemical at the end.

(e) Dendrites respond to stimulus by releasing a chemical within the cell that travels through the cell body. The axon generates an impulse in response, spreading it to the end, where it passes the signal on to the next cell.

4 Describe the location and function of intercalated discs.

_____

_____

_____

⑤ What tissue type is shown in Figure 4-3? _____

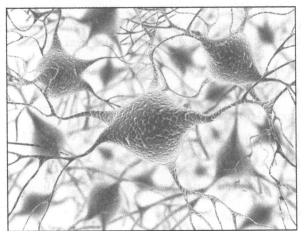

FIGURE 4-3:
Identify this
tissue.

Photo by ktsdesign

⑥ This thin, delicate tissue has the most "empty" space between cells; it helps hold organs in place and fills gaps between muscles.

(a) Reticular tissue

(b) Elastic cartilage

(c) Dense irregular tissue

(d) Dense regular tissue

(e) Areolar tissue

⑦ Transitional epithelial tissue

(a) primarily protects specific areas.

(b) has the ability to stretch.

(c) is constantly worn away and replaced.

(d) is tailor-made for absorption.

(e) is found in the salivary glands.

⑧ Which type of cell junction joins cells by linking filaments from their cytoskeleton?

(a) Desmosome

(b) Gap junction

(c) Tight junction

9 Which of the following are functions of the tissue illustrated in Figure 4-4? Choose all that apply.

(a) Mineral storage

(b) Thermoregulation

(c) Framework of body

(d) Protection of vital organs

(e) Manufacture of blood cells

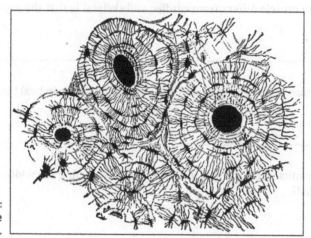

FIGURE 4-4:
Identify the
tissue type.

Illustration by Morphart Creation

10 What type of tissue makes up tendons?

(a) Elastic cartilage

(b) Dense regular tissue

(c) Areolar tissue

(d) Fibrocartilage

(e) Dense irregular tissue

For questions 11–15, choose true or false. When the statement is false, identify the error.

11 Nervous tissue is essentially formed by neurons connecting end to end with very little matrix.

_____

12　Between the bones of the knee is a cartilage pad called the meniscus that creates a cushion of fibrocartilage for shock absorption.

_____

13　Cilia are the fingerlike projections on a cell that serve to increase surface area for absorption and secretion.

_____

14　One way in which smooth muscle differs from cardiac and skeletal is that the myocytes are multinucleated.

_____

15　Structures made from dense regular connective tissue take a long time to heal because they are avascular.

_____

16　Both skeletal and cardiac muscle tissue have prominent lines spanning the width of the cell. What are these lines called?

(a) Fibers

(b) Multinucleates

(c) Lacunae

(d) Striations

(e) Intercalated discs

For questions 17–20, identify the epithelial tissues shown in Figure 4-5.

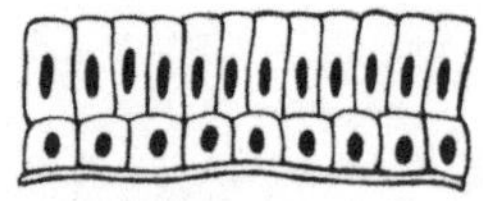

17. _____

18. _____

FIGURE 4-5:
Epithelial
tissues.　19. _____　　20. _____

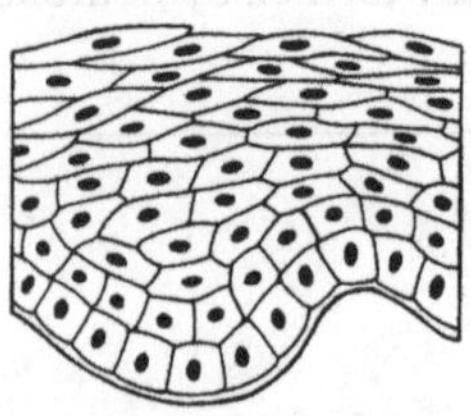

Illustration by Kathryn Born, MA

For questions 21–25, match the epithelial tissue with its location in the body.

(a) Urinary bladder

(b) Tubules of the kidney

(c) Digestive tract

(d) Epidermis of the skin

(e) Respiratory passages

21 _____ Simple columnar

22 _____ Stratified squamous

23 _____ Transitional

24 _____ Pseudostratified ciliated columnar

25 _____ Simple cuboidal

26 Cartilage heals slowly after an injury because

(a) It's dead rather than living tissue.

(b) The cells are packed together tightly.

(c) This tissue type is very complex.

(d) It has few, if any, blood vessels.

27 Which of the following contain smooth muscle tissue? Choose all that apply.

(a) The heart

(b) The small intestine

(c) The biceps

(d) The hamstrings

(e) The femoral artery

28 In what tissue is a basement membrane found, and what is its purpose?

_____

_____

_____

29   The epiglottis is the flap of tissue that closes off your trachea (windpipe) when you swallow. It is firm but also flexible. What type of cartilage is it made of?

(a) Hyaline cartilage

(b) Fibrocartilage

(c) Elastic cartilage

(d) Reticular cartilage

(e) Osseous cartilage

30   The appearance of the cells of the tissue shown in Figure 4-6 is a result of its role in storing what?

(a) Plasma

(b) Proteins

(c) Lipids

(d) Carbohydrates

(e) Calcium

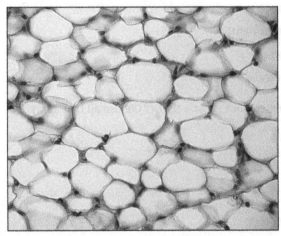

FIGURE 4-6: Identify this tissue type.

Photo by Jose Luis Calvo

Answers to Chapter 4 Quiz

1. **(b) To withstand stresses applied from any direction.** The irregular pattern also creates space for structures such as blood vessels to pass through.

2. **(c) contraction.** Only muscle tissue does that.

3. **(d) In response to stimulus, the dendrite generates an electrical signal that is passed to the cell body. The axon spreads an impulse down its length, releasing a chemical at the end.** An electrical signal can only be used to quickly travel the length of a single neuron - which can be inches (cm), even feet (m) long. At the end of the cell, though, that electricity can't be passed to another cell; a chemical must be used instead. Although both dendrites and axons can generate an electric signal, it is not termed "impulse" until the axon does it.

4. **Intercalated discs are found where cardiac muscle fibers meet end to end. These gap junctions form tunnels that allow cells to pass molecules to one another directly without expending energy on endo- and exocytosis.**

5. **Nervous tissue.** You can see the large cell bodies with numerous stringy projections: the dendrites and axon.

6. **(e) Areolar tissue.** With this tissue, you see mostly strings (protein fibers) and some cells scattered about, but no cells as big as the ones in nervous tissue.

7. **(b) has the ability to stretch.** That's why it looks like squamous cells when it's being pulled and cuboidal when it's relaxed.

8. **(a) Desmosome**

9. **(a) Mineral storage, (c) Framework of body,** and **(d) Protection of vital organs.** You didn't fall for my trick, did you? This is osseous tissue, which is found in bone. And bones, as organs, do make blood cells. But that doesn't happen in the osseous tissue, it happens in the marrow, which is reticular. Lastly, there's no way that bones could carry out thermoregulation.

10. **(b) Dense regular tissue**

11. **False.** Abundant space is visible between neurons dotted with glial cells and numerous "fibers," which are actually the ends of axons belonging to other neurons.

12. **True**

13. **False.** Microvilli are the structures that increase cell surface area. Cilia are much thinner and wave to move fluids past themselves.

14. **False.** Skeletal muscle is the only type that is multinucleate (having more than one nucleus).

15. **True**

(16) **(d) Striations.** Don't confuse them with the proteins you can see running end to end. Striations are perpendicular, running top to bottom.

(17) **Stratified columnar**

(18) **Pseudostratified**

(19) **Simple squamous**

(20) **Stratified squamous**

WARNING

Identifying epithelials can be a bit tricky. First, look for the basement membrane and the space; then look only at the cells in between. Simple cuboidal can appear to be stratified because the tissue forms tubes, and neighboring tubes connect at their basement membrane. When the tissue is stratified, you must look at the pattern of the cells throughout. Stratified squamous has cube-shaped cells along the bottom, but the bulk of the cells are clearly squamous. Transitional tissue is the most troublesome, but you will see both squamous and cuboidal throughout the width of the tissue.

(21) **(c) Digestive tract**

(22) **(d) Epidermis of the skin**

(23) **(a) Urinary bladder**

(24) **(e) Respiratory passages**

(25) **(b) Tubules of the kidney**

(26) **(d) It has few, if any, blood vessels.**

(27) **(b) The small intestine** and **(e) The femoral artery.** Smooth muscle is in the walls of our hollow organs except the heart, which has its own type. Though we associate blood vessels with the heart, they don't beat like the heart, so they aren't cardiac muscle.

(28) **The basement membrane is found only in epithelial tissues. These tissues are made entirely of cells, and they border empty space, which makes getting resources from blood flow rather difficult. Also, the cells need something to bind them to the connective tissue underneath them. So they secrete compounds that form the basement membrane, holding the epithelial tissue in place and controlling what can pass through the basement membrane in either direction.**

(29) **(c) Elastic cartilage.** Answers (d) and (e) aren't real things; fibrocartilage is the shock absorber; and hyaline cartilage is smooth to reduce friction.

(30) **(c) Lipids.** This tissue is adipose tissue. The cells look empty because on a slide, they are. The lipids (specifically, fats) that filled the cells get washed away in the process.

2

Weaving It All Together: The Structural Layers

In This Unit . . .

Chapter **5**

It's Just Skin Deep: The Integumentary System

Did you know that the skin is the body's largest organ? In an average person, its 17 to 20 square feet (5-6 square meters) of surface area represents 5 to 7 percent of the body's weight. Self-repairing and surprisingly durable, the skin is the first line of defense against the harmful effects of the outside world. It helps retain moisture; regulates body temperature; and hosts the sense receptors for touch, pain, and heat.

Skin is jam-packed with components. It's been estimated that every square inch of skin contains 15 feet (4.5 meters) of blood vessels, 4 yards (3.6 meters) of nerves, 650 sweat glands, 100 oil glands, 1,500 sensory receptors, more than 1,000 hair follicles, half a million melanocytes (pigment cells), and more than 3 million cells with an average life span of 26 days that are constantly being replaced. Refer to the "Skin (Cross Section)" color plate in the center of the book to see a detailed cross section of the skin.

Skin is made in layers, and there are layers within layers. New cells begin life in the lower levels and are gradually pushed to the surface to replace the old, dead ones. By the time the new cells reach the top, they've become hard and flat, like roof shingles. Eventually, they pop off like shingles blown from a roof in a strong wind.

Meet & Greet and Love the Skin You're In

The entity known as *you* is bounded by your *integument* — your skin. It mediates much of the interaction between *you* and *not-you* — the environment. Your integument identifies *you* to other humans, which is a very important function for members of the hypersocial human species. When we say the word *skin*, we're generally talking about the outermost layer: the *epidermis*. But there's far more to our integument than meets the eye.

Here's a look at the integument's important functions:

>> **Protection:** Skin protects the rest of the body by keeping out many threats from the environment, such as pathogens, damaging solar radiation, and nasty substances everywhere.

>> **Thermoregulation:** The skin and its accessories in particular support *thermoregulation* (maintaining a consistent body temperature) in several ways.

>> **Water balance:** The skin's outer layers are more or less impermeable to water, keeping water and salts at an optimum level inside the body and preventing excess fluid loss. A small amount of excess water and salts are eliminated through the skin.

>> **Incoming messages:** Many types of sensory organs are embedded in your skin, including receptors for heat and cold, pressure, vibration, and pain.

>> **Outgoing messages:** The skin and hair are messengers to the outside environment, mainly to other humans. People get information about your state of health by looking at them. Your emotional state is signaled by pallor, flushing, blushing, goose bumps, and sweating. The odors of sweat from certain sweat glands signal sexual arousal.

>> **Substance production:** *Sebaceous glands* in the skin, usually associated with a hair follicle, produce a waxy substance called *sebum* for waterproofing. Sweat glands in the skin make sweat. In fact, your skin has several types of glands, and each makes a specific type of secretion.

Some skin cells produce *keratin,* a fibrous protein that's an important structural and functional component of skin and essentially is the only component of hair and nails. Other skin cells produce *melanin,* our main pigment molecule, giving us our varieties of skin tone.

The integument wraps around the musculoskeletal systems, taking the form of your bones and muscles, and adding its own shape-forming structures. Although your skin feels tight, it's actually loosely attached to the layer of muscles below. In spots where muscles don't exist, such as on your knuckles, the skin is attached directly to the bone.

Without losing sight of the reality or being disturbed by the horror of the idea, it's helpful to imagine that you can unzip and remove your integument, spread the living skin out on a table, and look at it. What would you see, feel, and smell? (To be fair, without your *dermis* the answer to the last two questions would be "Nothing!")

One of the most obvious features you'd notice is that the skin, itself a thin layer, is made up of several layers. The layering is visible to the unaided eye because each layer is different from the others, and the transitions between layers appear to be relatively abrupt. The superficial (outermost) layer is the epidermis, followed by the dermis, which is where all the action is.

Deep to the dermis is the *hypodermis*, or *subcutaneous layer*. This layer is the point of attachment to the *fascia*, the fibrous connective tissue that surrounds all your internal organs.

Some people argue that the hypodermis isn't part of the skin. The argument is really about semantics; that is, it depends on how you choose to define *skin*, which isn't a proper anatomical term anyway. Skin or not, the hypodermis is certainly part of the integumentary system.

The epidermis is on top of the dermis. The prefix *epi* means *on top of*. Many anatomical structures are on top of other structures, so you see this prefix in many chapters throughout this book. Try to take notice of it and make the connections before digging into the term's definition. Other prefixes you may see are *endo* (within), *ecto* (outside), *peri* (around), and *hypo* (beneath or lower).

Now let's dig in and see what we find.

Digging Deep into Dermatology

Skin, together with its *appendages* — the hair, nails, and glands — composes the integumentary system. The name stems from the Latin verb *integere*, which means *to cover*. The relevant Greek and Latin roots include *dermato* and *cutis*, both of which mean *skin*. Check the color plate often as you go through the following sections.

Don't judge this book by its cover: The epidermis

The most familiar aspect of the integument is the *epidermis*; it's the part you see when you look in the mirror or at other people. The epidermis feels soft, slightly oily, elastic, resilient, and strong. In some places, the surface contains dense, coarse hairs; in other spots, it has a lighter covering of finer hairs; in a few places, it has no hairs at all. The nails cover the tips of the fingers and toes.

The epidermis is composed of *stratified squamous* tissue and is *avascular* (has no direct blood supply). Its nourishment diffuses through the *basement membrane* from the dermis below. (I mention in Chapter 4 that the basement membrane is the layer of glycoproteins that forms the base of all epithelial tissues.) As your skin cells age and get pushed farther away from the resource supply, they weaken and eventually die, giving the epidermis its layered appearance (not visible to the naked eye, though). Figure 5-1 illustrates the layers of the epidermis.

The healthiest, happiest layer of epidermal cells, called the *stratum basale* (or *stratum germinativum*), is along the basement membrane. This layer is the only one with cells that reproduce, because they get access to the resources filtering up through the dermis first. As new cells are made, older cells shift upward, becoming spindle-shape and forming the *stratum spinosum*, the thickest epidermal layer. Next is the *stratum granulosum*; cells in this layer are flattened, and the nuclei and organelles start to shrivel. The most superficial layer is the *stratum corneum*; the cells in this layer are hardened, dead, and full of keratin. Thick skin, such as on the palms of your hands, has an additional layer between the granulosum and corneum: the *stratum lucidum*, so named because the cells appear to be lucid (clear) under a microscope. This layer is built up for added protection in areas that receive a lot of wear and tear.

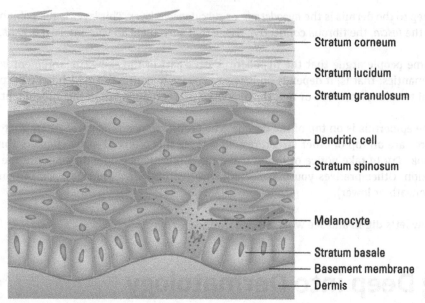

- Stratum corneum
- Stratum lucidum
- Stratum granulosum
- Dendritic cell
- Stratum spinosum
- Melanocyte
- Stratum basale
- Basement membrane
- Dermis

FIGURE 5-1:
The epidermis.

Shutterstock

REMEMBER

Cells are shed continuously from the top of the epidermis and replaced continuously by cells that are pushed up from the deeper layers. The entire epidermis is replaced approximately every six to eight weeks throughout most of your life, though it slows down with age.

The thin, impervious cover

Think of the stratum corneum as being a sheet of self-repairing fiberglass over the other layers of the epidermis. It's typically about 15 to 30 cells thick, though those cells are so flattened and tightly linked that the layer looks more like strands of proteins than cells (which, basically, they are).

As the cells are pushed toward the surface, they undergo *keratinization.* That is, their cytoplasm progressively fills with keratin, and they flatten to form the dense, relatively hard outer layer of the skin. Sebaceous glands in the dermis release sebum onto the surface of the stratum corneum for softening and waterproofing. The stratum corneum protects the entire body by making sure that some things stay in and everything else stays out.

One important thing that the stratum corneum doesn't seal out, however, is ultraviolet (UV) radiation. This form of energy goes right through the skin's surface and down to the layers below, where it stimulates the production of vitamin D. In high doses, it burns the skin and damages DNA, which can cause cells to become cancerous. Some exposure to UV radiation is necessary for our health, but too much UV radiation can clearly be problematic. Special cells called *melanocytes* in the stratum basale produce melanin, which absorbs harmful UV radiation and transforms the energy into harmless heat.

The transit zone

The *stratum lucidum*, found only on the palms of hands and soles of the feet (thick skin), the *stratum granulosum*, and the *stratum spinosum* lie in distinct layers below the stratum corneum. Most of the cells that comprise the epidermis, called *keratinocytes*, are within these three layers. These cells create structural proteins (such as keratin), lipids, and even some antimicrobial molecules. As they are pushed away from the nutrient source (blood vessels in the dermis), they become progressively more keratinized. These layers also contain *Langerhans cells* — immune cells that arrest microbial invaders (perform phagocytosis) and help activate our immune response.

The cell farm

The *stratum basale* (also called the *stratum germinativum*, or basal layer) is like a cell farm, constantly producing new cells and pushing them up into the layer above. This layer also contains the melanocytes. The pigment they produce (melanin) is released and taken in by surrounding cells; this pigment is released or recycled when the cells die.

REMEMBER

Everybody's stratum basale has roughly the same number of melanocytes — half a million to a million or more per square inch (2.5 square centimeters) of skin — but the amount of melanin produced varies depending mainly on genetics (heredity). Genetics leads to our beautiful variety of skin colors. The environment also plays a role in skin tone, because exposure to UV radiation stimulates increased production of melanin. Human groups living close to the equator have evolved genes that stimulate melanocytes to produce more melanin as protection from UV radiation. Without the melanin, the radiation can burn the skin, damage DNA, and ultimately cause skin cancer.

The basement

Deep to the stratum basale is the *basement membrane* of the epidermal tissue, which is thought to be created by the layer itself. In addition to anchoring the epidermal layer to the dermis below, it acts as a filter, controlling what gets up into this layer from below and what is allowed deeper. Because there is no blood flow in the epidermis, the only way resources can get into tissue is through the basement membrane. Interstitial fluid seeps through from the dermis, carrying resources with it. Cells that aren't along the basement membrane get their resources from the limited amount of interstitial fluid in epithelial tissue or from their neighboring cells. This is why it gets progressively harder for cells to survive as they get pushed farther up.

TIP

To remember the layers of the epidermis, try this phrase: "Come, let's get sun burned." Conveniently, each layer starts with a different letter, and this little phrase even corresponds to one of the functions. You can remember that you're going from the outside in because the deeper the sun goes, the worse the burn is. Plus, *b* is for *basale*, which is like *base*, which is along the *basement* membrane.

Exploring the dermis

Below the layers of the epidermis and several times thicker is the *dermis*. The dermis itself is made up of two layers: the papillary region and the reticular region.

Under the basement

The *papillary region* contains areolar connective tissue that projects into the epidermis. The loosely packed fibers of the areolar tissue leave plenty of space to bring small blood vessels and nerve endings closer to the surface. These projections, called the *dermal papillae* or *dermal pegs*, are so pronounced in your hands and feet that they form *friction ridges* on the surface. When you dip these structures in some ink and then press them onto something, you get to see the unique pattern of your friction ridges: your fingerprint. Their purpose, though, is to aid in grasping.

REMEMBER

The papillary region is an example of a common anatomical strategy for increasing the surface area between two structures. More areas of direct contact leads to more chances for molecules to travel from one side to the other. Think about the difference between leaving a crowded parking lot after a concert with only 2 exits and one with 20 exits. I discuss another prominent example of the benefits of high surface area — the intestines — in Chapter 14.

The manufacturing site

The *reticular region* is, annoyingly, made of dense irregular (as opposed to actual reticular) connective tissue. It is chock-full of protein fibers and is complex and metabolically active. The dermis contains interlacing bundles of collagenous and elastic fibers that form the strong yet flexible layer. The various skin "accessories" — blood vessels, nerves, glands, and hair follicles — are embedded here. Adipocytes (fat cells) are found at the bottom of the dermis, which transitions into the hypodermis below.

The blood vessels in the dermis provide nourishment and waste removal from its own cells as well as from the epidermis above. The blood vessels in the dermis dilate (become larger) when the body needs to lose heat and constrict to keep heat in. They also dilate and constrict in response to your emotional state, brightening or darkening skin color, thereby functioning as social signaling.

Getting under your skin: The hypodermis

The *hypodermis* (or *subcutaneous layer* or *superficial fascia*) is the layer of tissue directly underneath the dermis. It is composed mainly of connective and *adipose* (fat) tissue. The blood vessels, lymphatic vessels, and nerve fibers here are larger than those in the dermis. The loosely arranged elastic fibers of the hypodermis anchor it to the *deep fascia* of the muscle below.

The physiological functions of the hypodermis include insulation, cushioning, and energy storage (in the form of lipids). Recently, researchers have found that adipose tissue also plays a very active role in the endocrine system. All the hormones produced by the adipocytes (fat cells) and their functions are still being sorted out, but there is a definite connection to insulin sensitivity, appetite suppression, and likely much more.

PRACTICE

See if you've got the skinny on skin so far.

For questions 1–8, label the structures of the epidermis shown in Figure 5-2.

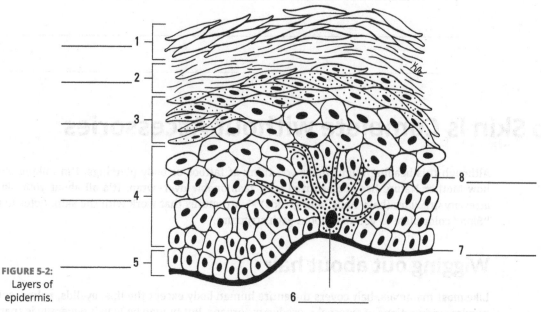

_____ 1

_____ 2

_____ 3

_____ 4

8 _____

7 _____

_____ 5

6 _____

FIGURE 5-2:
Layers of
epidermis.

Illustration by Kathryn Born, MA

9 What does the papillary layer of the dermis do? Choose all that apply.

(a) Filter out microbes

(b) Extend into the epidermis

(c) Carry the blood and nerve endings close to the epidermis

(d) Aid in holding the epidermis and dermis together

(e) Contain cells sensitive to touch

10 What do the epidermal ridges on the fingers do?

(a) Provide a means of identification

(b) Increase the friction of the epidermal surface

(c) Decrease water loss by the tissues

(d) Aid in regulating body temperature

(e) Prevent bacterial infection

11 What do the keratinocytes in the spinous layer do? Choose all that apply.

(a) They metabolize nutrients to supply the epidermis.

(b) They connect the cells of the layer.

(c) They transform other cells that have lost their nuclei and cytoplasm.

(d) They move nutrients from the blood into the epidermis.

(e) They build keratin proteins.

12 Briefly outline the steps of epidermal growth.

_____

_____

_____

No Skin Is Complete without Accessories

Although this heading might make you think of tattoos or body piercings, I'm talking about how Mother Nature has accessorized your fashionable over-wrap. It's all about your skin's *accessory structures* (hair, nails, and glands) — structures that work with the skin. Refer to the "Skin" color plate as you read about each structure.

Wigging out about hair

Like most mammals, hair covers the entire human body except the lips, eyelids, palms, soles, nipples, and portions of external reproductive organs. But human body hair generally is sparser and much lighter in color than that sported by most other mammals. Animals have hair for protection and temperature control. For humans, however, body hair is largely a secondary sex characteristic.

A thick head of hair protects the scalp from exposure to the sun's harmful rays and limits heat loss. Eyelashes block sunlight and deflect debris from the eyes. Hair in the nose and ears prevents airborne particles and insects from entering. Touch receptors connected to hair follicles respond to the lightest brush.

The average adult has about 5 million hairs, with about 100,000 of those growing from the scalp. Normal hair loss from an adult scalp is anywhere from 40 to 100 hairs each day, although baldness can result from genetic factors, hormonal imbalances, scalp injuries, disease, dietary deficiencies, radiation, or chemotherapy.

Hairs grow out from a *hair follicle*, a small tube made up of epidermal cells that extends down into the dermis to take advantage of its rich blood supply. Nerves reach the hair at the follicle's expanded base, called the *bulb*, where a pinch of connective tissue called the *papilla* contains capillaries that provide nutrients to the growing hair. Epithelial cells in the bulb divide to produce the hair's *root*. Like cells in the epidermis, epithelial cells die as they're pushed away from the source of nutrients. By the time the hair reaches the scalp's surface, now referred to as the *shaft*, the cells have long been dead. The shape of a hair's cross section can vary from round to oval or even flat, influenced by the follicle's shape. Oval hairs appear to be wavy or curly, flat hairs appear to be kinky, and round hairs grow out straight. Each scalp hair grows for 2 to 3 years at a rate of about ⅓ to ½ millimeter (0.01-0.02 inch) per day, or 10 to 18 centimeters (4-7 inches) per year. When mature, the hair rests for three or four months before slowly losing its attachment. Eventually, it falls out and is replaced by a new hair.

Hair pigment (which is mostly melanin, just as in the skin) is produced by melanocytes in the follicle and transferred to the cells. Three types of melanin — black, brown, and yellow — combine in different quantities for each person to produce different hair colors ranging from light blonde to black. Gray and white hairs grow in when melanin levels decrease and air pockets form where the pigment used to be.

Nailing nails

Your finger and toenails are also made of keratin, though much more tightly packed, making them hard. The nails function as an aid to grasping, as a tool for manipulating small objects, and as protection again trauma to the ends of fingers and toes.

Your fingernails and toenails lie on a *nail bed* (not to be confused with a bed of nails). At the back of the nail bed is the *nail root*. Just like skin and hair, nails start growing near the blood supply that lies under the nail bed, and the cells move outward at the rate of about 1 millimeter (0.04 inch) per week. As they move out over the nail bed, they become keratinized. (See Figure 5-3.)

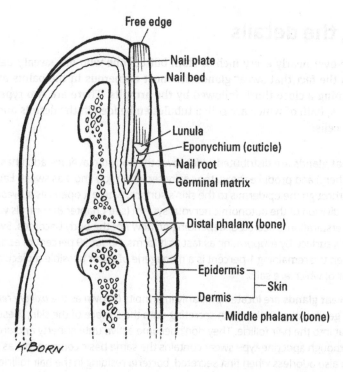

Free edge

Nail plate

Nail bed

Lunula

Eponychium (cuticle)

Nail root

Germinal matrix

Distal phalanx (bone)

Epidermis

Skin

Dermis

Middle phalanx (bone)

FIGURE 5-3: A nail bed.

K. BORN

Illustration by Kathryn Born, MA

Since the area of growth, the *germinal center,* is so close to the surface, a protective bit of skin folds back over the top, forming the *cuticle,* or *eponychium.* Above that structure is a white, half-moon-shaped area called the *lunula.* The lunula is white because it's the area of cell growth. In the nail body, the nail appears to be pink because the blood vessels lie underneath the nail bed. The area of growth contains many more cells, however, so this layer is thicker, which means that you see white instead of pink.

The oil slick

The dermis is also home to our *sebaceous glands,* which are most often associated with a hair follicle. They secrete a fatty substance called *sebum* (which you know as oil) in response to hormones. Because the sebaceous gland is attached to the follicle, when the *arrector pili* contract just the tiniest bit, the pressure put on the gland can cause it to release its oil. The sebum is made up of cholesterol, fats, and other oily substances, and though it's annoying, it does have important functions. It softens the hairs and skin surface, making them both more pliable; it also serves to reduce friction and provide waterproofing.

REMEMBER

Ever wonder why it seems like your hair gets dirtier faster than anything else on your body? This is the result of the sebaceous gland ducts leading into a hair follicle instead of your skin's surface. It's also why excess oils cause acne; a follicle becomes clogged with oil and gives the bacteria there an infinite buffet. And those little *arrector pili?* Well, those tiny strips of smooth muscle can contract enough to straighten the follicle, which in turn straightens the hair shaft, elevating the skin in a pattern called *goose bumps.* In other mammals, this response traps heat to keep them warm. We, however, do not have enough hair to enjoy that benefit, but we certainly still get goose bumps!

Sweating the details

Humans perspire over nearly every inch of skin, but anyone who has sweaty palms or smelly feet can attest to the fact that sweat glands are most numerous in the palms and soles, with the forehead running a close third, followed by the armpits. There are two types of sweat, or *sudoriferous,* glands, both of which are coiled tubules embedded in the dermis and composed of simple columnar cells:

>> **Eccrine sweat glands** are distributed widely over the body (an average adult has roughly 3 million of them) and produce the watery, salty secretion you know as sweat. Each gland's duct passes through the epidermis to the skin's surface, where it opens as a *sweat pore.* The sympathetic division of the autonomic nervous system (see Chapter 8) controls when and how much perspiration is secreted, depending on how hot the body becomes. Sweat helps cool the skin's surface by evaporating as fast as it forms. About 99 percent of eccrine-type sweat is water; the remaining 1 percent is a menagerie of other possible molecules, the most familiar of which are salts.

>> **Apocrine sweat glands** are located primarily in armpits (known as the *axillary region)* and the external genital area. Rather than secreting onto the surface of the skin, these glands release sweat into the hair follicle. They don't become active until puberty and are triggered by stress. Although apocrine-type sweat contains the same basic components as eccrine sweat and is also odorless when first secreted, bacteria residing in the hair follicle quickly begin to break down its additional fatty acids and proteins, which explains the lovely post-exercise underarm stench. In addition to exercise, sexual and other emotional stimuli can cause contraction of cells around these glands, releasing sweat.

Grateful for glands

There are many more types of glands embedded in the dermis all over the body, releasing their specialized secretions based on their location. Two of them are worth mentioning:

>> **Mammary glands:** Breast tissue is formed from extended dermis and hypodermis. Fat is deposited here in females to serve as raw material for milk production. The mammary glands that produce the milk are very similar to the apocrine sweat glands.

>> **Ceruminous glands:** The proper term for earwax is *cerumen,* which is the combination of the secretion of these glands mixing with sebum, forming the waxy substance that serves to trap particles before they reach the eardrum.

Think you've got a grip on the skin's accessories? Let's find out.

PRACTICE

13 The cause of graying hair is

(a) production of melanin in the shaft of the hair.

(b) production of carotene in the shaft of the hair.

(c) decrease in blood supply to the hair.

(d) lack of melanin in the shaft of the hair.

(e) parenthood.

14 Trimming back the cuticles has become part of many people's nail-care routine. Explain how this practice can lead to problems (because cuticles actually serve an important function).

_____

_____

_____

15 The nails are modifications of which epidermal layers?

(a) Corneum and lucidum

(b) Lucidum and granulosum

(c) Granulosum and spinosum

(d) Spinosum and basale

(e) Lucidum and spinosum

16 What does the bulb at the base of a hair follicle do?

(a) Prevents dirt and debris from becoming embedded in the skin

(b) Establishes additional thermal protection

(c) Provides nutrients to the growing hair

(d) Regulates sweat production

(e) Injects melanin into the hair

17 When the arrector pili contract, which gland could incidentally release its contents?

(a) Apocrine sweat gland

(b) Eccrine sweat gland

(c) Ceruminous gland

(d) Sebaceous gland

(e) Endocrine gland

18 What's the difference between apocrine and eccrine sweat glands?

_____

_____

_____

Your Sensational Skin

Have you ever wondered how you feel? Not emotionally, but how you actually *feel?* How does your body know when it's cold or hot? How do you know when you get a cut or splinter? How can you tell the difference between being tickled with a feather and punched with a fist? The answer lies in the dermis. This layer contains numerous nerve endings that serve as specialized receptors for touch, pressure, hot and cold, pain, and more. These nerve endings are sprinkled throughout the dermis all over your body and are connected to the nerves that run through the dermis and hypodermis on their way to your central nervous system.

TECHNICAL STUFF

Nerves are the roads that travel to and from the central nervous system (the brain and spinal cord). *Neurons* are the individual cells for nervous communication. Neurons have two key parts: the *dendrites*, which they use to receive stimulus, and the *axons*, which they use to send information. Receptors are usually bundles of connective tissue situated around the end of a dendrite or a dendrite arranged in a specific way. I cover how receptors work and their role in nervous communication in Chapter 8.

Receptors are designed to respond to a specific stimulus or change. There are no *temperature* receptors, rather there are *hot thermoreceptors* and *cold thermoreceptors*. Hot receptors communicate an increase in temperature, while the cold receptors signal a decrease. You could set a cold thermoreceptor on fire, and it won't send a message to the brain. The combination of input from all your related receptors is what gives you sensations. Touch is no exception.

Numerous receptors are involved in creating the sensation of touch, including the following:

>> **Meissner's (tactile) corpuscles:** These light-touch mechanoreceptors lie within the dermal papillae (see the "Skin" color plate). They're small, egg-shaped capsules of connective tissue surrounding the spiraled end of a dendrite. They are most abundant in sensitive skin areas such as the lips and fingertips.

>> **Pacinian (lamellated) corpuscles:** These deep-pressure mechanoreceptors are dendrites surrounded by concentric layers of connective tissue. Located deep within the dermis and hypodermis, they respond to deep or firm pressure and vibrations.

>> **Ruffini's corpuscles:** These capsules of collagen surround the ends of dendrites. Located throughout the reticular layer of the dermis, they provide information about stretch.

>> **Hair plexus:** These nerve endings wrap around the hair follicle and function as mechano-receptors. When the hair is moved, the movement stimulates the nerve endings, creating a very light touch sensation.

Practice Questions Answers and Explanations

(1) **Stratum corneum**

(2) **Stratum lucidum**

(3) **Stratum granulosum**

(4) **Stratum spinosum**

(5) **Stratum basale**

(6) **Melanocyte**

(7) **Basement membrane**

(8) **Keratinocyte**

(9) **All the answers are functions except (a) Filter out microbes.** Papillae are busy little finger-like projections, but they're not filters. That function is carried out by the basement membrane of the epidermis.

(10) **(b) Increase the friction of the epidermal surface.** These ridges are Mother Nature's way of helping you cling to tree branches or grab food. Although answer (a) seems to be reasonable, it's a perk, not a function.

(11) **(b) They connect the cells of the layer** and **(e) They build keratin proteins.** Since they are located in the epidermis itself, they can't be the ones carrying nutrients there. Cells are replaced by cell division along the basement membrane, which is also where nutrients enter.

(12) Your answer should include these points:

- Along the basement membrane, cells in the stratum basale undergo mitosis. As a result, older cells are pushed upward.

- Cells are pushed into the stratum spinosum, where they spend most of their time creating proteins and fibers for desmosomes, which hold all the cells tightly together. They also take up melanin produced by melanocytes in the basal layer.

- Because nutrients must pass through the basement membrane, cells receive less and less access to them as they're pushed up. When they reach the stratum granulosum, they're usually dead. The keratohyalin contained in the cells is the precursor of keratin.

- By the time the cells reach the stratum corneum, they're unrecognizable — completely flattened and made mostly of keratin.

  The hair and nails are formed by this same general process, so you have to memorize it only once!

TIP

(13) **(d) lack of melanin in the shaft of the hair.** Despite the medical cause, people often suspect that answer (e) (parenthood) has a lot to do with graying hair.

14. The cuticle folds over the top of the germinal center, where cells are actively dividing. This means they're right on top of the basement membrane, and interstitial fluid is just under that. Removing the cuticles increases the chance that something could get down there, which could lead to infection.

15. (a) Corneum and lucidum. These are the two upper layers.

16. (c) Provides nutrients to the growing hair. The bulb is where connective tissue and capillaries come together to provide those nutrients.

17. (d) Sebaceous gland. None of the other glands is that closely associated with a hair follicle. An apocrine sweat gland does secrete into the follicle, but only the duct is near the follicle.

18. Although both types of glands secrete sweat that is mostly water, their triggers are different. Eccrine sweat is purely for thermoregulation. You get hot, you sweat, and that's it. Apocrine sweat is triggered by stress, like when you're scared or anxious. Exercise is stress too, as is being hot. Also, eccrine ducts release directly onto the skin, whereas apocrine ducts release into a hair follicle. This is why apocrine sweat is the smellier variety because the bacteria get a big meal every time and release gasses as a result.

If you're ready to test your skills a bit more, take the following chapter quiz, which incorporates all the chapter topics.

Whaddya Know? Chapter 5 Quiz

It's quiz time! Complete each question to test your knowledge of the various topics covered in this chapter. You can find the solutions and explanations in the next section.

1. Why is keratin important to the skin?

   (a) It makes the stratum corneum thick, tough, and water-repellant.

   (b) It keeps the stratum lucidum moisturized.

   (c) It helps nourish the stratum granulosum.

   (d) It maintains connections between the cells of the stratum spinosum.

   (e) It distributes sweat evenly through the epidermal layer.

2. The _____ glands form perspiration.

   (a) sebaceous

   (b) ceruminous

   (c) dendritic

   (d) Langerhans

   (e) sudoriferous

3. Sebaceous glands

   (a) produce a watery solution called sweat.

   (b) produce an oily mixture of cholesterol, fats, and other substances.

   (c) produce a waxy secretion called cerumen.

   (d) accelerate aging.

   (e) are associated with endocrine glands.

4. Differentiate among the four receptors involved in the sensation of touch.

   _____

   _____

   _____

   _____

5   What is the name of a layer of dense irregular connective tissue containing interlacing bundles of collagenous and elastic fibers?

(a) Basale layer of the epidermis

(b) Reticular layer of the dermis

(c) Outer layer of the hypodermis

(d) Papillary layer of the dermis

(e) Inner layer of the hypodermis

6   What do ceruminous glands secrete?

(a) Oil

(b) Wax

(c) Tears

(d) Sweat

(e) Milk

For questions 7–11, answer true or false. When the statement is false, identify the error.

7   People with darker skin tones have more melanin-producing cells.

_____

8   Cells in the stratum spinosum can be stimulated to make vitamin D.

_____

9   The root of a hair is the portion with cells that are still alive.

_____

10   Arrector pili contraction give us goose bumps by pushing the hair up from the bottom.

_____

11   The stratum lucidum is found only in thick or callused skin.

_____

12   What is another name for the cuticle?

(a) Lunula

(b) Hyponychium

(c) Eponychium

(d) Nail matrix

(e) Perionychium

13 These glands secrete water with fatty acids and proteins in response to stress. The secretion acquires an unpleasant odor when bacteria break down the organic molecules.

(a) Apocrine sweat glands

(b) Sebaceous glands

(c) Ceruminous glands

(d) Eccrine sweat glands

(e) Mammary glands

14 When a new cell is created in the epidermis, it goes through the following layers before it's eventually brushed off.

_____

_____

_____

15 Meissner's corpuscles play a role in which function?

(a) The sensation of heavy pressure

(b) The skin's ability to rebound into shape after pressure is applied

(c) The immediate withdrawal response from intense heat

(d) The sense of motion

(e) The sensation of soft touch

16 What are the four main functions of the hypodermis?

_____

_____

_____

_____

Answers to Chapter 5 Quiz

1. **(a) It makes the stratum corneum thick, tough, and water-repellant.** Associate the words *corneum* and *keratin*, and you're in great shape.

2. **(e) sudoriferous.** These glands form perspiration. The Latin word *sudor* means *sweat*. It also looks like the word *odor* is in there.

3. **(b) produce an oily mixture of cholesterol, fats, and other substances.** That secretion is called *sebum* — hence the name *sebaceous glands*.

4. **The three corpuscles are Meissner's (light pressure), Pacinian (heavier pressure), and Ruffini's (stretch.) The hair plexus is stimulated when a hair bends, providing the sensation of a very gentle touch.**

5. **(b) Reticular layer of the dermis.** The description in this question sounds like a tough structure, so it may help you to remember that the reticular layer is used to make leather from animal hides.

6. **(b) Wax** (earwax, to be specific)

7. **False.** The number of melanocytes is fairly consistent across all people. It's their production levels, pigment sizes, and pigment colors that differ.

8. **True.** The cells are stimulated by UV light, and we need that vitamin D to absorb calcium from the foods we eat.

9. **False.** That portion is the bulb. The root is simply the portion of the hair that lies below the surface.

10. **False.** The arrector pili cause goose bumps because the follicle is pulled straight up and down, making the hair do the same.

11. **True**

12. **(c) Eponychium.** The prefix *ep* refers to "upon" or "around," whereas the prefix *hypo* refers to "below" or "under." The cuticle is around the base of the nail, so it's the eponychium, not the hyponychium.

13. **(a) Apocrine sweat glands.** These glands are the truly stinky sweat glands.

14. The sequence of layers in the epidermis from the basement membrane outward, which is where cell division occurs, is basale, spinosum, granulosum, lucidum, corneum.

15. **(e) The sensation of soft touch.** Although it's true that several different nerves are involved in the overall sense of touch, Meissner's corpuscles are the most responsive to touch.

16. **Insulation: keep the warmth of your body in there, cushioning: absorb forces from movement and falls so the organs don't, energy storage: hypodermis is adipose, which stores lipids, Chemical signaling: Fat cells (adipocytes) secrete numerous hormones.**

Chapter **6**

More than a Scaffold to Build On: The Skeletal System

When it comes to your internal anatomy, the skeleton is the most familiar. Odds are that at this point in your life, you've bruised or broken a bone or had a sprain or strain. As Halloween approaches, skeletons start popping up everywhere. You're familiar with the structure and probably know the names of a couple of bones. Those spooky, scary skeletons that start populating store shelves in September, however, aren't exactly anatomically correct.

Meet & Greet Your Body's Foundation

Because most of us don't have anatomically accurate skeletons in our closets, we have to make do with diagrams. Flip to the "Major Bones of the Skeleton" color plate in the center of this book. Try to think about where each bone is on your body, connecting the 2D picture to the 3D structure. We will take a tour of the skeletal sections near the end of this chapter, but it's important to start with a context. So take a moment to greet each of the 206 bones and address them by name!

Also included in the skeletal system, but rarely indicated in a skeleton diagram, are all the joints that connect your bones. These joints often include cartilage and may be encased in a capsule of connective tissue. There are also numerous bands of dense regular connective tissue: the *ligaments* that connect two bones and *tendons* that connect a bone to a muscle.

You could likely tell me right now, before reading any further, a couple of functions of the skeletal system; but could you get all five? Yes, I said five! We tend to forget about the last two on this list, and the third one is ripe for a trick question, so don't just glance over the list below because it seems familiar. Besides, you already know there's a quiz at the end!

So here goes. The skeletal system as a whole serves five key functions:

>> **Shape and support:** The bones of the skeleton create the framework onto which the rest of our body is built, giving us our characteristic human shape. The density of the bones creates the rigidity needed to support all that tissue. The leg and foot bones are especially important for bearing the weight of our upright posture, as is the curvature of our backbone.

>> **Protection:** The skeleton encases and shields delicate internal organs that might otherwise be damaged during motion or crushed by the weight of the body itself. The skull bones, for example, create the *cranium* which houses the brain, and the ribs and sternum of the thoracic cage protect the heart and lungs.

>> **Allow for movement:** By providing attachment sites and a scaffold against which muscles can contract, the skeleton makes motion possible. The bones act as levers, the joints are the fulcrums, and the muscles apply the force. When the biceps muscle contracts, for example, the radius and ulna bones of the forearm are lifted toward the humerus bone of the upper arm.

>> **Mineral storage:** Our bodies can use many minerals to create bone, including phosphorus and magnesium, but calcium is by far the most prevalent. Building our structural compounds with these minerals serves a dual purpose, particularly calcium compounds. Besides creating an incredibly strong matrix, calcium ions are essential for both muscle contraction and nervous communication. A mere 35 percent decrease in blood calcium can cause symptoms such as depression, memory loss, and convulsions. Good thing we can stockpile it in our bones!

>> **Blood cell formation:** The creation of new blood cells, called *hematopoiesis,* begins with stem cells in the *bone marrow.* All blood cell types are made here: *erythrocytes* (red blood cells), *leukocytes* (white blood cells), and *thrombocytes* (platelets). Some of these cells, however, will complete *differentiation* in other organs. Stem cells, for example, may mature into *lymphocytes* (a type of white blood cell) in the marrow and then enter blood flow. But at this point, the cells are not yet functional. It's like they're all dressed for the ball but still need to have their hair and makeup done, as well as any number of crucial-to-fashion accessories. Adding functionality (accessorizing the outfit) can occur in other organs. The thymus, for example, adds the final touches to create T cells, a crucial part of our immune system. (B cells, yet another crucial element of our immune system, complete differentiation while still in the marrow.)

Blood cell functions pop up all over this book. To review differentiation, flip back to Chapter 3. Blood as a whole is covered in Chapter 10, and white blood cells are covered in Chapter 12.

WARNING

Condensing information is a great study skill (essential, in my opinion). Consider the question "What are the functions of the skeletal system?" The answer isn't the entire box of information you just read. It's only the bold words. You must be aware, though, of what the terms you choose mean to you. *Movement* is a great example. What does this word mean to you? Does it cue you to talk about how movement is generated? Because the skeletal system most certainly does not cause or create movement in or of the body. It allows for movement through the joints, but that power is generated by the muscles.

Boning Up on Bone Structures

I like to think of a bone as being like a jelly doughnut. Most of the doughnut is the bready part, just as most of the bone is the bone part. Inside is the jelly, goopy and delicious. Without it, there's just a hole. And a jelly doughnut just isn't right without the coating of powdered sugar now, is it? (Clearly, I was hungry while writing this chapter.) Furthermore, who eats just one jelly doughnut anyway? And let's not forget that sharing is caring! So like the skeleton, we need a box full of jelly doughnuts. But if you just toss them in, they'll get squished and stuck together. So the box is packed neatly, with a piece of tissue paper between doughnuts.

Now that you're daydreaming of doughnuts and not bones, let me connect the dots for you. Bones, like all our organs, have multiple layers and different tissues comprising their overall structure. The powdered sugar is the outer protective covering of the bone, called the *periosteum*. When two bones come into contact, called an *articulation*, there is a layer of tissue between them. Then you have bone, the actual *osseous tissue*, surrounding the jelly. The analogy works even better when you think about the *bone marrow*, which is what the jelly represents; it's very much like jelly in appearance.

Classifying bones

The 206(ish) adult bones are made of 35 percent protein, 65 percent minerals, and a small amount of water. The protein, a specialized type of collagen called *ossein*, creates the shape, or framework. The minerals give the bone strength and hardness. The different shapes of the bones relate to their specific functions and are divided into four categories, as shown in Table 6-1. For the most part, the classification matches what the bone looks like.

WARNING

Though they aren't very long at all (some people would even say they're short), the *phalanges* (finger bones) are in fact long bones. Remember that short versus long is about the proportion of length to width, not the overall length of the bone!

What makes the classification of bones a bit tricky are all the other bumps, lines, knobs, and valleys that are on there too. Let's look at the most familiar bone as an example: the femur. (You can see one in Figure 6-1 in the next section.) It's a long bone, as it's clearly much longer than it is wide. But the odd shapes on the ends look awfully irregular. First off, keep in mind that the classification is addressing the overall shape of the bone without the added details. Even flat bones like the shoulder blade have an odd cup shape attached to them, but the rest of the bone is clearly flat. Second, bones have numerous external features, collectively called *processes*. These processes serve a variety of purposes, such as providing a point of attachment for tendons and ligaments and creating surfaces for articulations (joints). And although the terms of these ridges and valleys correspond to the shape, such as a *spine* (pointed projection, like a spike) or a *foramen* (hole, like the one at the bottom of your skull), they are used in conjunction with specific bones. As such, they're beyond the scope of this book.

Table 6-1   Characteristics of Bone Types

Bone Type	Characteristics	Example Location in the Body	Functions
Flat	Thin, flat, slightly curved	Shoulder blades, ribs, sternum, pelvic bones, outer skull bones	Like plates of armor, flat bones protect the soft tissues of the brain and organs in the thorax and pelvis.
Long	Longer than they are wide	Arms and legs	Like steel beams, these weight-bearing bones provide structural support.
Short	Blocklike, length and width about the same	Wrists and ankles	Connect with other short bones to allow for a wider range of movement than larger bones.
Irregular	Complex, no general shape	Vertebral column; facial and internal skull bones	The variety of shapes create surfaces that tendons and ligaments can attach to.

Bone structures

All of our bones, regardless of category, are covered in the fibrous periosteum. Directly underneath that is the osseous tissue, followed by the marrow. Here is where the four categories begin to differ internally. They all contain the same types of cells, the two types of osseous tissue, and bone marrow, but their internal organization varies.

REMEMBER

Keep in mind the difference between *bone tissue* and a specific, named bone. Both the *femur* (thigh bone) and the *scapula* (shoulder blade) contain bone tissue, but each bone has its own specialized configuration of the components of bone tissue.

Parts of a long bone

Long bones such as your thigh bone (femur) or forearm bone (radius) are the type of bones people usually think of first. And in fact, they make a good illustration of the general anatomy and physiology of bone tissue. (See Figure 6-1.)

The *periosteum* is made mostly of a fibrous connective tissue (often dense irregular connective tissue) and encases the entire bone. In addition to forming a protective barrier, it assists in routing blood vessels deep into the bone. The periosteum also anchors the cord-like tendons and ligaments. The collagen fibers in each of these intertwine, creating a strong connection.

This makes the periosteum effectively continuous with the ligaments and tendons because there's no real separation between the "sheet" and the "cords." This prevents them from letting go of the bone; in fact, you're more likely to tear them than to pull them off the bone.

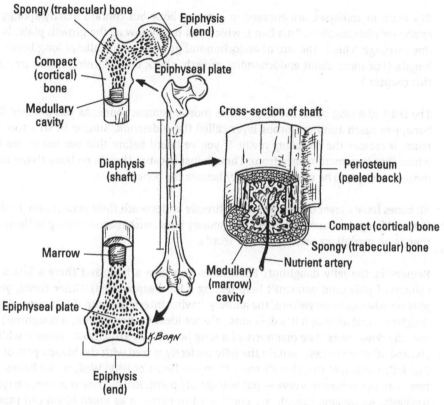

FIGURE 6-1:
Long bone structure.

Illustration by Kathryn Born, MA

REMEMBER

Dense connective tissues (regular and irregular) have very little direct blood flow, which is why they take so long to heal.

The ends of a long bone, the *epiphyses* (singular is *epiphysis*), have a pad of *hyaline cartilage* to reduce friction where it articulates (joins) with another bone to allow for movement. The protein fibers in the matrix of hyaline cartilage are so fine that the tissue has a smooth, glassy appearance. This makes it particularly well suited for forming joints where bones will regularly be moving past each other.

TECHNICAL STUFF

Another type of cartilage, *fibrocartilage*, is found between other bone types, such as between the vertebrae or the ribs and sternum. Its matrix has bulkier protein fibers, making it an effective shock absorber. *Elastic cartilage*, the third type, forms flexible structures (like your outer ears) but is not found in conjunction with bones.

The two types of osseous tissue are *spongy* (*trabecular* is the fancy term) and *compact* (or *cortical*, if you want to be equally fancy). The difference between the two lies in the matrix, whether or not it is packed completely full of mineral compounds. (For ease of understanding, I'm going to say just *calcium compounds* from now on since calcium is the major component.) Compact bone has a matrix with virtually no empty space; it's packed to the gills with calcium compounds. Spongy bone has pockets of empty space (hence the name *spongy*, because that's exactly what it looks like!). The *histology* — the fancy word for the microscopic structure of a tissue — of bone tissue is covered in the following section.

The ends, or *epiphyses*, are encased in compact bone but contain mostly spongy bone and the *epiphyseal plate/disc/line*. This band, which you may know as the growth plate, is made of hyaline cartilage. This is the site of endochondral growth, which allows long bones to increase in length. (For more about endochondral growth, check out "Turning Bones into Bone" later in this chapter.)

The shaft of a long bone, the *diaphysis*, is mostly compact bone. As you proceed deeper into the bone, you reach another fibrous layer called the *endosteum*. Similar in structure to the periosteum, it creates the *medullary cavity*. If you've heard before that our bones are hollow, this is where that statement comes from. Though just because there's no bone tissue in there doesn't mean it's empty. The medullary cavity houses the *bone marrow*.

REMEMBER

All bones have a layer of compact bone directly underneath their periosteum. From there, short, flat, and irregular bones are filled with spongy bone, with marrow filling in those empty spaces; there's no medullary cavity, in other words.

Remember the jelly doughnut? Sometimes you take a bite, and there's just a narrow, little column of jelly (and you can't help feeling a bit disappointed). Other times, you bite in, and jelly squishes out everywhere, the amount having interfered with the structural integrity of the doughnut (and although it's delicious, it's not ideal for being, well, a doughnut). All our bones are jelly doughnuts. The diaphysis of a long bone has a structured column; while the epiphysis, and all other bones, contain the jelly perfectly mixed with the bready part of the doughnut. And let's not forget that there's more than one flavor of jelly! Well, in our bones, there are only two — red or yellow marrow — but you get my point. *Red marrow* is active, carrying out hematopoiesis. As we age, though, we don't need to perform as much blood cell production. Thus, red marrow gets deactivated by being signaled to store lipids (fat), which is why the inactive form is called *yellow marrow*.

TECHNICAL STUFF

When we're young, we need to replace old and broken blood cells as well as create new ones because we're making more space within our ever-growing bodies. Nearly all the marrow in infants and young children is red. In adults, it's mostly yellow; red marrow remains only in the ribs, sternum, and pelvis (plus a few other flat bones). But if blood cell levels drop too low — after blood loss, for example — the yellow marrow can be reactivated and again perform hematopoiesis.

PRACTICE

Let's see how well you're boning up on the skeletal structures.

 1   Which of the following are functions of the skeleton? Choose all that apply.

    (a) Support of soft tissue

    (b) Hemostasis

    (c) Production of red blood cells

    (d) Movement

    (e) Mineral storage

2 Which type of bones form the weight-bearing part of the skeleton?

    (a) Flat bones

    (b) Irregular bones

    (c) Long bones

    (d) Short bones

For questions 3–8, label Figure 6-2.

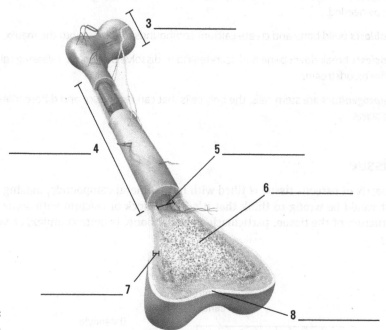

FIGURE 6-2:
The long bone.

9 What is the function of the endosteum?

    (a) It ensures circulation into even the hardest bone structures.

    (b) It allows bones to connect with ligaments and tendons.

    (c) It creates a hollowed-out space inside the bone to house marrow.

    (d) It controls calcification over time.

    (e) It allows flexibility in infant bone structures that is needed for proper development.

Histology

When we start looking at bone *tissue* histologically (*histology* being the microscopic study of tissues and cells), my jelly-doughnut analogy falls apart. The bread part of the doughnut does represent the osseous tissue; the problem is that there are two types. Imagine having a layer of cake doughnut underneath the powdered sugar, but before the standard yeast doughnut with the jelly. And although that does sound delicious, it's getting far too complicated to be useful.

Though it seems more like a rock than living cells, osseous tissue is physiologically very active. It's constantly generating and repairing itself and has a generous blood supply all through it.

Bone cells

Bone contains four specialized types of cells: osteocytes, osteoblasts, osteoclasts, and osteoprogenitor (or osteogenic) cells. The skeletal system's functions depend on the work of these specialized cells within the bone tissue:

>> *Osteocytes* manage the *matrix* (the space between the cells), signal osteoblasts, and osteoclasts as needed.

>> *Osteoblasts* build bone and create calcium compounds to deposit into the matrix.

>> *Osteoclasts* break down bone and secrete acid to dissolve the matrix, releasing minerals into the bloodstream.

>> *Osteoprogenitors* are stem cells, the only cells that can reproduce and differentiate into osteoblasts.

Bone tissue

Yes, the matrix of osseous tissue is filled with hard mineral compounds, making it very rock-like, but it would be wrong to think that it's just a block of calcium with some cells thrown in. The structure of the tissue, particularly compact bone, is quite complex, as you can see in Figure 6-3.

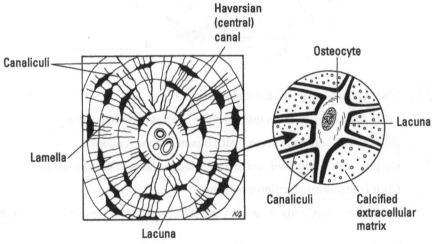

FIGURE 6-3:
Compact
osseous tissue
structure.

Illustration by Kathryn Born, MA

The first things you notice when looking at osseous tissue are the big holes. Then you see that everything seems to be arranged in circles around those holes. One set of these circles is called an *osteon*; these structures are repeated and glued together to form compact bone. The hole in the middle of each osteon, the *Haversian* (or central) *canal*, creates a path for nerves and blood vessels to run throughout all areas of the bone.

Each ring encircling the Haversian canal is called a *lamella*. These rings are formed as calcium compounds (such as calcium phosphate and calcium carbonate) are deposited into the matrix, one layer at a time. The seed-shaped spots along the lamellae are little caves in the matrix called *lacunae*. Each lacuna houses a single osteocyte.

Because osteocytes are immobile (not that they could go very far through the solid matrix), we have a resource problem. Like all other cells, they need access to the resources delivered by blood flow. That's where all those squiggly lines come in. Spreading through the lamellae are tiny tunnels called *canaliculi*, which connect the lacunae. Osteocytes are able to form gap junctions (see Chapter 4) through the canaliculi, allowing resources to be passed cell to cell to those far away from the Haversian canal (much like a game of Telephone). The Haversian canals of each osteon are connected to each other by *Volkmann's canals*, carrying the vessels deeper into the bone.

Spongy bone is much less organized (which means you have fewer things to label!). Rather than the matrix being built into rings, creating numerous osteons, the calcium compounds form bony, bridgelike structures called *trabeculae*. While they do still have canaliculi (which in spongy bone are just tiny holes) and lacunae (which are far roomier), there are no lamellae nor canals.

Are you making sense of the microscopic structures? How about a quick check?

PRACTICE

10 What is the role of progenitor cells?

(a) They undergo mitosis to lay down new matrix to create bone.

(b) They destroy old bone cells to make room for new, healthy ones.

(c) They spread throughout the bone and maintain the matrix.

(d) They build and break calcium compounds for cells to use.

(e) They differentiate into bone-building cells.

11 Compare and contrast Haversian and Volkmann's canals.

_____

_____

_____

For questions 12–17, provide the name of the structure described, followed by the corresponding letter from Figure 6-4. (Not all the letters will be used.)

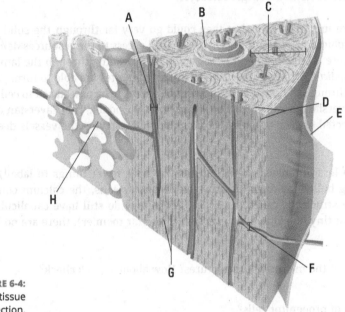

FIGURE 6-4:
Bone tissue cross section.

12 Carries blood vessels deeper into the bone: _____ letter: _____

13 Area where osteocytes form gap junctions: _____ letter: _____

14 Structure created by a single osteoblast depositing calcium compounds:
_____ letter: _____

15 Group of cells receiving nutrients from the same blood vessels:
_____ letter: _____

16 Organizes nerves and vessels before routing them into the matrix:
_____ letter: _____

17 Bony plates formed when fibers become calcified:
_____ letter: _____

Turning Bones into Bone

At birth, bones are soft and pliable because they're composed mostly of cartilage. The softer cartilage allows the fetus to bend into poses that would make a yoga instructor beam with pride. The shape of the bone is determined by the shape of the cartilage, so it serves as a template.

As the body grows, the cartilage is gradually replaced by hard bone tissue. In fact, infants start life with about 300 distinct bones, which will fuse into 206 by adulthood.

TECHNICAL STUFF

During fetal development, the flat bones form from fibrous connective tissue rather than cartilage. They are built layers at a time, forming around the developing blood vessels. Most other bones use a template made from hyaline cartilage. Either way, at birth, bones are far from done forming, having an increasing amount of connective tissue to turn into bone (until we're done growing as teenagers).

Observing ossification

Ossification is the process of turning connective tissue into bones. That's the first thing to wrap your head around to understand the physiology of bone growth: we don't just make bone, we turn other tissue into bone. We need the space to already exist, and that space needs to have a scaffold to build on. This is why cartilage and fibrous connective tissues are used; they contain numerous collagen fibers to serve as the skeleton for our bones. (Pun totally intended!) Regardless of the origin, the ossification process begins with *osteoid* — the area of tissue that is effectively fibers and space. *Osteoblasts*, the bone-building cells, will be sent to the area to begin *calcifying* the osteoid.

TIP

Think of it this way: Osseous tissue, the actual bone tissue, is comprised of the bone cells and their matrix (the space between the cells that is still part of the tissue). The matrix of all connective tissues contains fibers, which are just long strings of proteins. This bundle of tissues is the osteoid — the location where the osteoblasts will pile up the calcium compounds (*calcify*, in other words), thus creating the hardened matrix that we call bone.

Osteoblasts take in calcium ions from their surroundings (which would be the interstitial fluid, but it's more like gelatin than water here) and build them into compounds such as *hydroxyapatite*. This specific variation of calcium phosphate is the most common component of the bone matrix. The cells release the mineral compounds into the space, where they will attach to the fibers. Eventually, the fibers become completely encased, and voilà: spongy bone!

TIP

I find it helpful to picture the areolar (loose) connective tissue and all of its messy, crisscrossing fibers. Ignore anything else present in the tissue (cells, blood vessels, and so on), and focus on the lines (the fibers). Are you picturing the mess of blurry pink lines from the histology unit? (See Chapter 4.) Now imagine coating them in white paint — layers and layers of thick, white paint that hardens when it dries. After a while, you won't see the fibers anymore, just these bridges, ropes, and plates hardened and rigid. Those are the trabeculae of spongy bone.

Eventually, the osteoblasts are signaled to stop, and they begin their life in retirement as osteocytes. They maintain the matrix around them in their spacious lacunae as they live out their days. If spongy bone is the osseous tissue that's desired in that area, that's where the story ends. Ossification is complete. If compact bone is the needed bone type, however, there's another chapter to the story. Unfortunately, the osteocytes in that area won't be around to see the ending.

When compact bone is built, remember that spongy bone is built first. Those osteoblasts will retire and differentiate into osteocytes, which is a one-way road. Osteocytes cannot be reactivated to begin building bone again. New osteoblasts are sent to the area around a blood vessel to begin the process anew: absorb calcium, build compounds, deposit onto trabeculae (because the fibers are already covered, but they're still in there). These new osteoblasts continue their work until they've walled themselves off, creating a lacuna that isn't much bigger than themselves. This process creates the first ring, or lamella, of an osteon and thus creates the Haversian canal (Refresh your memory of what this looks like by flipping back to Figure 6-2.) Another round of osteoblasts is sent in to fill in the space around the newly built ring. They grab on to the "arms" of the newly retired osteocyte and do the same thing: wall themselves off into a lacuna. They get their needed resources from the osteocyte that is closer to the blood vessel enclosed in the Haversian canal. This process creates the next lamella as well as the little tunnels (the canaliculi) that connect the cells.

REMEMBER

The number of rings that can be built around a single blood vessel is limited because cells will only pass on the resources they aren't using. Eventually, not enough molecules like glucose will reach a cell to build another ring. Fortunately, blood vessels are branching throughout the bone, and osteoblasts will build rings around each of them as well. As a result you see many sets of rings encircling a hole — the lamellae surrounding each Haversian canal, and each set of rings is an osteon.

Getting a grip on growth

I talk about how bone tissue is built in the previous section, so now it's time to look at how bones, as organs, get bigger. There are two ways the body accomplishes this goal, both of which lead to the ossification process. The difference lies in how the new space is created. Most of our bones — the short, flat, and irregular ones — need to grow proportionally. That is, as the body as a whole grows, the bones keep the same shapes; they just get bigger. The body accomplishes this task by performing *intramembranous growth*. Long bones, on the other hand, cannot grow proportionally. If our leg bones grew in width as much as they did in length, our legs would be far too heavy to move, which basically defeats their purpose. Thus, long bones use *endochondral growth* to grow in length. (For more on intramembranous versus endochondral growth, see the next sections.)

REMEMBER

Long bones do still use intramembranous growth as well. If they grew only in length, not width, we wouldn't even be able to stand, let alone walk. Our leg bones would be too thin to support our weight.

Intramembranous growth

Intramembranous growth occurs in all bones; it's how short, flat, and irregular bones grow, and it's how long bones increase their width. A sheet of connective tissue is formed beneath the periosteum, all the way around the bone, by sending fibroblasts to the area, which start building protein fibers. This effectively pulls the periosteum off of the osseous tissue, creating the needed new space full of fibers. Then osteoblasts come in and complete ossification. The result is a proportionally bigger bone.

Endochondral growth

To increase their length, long bones use *endochondral ossification*. If you flip back to Figure 6-1, you'll see an *epiphyseal plate* (line) on each end. Rather than being osseous tissue, the epiphyseal line (or plate) is made of hyaline cartilage. When stimulated by growth hormones (see

Chapter 9), the *chondrocytes* (cartilage cells) begin to copy themselves. The new cells begin to enlarge, pushing the end of the bone farther away from the shaft. The cells become so large that they eventually burst. *Phagocytes*, the body's clean-up cells (picture little cells running around with vacuum cleaners, sucking up all the broken bits), come in and clean up the debris, leaving only the fibers, ready for osteoblasts to begin depositing calcium compounds.

REMEMBER

When your long bones increase in length, you increase in height. Eventually, around age 18, your chondrocytes stop dividing. The entire epiphyseal plate is ossified (turned into bone). At this stage, the growth plates are closed.

Is your bone growth knowledge growing? Let's check.

PRACTICE

18   Explain how a network of fibers becomes calcified.

_____

_____

_____

19   Which statement best explains how an osteon is created?

(a) Numerous osteoblasts gather around the blood vessels in the connective tissue and lay down calcium compounds until all the space between them is filled.

(b) Osteoblasts migrate to spongy bone, gather around blood vessels one ring at a time, and deposit matrix until they've walled themselves off.

(c) An osteoblast moves to a blood vessel and builds the matrix outwards in a ring, until it can't access enough resources to continue.

(d) Osteoblasts build calcium compounds and deposit them onto collagen fibers that formed in rings around a blood vessel.

20   Identify the differences between intramembranous and endochondral growth.

_____

_____

_____

Remodel & repair

Though you may be done growing at this point in your life, your bones never really stop developing. The bones' mineral content is constantly renewed, refreshing entirely about every nine months. Through the process called *remodeling*, our bones adjust to the forces we apply on them to remain strong and supportive.

As we apply forces through exercise, or just our daily activities, the bone matrix weakens. (Picture a brand-new brick wall compared with an old, weathered one.) Osteocytes will communicate the need to strengthen the matrix and osteoclasts are signaled to begin the remodeling process. They migrate to the area and secrete acid, dissolving the matrix. The calcium compounds are broken apart, and the calcium ions return to the bloodstream. Conveniently, though, the collagen fibers are left intact. The osteoclasts will then clean up the area via phagocytosis, leaving a clean framework. Osteoblasts move in and deposit new calcium compounds, thereby strengthening the bone.

Think of this process like road construction on a smaller scale. The roads wear out with continued use and need to be refreshed. But you can't do all the roads, or even a single road in its entirety, at the same time; if you did, people wouldn't be able to get anywhere. (Some cities haven't quite figured this part out yet.) So you do the work in small sections at a time, all the time. Luckily, bone remodeling doesn't cause the headaches that road construction does!

Remodeling is also under hormonal control. Because the bones serve as a reservoir for minerals (calcium in particular), the body can trigger osteoblasts or osteoclasts, depending on the need. When calcium in the blood is in abundance, the thyroid gland releases *calcitonin,* triggering the osteoblasts to create calcium compounds and deposit them into the matrix. When blood calcium levels are low, the parathyroid gland releases *parathyroid hormone* (PTH) to signal the osteoclasts to break down the matrix. Calcium ions are released into the bloodstream so normal functioning can continue; calcium ions are particularly important for muscle contraction as well as nervous communication.

An important distinction to make is that the remodeling process is not the same as healing; it's a response to normal wear and tear. When a bone is broken, or fractured, blood vessels break along with the matrix. As a result, the first step in the healing process is the formation of a *hematoma,* or blood clot. Osteoclasts come to clear up the debris, leaving behind the fibers. (The clotting process involves the creation of several fibers, not just platelets.) From there, a *callus* — or thickened layer of cartilage — forms, and the area slowly becomes ossified again.

Technically speaking, your body doesn't require medical intervention to heal a broken bone. If you want it to heal properly, however, you might need to get the bone realigned. Immobilizing it is probably a good idea as well, because it keeps the bone from falling back out of line or breaking further.

How about a couple more questions before we switch gears and start learning which bones are which?

21   Endochondral growth is triggered by which hormone?

    (a) PTH

    (b) Calcitonin

    (c) GH

    (d) Ossein

22   How is it that calcitonin and parathyroid hormone are said to oppose each other?

_____

_____

The Axial Skeleton

Just as the Earth rotates around its axis, the axial skeleton lies along the midline, or center, of the body. Think of your vertebral column and the bones that connect directly to it: the rib (thoracic) cage and the skull. The bones of the axial skeleton support the head and trunk of the body and serve as an anchor for the pelvis.

It's good to be hardheaded

Rather than being one big bone, like a huge cap that fits over the brain, the skull consists of multiple cranial and facial bones. Of the 80 named bones in the axial skeleton, 29 are in (or very near) the skull. In addition to the hyoid bone, 8 bones form the cranium to house and protect the brain, 14 form the face, and 6 bones make it possible for you to hear.

TECHNICAL STUFF

The tiny *hyoid* bone, which lies just above your larynx, or voice box, is part of the axial skeleton but is not connected to your skull or backbone. In fact, it's not directly connected to any other bone at all, making it unique from all of your other bones. The hyoid bone plays a key role in swallowing as it is the anchoring point for the tongue.

Cracking the cranium

The eight bones of your cranium protect your brain and have immovable joints between them called *sutures*. (These joints look a lot like the sutures, or stitches, that you may receive to close a wound.) The bones of the cranium that are joined by sutures include the following:

>> **Frontal bone:** Gives shape to the forehead and part of the eye sockets.

>> **Parietal bones:** Form the roof and sides of the cranium.

>> **Occipital bone:** Forms the back of the skull and the base of the cranium. The *foramen magnum*, an opening in the occipital bone, allows the spinal cord to pass into the skull and join the brain.

>> **Temporal bones:** Form the sides of the cranium near the temples. The temporal bone on each side of your head has the following processes (features):

    • *External auditory meatus:* Serves as the opening to your ear canal

    • *Mandibular fossa:* Joins with the mandible (the lower jaw)

- *Mastoid process:* Provides a place for neck muscles to join your head

- *Styloid process:* Serves as an attachment site for muscles of the tongue and larynx (voice box)

>> **Ethmoid bone:** Contains several sections called *plates*. This bone forms the medial (inside) part of the eye sockets and much of the nasal cavity.

>> **Sphenoid bone:** Forms the floor of the cranium and the back and lateral sides of the eye sockets (orbits). This bone is shaped like a butterfly or a saddle, depending on how you look at it. A sunken central portion of the sphenoid bone called the *sella turcica* shelters the pituitary gland. (The importance of this gland is covered in Chapter 9.)

Facing the facial bones

The bones that form facial structures are

>> **Lacrimal bones:** Two tiny bones on the inside walls of the orbits, at the inside corners of your eyes. Tears flow across the eyeball and through that canal into your nasal cavity.

>> **Mandible:** The lower jaw and the only movable bone of the skull.

>> **Maxillae:** Two bones that form the upper jaw, part of the hard palate (roof of your mouth), and bottom of the orbits.

>> **Nasal bones:** Two rectangular bones that form the bridge of your nose. The lower, movable portion of your nose is made of cartilage.

>> **Palatine bones:** Form the back portion of the hard palate and the floor of the nasal cavity.

>> **Vomer bone:** Joins the ethmoid bone to form the nasal septum — that part of your nose that can be deviated by a strong left hook.

>> **Inferior turbinate bones:** Form the inner walls of the nasal cavities.

>> **Zygomatic bones:** Form the cheekbones and the lateral (outer) sides of the orbits.

Sniffing out the sinuses

Encased within the frontal, sphenoid, ethmoid, and maxillary bones of the skull are several air-filled, mucus-lined cavities called *paranasal sinuses* (named for the bones in which they are contained). Although you may think their primary function is to drive you crazy with pressure and infections, the sinuses actually lighten the skull's weight. They make it easier to hold your head up high; they warm and humidify inhaled air; and they act as resonance chambers to prolong and intensify the reverberations of your voice. (Unfortunately, they also drain into the nose and cause trouble when you cry or have a cold.) The temporal bone contains the *mastoid sinuses*, which drain into the middle ear (hence, the earache referred to as *mastoiditis*).

Stand up straight

The spinal column (see Figure 6-5) begins within the skull and extends down to the pelvis. It's made up of 33 bones in all: 24 separate bones called *vertebrae* (singular, *vertebra)*, plus the fused bones of the sacrum and the coccyx. Each vertebra has its core, stacked section called the *body*. Posterior to (behind) that is the *vertebral foramen*, the hollow space that encases and protects the spinal cord. The other side includes numerous processes, including the *spinous process*, which you can see stack down the middle of someone's back. These features allow for attachment to other bones and muscles, as well as create passageways for nerves and blood vessels.

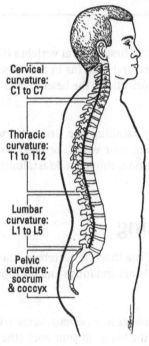

Cervical curvature: C1 to C7

Thoracic curvature: T1 to T12

Lumbar curvature: L1 to L5

Pelvic curvature: socrum & coccyx

FIGURE 6-5: The spinal column, side view.

Illustration by Kathryn Born, MA

Between vertebrae is an *intervertebral disc* made of fibrocartilage for shock absorption. Your spinal column is the central support for the upper body, carrying most of the weight of your head, chest, and arms. Together with the muscles and ligaments of your back, your spinal column enables you to walk upright.

If you look at the spine from the side, you notice that it curves four times: inward, outward, inward, and outward. The curvature of the spine helps it absorb shock and pressure much better than if the spine were straight. A curved spine also affords more balance by better distributing the weight of the skull over the pelvic bones, which is necessary for you to walk upright. A curved spine keeps you from being top-heavy. Each curvature spans a region of the spine: cervical, thoracic, lumbar, and pelvic (or sacral). The number of vertebrae in each region and some important vertebral features are listed in Table 6-2.

Table 6-2   Regions of the Vertebral Column

Region	Number of Vertebrae	Features
Cervical	7 (known officially as C1-C7)	The skull attaches at the top of this region to the first vertebra
Thoracic	12 (known officially as T1-T12)	Each one of these has a rib attached.
Lumbar	5 (known officially as L1-L5)	Commonly referred to as the small of the back, it takes the most stress.
Pelvic	Sacrum (5 bones fused)	The sacrum forms a joint with the hip bones and the last lumbar vertebra.
	Coccyx (4 bones fused)	Also known as the tailbone, the coccyx absorbs the shock of sitting.

The skull is attached to the top of the cervical spine. The first cervical vertebra (C1) is the *atlas*, which supports the head and allows it to move forward and back (as in the "yes" movement). The second cervical vertebra (C2) is called the *axis*, which allows the head to pivot and turn side to side (the "no" movement).

TIP You can differentiate these two important bones by recalling the Greek story about Atlas, who held the world on his shoulders. Your atlas holds your world, or head, on your shoulders. To remember the number of vertebrae in each region, think breakfast, lunch, and dinner: breakfast at 7, lunch at 12, and dinner at 5.

Being caged can be a good thing

REMEMBER The rib cage (also called the *thoracic cage*) consists of the thoracic vertebrae, the ribs, and the sternum. The rib cage is essential for protecting your heart and lungs and for providing a place for the *pectoral girdle* (the shoulders) to attach.

You have 12 pairs of bars in your cage. Some of your ribs are *true* (seven), some are *false* (three), and some are *floating* (two). All ribs are connected to the bones in your back (the thoracic vertebrae). In the front, true ribs are connected to the *sternum* (breastbone) by individual *costal cartilages* (*cost-* means *rib*); false ribs are connected to the sternum by a single, branched costal cartilage. The last two pairs of ribs are called *floating ribs* because they remain unattached in the front. The floating ribs protect abdominal organs without hampering the space in your abdomen for the intestines.

The sternum (breastbone) has three parts: the *manubrium*, the shield-shape top bone; the *body*, the long middle portion; and the *xiphoid* (pronounced *zi-foid*) *process*. The notch that you can feel at the top center of your chest, in line with your collarbones (the *clavicles*), is the top of the manubrium. The middle part of the sternum is the body, and the lower part of the sternum is the xiphoid process.

PRACTICE For questions 23–32, label Figure 6-6.

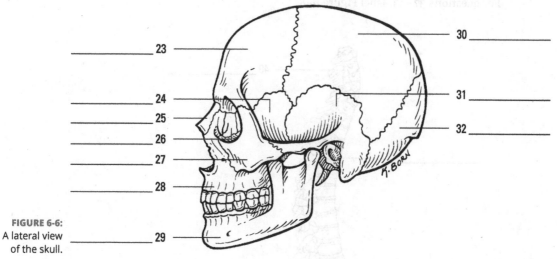

FIGURE 6-6:
A lateral view
of the skull.

23 _____

24 _____

25 _____

26 _____

27 _____

28 _____

29 _____

30 _____

31 _____

32 _____

Illustration by Kathryn Born, MA

Use the terms that follow to identify questions 33–36 in Figure 6-7.

(a) Sphenoid sinus

(b) Frontal sinus

(c) Maxillary sinus

(d) Ethmoid sinus

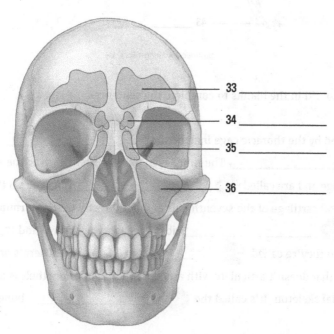

33 _____

34 _____

35 _____

36 _____

FIGURE 6-7:
Sinus view of
the skull.

Illustration by Imagineering Media Services Inc.

For questions 37–43, label Figure 6-8.

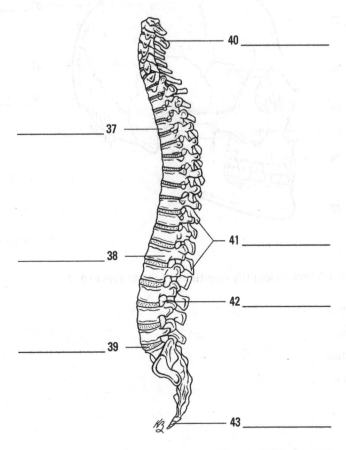

37 _____

38 _____

39 _____

40 _____

41 _____

42 _____

43 _____

FIGURE 6-8:
Vertebral
column.

Illustration by Kathryn Born, MA

For questions 44–49, fill in the blanks to complete the sentences.

The organs protected by the thoracic cage include the ⓐ⒋⒋ _____
and the ⒋⒌ _____. The first seven pairs of ribs attach to the sternum
by the costal cartilage and are called ⒋⒍ _____ ribs. Pairs 8 through
10 attach to the costal cartilage of the seventh pair and not directly to the sternum,
so they're called ⒋⒎ _____ ribs. The last two pairs, 11 and 12, are unattached anteriorly, so they're called ⒋⒏ _____ ribs. There's one bone in
the entire skeleton that doesn't articulate with any other bones but nonetheless is considered
to be part of the axial skeleton. It's called the ⒋⒐ _____ bone.

The Appendicular Skeleton

Whereas the axial skeleton lies along the body's central axis, the appendicular skeleton's 126 bones include those in all four appendages — arms and legs — plus the two primary girdles to which the appendages attach, which are the pectoral (chest) girdle and the pelvic (hip) girdle.

Wearing girdles: Everybody has two

REMEMBER

The word *girdle* is a verb than means *to encircle*. It has nothing to do with that funny undergarment that all proper women wore in the early 20th century (other than that it encircled the waist).

The body contains two girdles: the *pectoral girdle*, which encircles the vertebral column at the top, and the *pelvic girdle*, which encircles the vertebral column at the bottom. The girdles serve to attach the appendicular skeleton to the axial skeleton. Refer to the "Major Bones of the Skeleton" color plate to see the individual parts of each girdle.

The pectoral girdle consists of the two *clavicles* (collarbones) and the two triangle-shaped *scapulae* (shoulder blades). The scapulae provide a broad surface to which arm and chest muscles attach. The clavicles are attached to the sternum's *manubrium*. Significantly, this is the only point of attachment of the pectoral girdle and the axial skeleton, giving the shoulders a wide range of motion.

The pelvic girdle (see Figure 6-9) is formed by the hip bones (called *coxal bones*, which are actually each three fused bones). The hip bones, along with the sacrum and the coccyx, form the pelvis. In addition to providing the attachment point for the lower limbs, the pelvis bears the weight of the body.

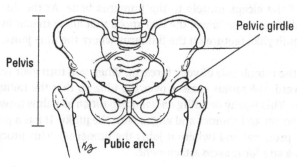

FIGURE 6-9:
The bones of
the pelvis.

Illustration by Kathryn Born, MA

The three bones that fuse into the coxae are the *ilium*, *ischium*, and *pubis*. The ilia are what you probably think of as your hip bones; they're the large, flared parts that you can feel your sides. The part that you can feel at the tip of the ilium is the *iliac crest*. In your lower back, the ilium connects with the vertebral column at the sacrum. The joint that's formed is appropriately called the *sacroiliac joint* — a point of woe for many people with lower back pain.

The *ischium* is the bottom part of your hip. You have an ischium on each side, within each buttock. You're most likely sitting on your *ischial tuberosity* right now. (These bones are what some people call your sitz bones because, you know, they allow you to sit.)

The *pubis* bones join the right and left sides, connected by a piece of fibrocartilage called the *pubic symphysis*. Pelvic floor muscles attach to the pelvic girdle at the pubis. A defining difference between that male and female pelvis is the angle of the *pubic arch*, or the *subpubic angle* (indicated in Figure 6-9). In males, the joining of the two pubic bones forms an acute angle. In women, as with most aspects of the pelvis, the angle of the pubic arch is obtuse.

Going out on a limb

Your arms and legs are *limbs* or *appendages*. The word *append* means to attach something to a larger body. Your appendages are attached to the axial skeleton by the girdles.

Giving you a hand with hands (and arms and elbows)

Your upper limb or arm is connected to your pectoral girdle. The bones of your upper limb include the *humerus* (upper arm); the *radius* and *ulna* (forearm); the *carpals* of the wrist; and the *metacarpal bones* and *phalanges*, which make up the hand.

The head (ball at the top) of the humerus connects to the scapula (shoulder blade) at the *glenoid cavity*. Muscles that move the arm and shoulder attach to the *greater* and *lesser tubercles*, two points near the head. Between the greater and lesser tubercles is the *intertubercular groove*, which holds the tendon of the biceps muscle to the humerus bone. At the distal end of the humerus (the opposite end) are the *capitulum*, the two knobs that allow the radius to articulate (join), and the *trochlea*, a pulleylike feature on the humerus where the ulna joins.

The radius is the bone on the thumb side of your forearm. When you turn your forearm so that your palm is facing backward, the radius crosses over the ulna so that the radius can stay on the thumb side of your arm. This is why defining *anatomical position* as palms forward is important. The ulna is slightly shorter and thinner and aligns with the pinky. It has a pointed process called the *olecranon* on its proximal end (where it joins the humerus). This process forms the point of your elbow and is a site for muscle attachment.

Both the radius and ulna connect with the bones of the wrist. The wrist contains eight short bones called the *carpal bones*. The ligaments binding the carpal bones are very tight, but the numerous bones allow the wrist to flex easily. The eight carpal bones are arranged in two rows. From thumb to pinky, the proximal row (farthest from your fingertips) contains the *scaphoid*, *lunate*, *triquetrum*, and *pisiform*. (Note that *pisiform* and *pinky* both start with *p*.) The distal row (also from thumb to pinky) contains the *trapezium*, *trapezoid*, *capitate*, and *hamate*.

The palm of your hand contains five bones called the *metacarpals*. When you make a fist, you can see the ends of the metacarpals as your knuckles. Your fingers are made up of bones called the *phalanges*. Each finger has three phalanges (*phalanx* is singular): the *proximal phalanx*, which joins your knuckle; the *middle phalanx*; and the *distal phalanx*, which is the bone in your fingertip. The thumb, though, has only two phalanges, so some people like to argue that it's not considered to be a true finger. So you may have eight fingers and two thumbs or ten fingers, depending on how you look at it. Regardless, on each hand, the thumb is referred to as the first digit.

Getting a leg up on your lower limbs

Your lower limb consists of the *femur* (thigh bone), the *tibia* and *fibula* of the leg, the bones of the ankle (the *tarsals)*, and the bones of the foot (*metatarsals* and *phalanges*).

The term *phalanges* refers to the finger bones *and* the toe bones.

The femur is the strongest bone in the body; it's also the longest. The *head* of the femur fits into a hollowed out area of the hip bone called the *acetabulum* (located where all three bones join). The *greater* and *lesser trochanters* on each side of the head are surfaces to which the muscles of the legs and buttocks attach. The *linea aspera* is a ridge along the back of the femur where several muscles attach.

The femur joins the *tibia*, or shinbone, to create the knee. The *patella* (commonly known as the kneecap) sits atop this joint. The femur also has knobs (*lateral* and *medial condyles)* that articulate with the top of the tibia (at its own condyles). The ligaments of the patella attach to the *tibial tuberosity*. The bottom, inner end of the tibia has a bulge called the *medial malleolus,* which forms part of the inner ankle.

The fibula, located on the lateral (outside) part of the lower leg, is thinner but about the same length as the tibia. The bottom, outside end of the fibula is the *lateral malleolus,* which is the bulge on the outside of your ankle.

Your foot is designed in much the same way as your hand. The ankle, or *tarsus*, which is akin to the wrist, consists of seven tarsal bones. But only one of those seven bones — the *talus* — is part of a joint with a great range of motion. The talus bone joins to the tibia and fibula and allows for the movements of your ankle. Below the talus is the largest tarsal bone, the *calcaneus,* which is the heel bone. The calcaneus and talus help support your body weight. The remaining tarsals are the *cuboid* on the outside, the *navicular* on the inside, and the lateral, intermediate, and medial *cuneiforms* just below the navicular.

The instep of your foot is akin to the palm of your hand, and, just as the hand has metacarpals and phalanges, the foot has metatarsals and phalanges. The ends of the metatarsals on the bottom of your foot form the ball of your foot. As such, the metatarsals also help support your body weight. Together, the tarsals and metatarsals (held together by ligaments and tendons) form the arches of your feet.

Your toes are also called phalanges, just like your fingers. And just as your thumbs have only two phalanges, your big toes have only two phalanges. But the rest of your toes have three: proximal, middle, and distal.

PRACTICE

For questions 50–55, use the letters in Figure 6-10 to identify the given bone. Not all the letters will be used.

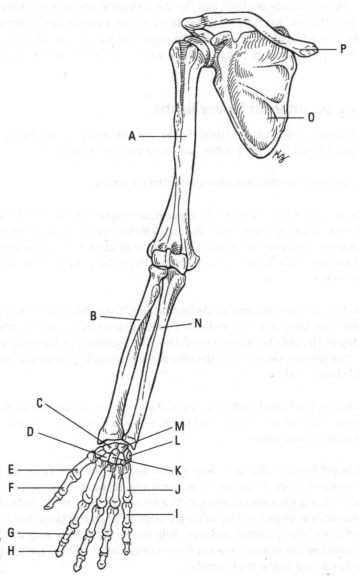

FIGURE 6-10:
The bones of
the upper limb
and hand.

Illustration by Kathryn Born, MA

50 distal phalange _____

51 clavicle _____

52 radius _____

53 pisiform _____

54 first metacarpal _____

55 trapezoid _____

For questions 56–66, label Figure 6-11.

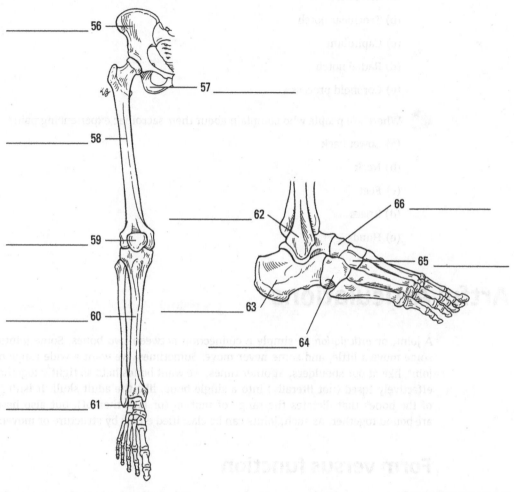

56 _____

57 _____

58 _____

62 _____

66 _____

59 _____

65 _____

63 _____

60 _____

64 _____

61 _____

FIGURE 6-11:
The bones of
the lower limb
and foot.

Illustration by Kathryn Born, MA

 67 Where do the clavicles articulate with the scapulae?

(a) At the acromion process

(b) Below the glenoid cavity

(c) Behind the coracoid process

(d) At the acetabulum

(e) Along the upper spine

68 What is the formal term for the prominence commonly referred to as the elbow?

(a) Olecranon

(b) Trochlear notch

(c) Capitulum

(d) Radial notch

(e) Coronoid process

69 Where are people who complain about their sacroiliac experiencing pain?

(a) Lower back

(b) Neck

(c) Feet

(d) Knees

(e) Hands

Artful Articulations

A joint, or *articulation*, is simply a connection between two bones. Some joints move freely, some move a little, and some never move. Sometimes, we want a wide range of motion in a joint, like at our shoulders. At other times, we want bones held so tightly together that they're effectively fused (not literally) into a single bone, like the adult skull. It isn't just the shape of the bones that dictates the range of motion (or lack thereof), but also how those bones are bound together. As such, joints can be classified either by structure or movement.

Form versus function

Classifying joints by the amount of movement they allow leads to three categories:

>> **Synarthrotic:** Immovable

>> **Amphiarthrotic:** Slightly movable (they have "give" or allow compression)

>> **Diarthrotic:** Highly movable

Although these categories seem pretty straightforward, joints of the same structure can allow different categories of movement. Most often, you see those terms used as descriptors, giving sublevels of the structural classifications, which are as follows:

>> **Fibrous:** These joints are characterized by the fibrous connective tissue that joins the two bones. Most of these are synarthrotic and come in three forms:

● *Sutures* occur when the bones are joined by a piece of dense connective tissue. Sutures can be found between skull bones.

- *Gomphoses* are the bundles of collagen fibers found where the teeth attach to the sockets of the mandible and maxilla (although teeth themselves aren't actually bones).

- *Syndesmosis* is the only amphiarthrotic fibrous joint. Bones are joined by a sheet of connective tissue, like that of the articulations between the tibia and fibula. It allows for stretch and even some twisting.

>> **Cartilaginous:** This type of joint is found in two forms:

- *Synchondrosis* is formed from rigid, hyaline cartilage that allows no movement. The most common example is the epiphyseal plate of the long bone. Other examples are the joints between the ribs, costal cartilage, and sternum.

- *Symphysis* occurs where fibrocartilage pads connect bones. Since the pads are flexible, these joints are amphiarthrotic. (Think about how you can arch your back.) Examples include the intervertebral discs and the pubic symphysis.

>> **Synovial:** All synovial joints are diarthrotic but to varying degrees. They are surrounded by a *joint capsule* of fibrous connective tissue. This capsule isolates the articulation, leaving space that is filled with *synovial fluid* to reduce friction. Each adjoining bone is capped with hyaline cartilage called the *articular cartilage* and stabilized with ligaments. Some joints, such as the knee and wrist, have an additional pad of fibrocartilage called a *meniscus* between the bones for shock absorption. Many synovial joints also contain *bursae,* which are fluid-filled sacs that reduce friction as tendons rub past the bones during movement. Inflammation of the bursae, *bursitis,* results from repetitive use of a joint and is a common culprit for nagging joint pain.

There are six classifications of synovial joints:

>> **Gliding (or planar):** Curved or flat surfaces slide against one another, such as between the carpal bones in the wrist or between the tarsal bones in the ankle. Allows movement in two planes and rotation is possible but limited by their small size and ligament connections.

>> **Hinge:** A convex surface joint with a concave surface, allowing right-angle motions in one plane, such as elbows, knees, and joints between the phalanges.

>> **Pivot:** One bone rotates (pivots) around a stationary bone, such as the atlas rotating around the axis at the top of the vertebral column and the radioulnar joint. Rotation is the only type of movement allowed.

>> **Condyloid:** The oval head of one bone fits into a shallow depression in another, as in the carpal–metacarpal joint at the wrist and the tarsal–metatarsal joint at the ankle. Movement is allowed through all three planes, but there's no rotation.

>> **Saddle:** Each of the adjoining bones is shaped like a saddle, as in the carpometacarpal joint of the thumb (where the thumb meets the wrist). Movement is allowed through two planes: there's no rotation.

>> **Ball-and-socket:** The round head of one bone fits into a cuplike cavity in the other bone, allowing movement in all planes of motion plus rotation. These joints include your two most movable joints: shoulders and hips.

Can you articulate what you've learned about joints so far?

PRACTICE

For questions 70–75, complete the following minitable:

	Structure Category	Movement Category
70. between skull bones		
71. between carpals		
72. knee joint		
73. pubic symphysis		
74. epiphyseal plate		
75. between vertebrae		

For questions 76–78, provide the name of the structure described, followed by the corresponding letter from Figure 6-12. (Not all the letters will be used.)

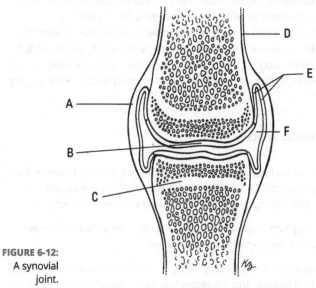

FIGURE 6-12:
A synovial
joint.

Illustration by Kathryn Born, MA

76 The fibrous layer that isolates an articulation: _____

letter: _____

77 A band of cartilage where a bone can grow in length: _____

letter: _____

78 Layer that produces a serous fluid to reduce friction: _____

letter: _____

79 What type of joint would you find between the trapezium and the first metacarpal?

(a) condylar

(b) saddle

(c) gliding

(d) hinge

(e) ball-and-socket

80 Choose an example of a pivot joint.

(a) Between the radius and the ulna

(b) Between the phalanges

(c) Between the mandible and the temporal bone

(d) Between the ribs and the sternum

(e) Between the tibia and the tarsals

Jump around with your joints

You're well aware that certain types of joints (which you now know are called *synovial joints*) can perform certain kinds of movements. The movement of a body part — say, raising your hand — often has an opposing movement to return it to its original position, such as putting your hand down in frustration when you don't get called on. Here's a quick overview of those special movements:

>> **Abduction:** Moves a body part to the side, away from the body's middle. When you make a snow angel, moving your arms and legs out and up, that's abduction.

>> **Adduction:** Moves a body part from the side toward the body's middle. When you're in snow-angel position and move your arms and legs back down, that's adduction; you're "adding" your body back together.

>> **Flexion:** Decreases the joint angle. When you flex to show off your biceps, you move your wrist to your shoulder, decreasing the angle at the elbow.

>> **Extension:** Makes the angle larger. Returning your arm from the flexed position increases the angle and the elbow and, thus, is extension. Hyperextension occurs when the body part moves beyond a straight line (180 degrees), such as tilting your head back in exasperation.

>> **Elevation:** The upward movement of a body part, such as shrugging your shoulders.

>> **Depression:** The downward movement of a body part, such as the downward movement after shrugging your shoulders.

>> **Eversion:** Happens only in the feet when the foot is turned so that the sole is facing outward.

>> **Inversion:** Happens only in the feet when the foot is turned so that the sole is facing inward.

>> **Supination:** Happens only in the arm when the forearm is rotated to make the palm face upward or forward. (Think about holding a bowl of soup.)

>> **Pronation:** Happens only in the arm when the forearm is rotated to make the palm face down or backward.

>> **Rotation:** The movement of a body part around its own axis, such as shaking your head to answer "No." The partnered motions are *medial rotation* (movement toward the midline) and *lateral rotation* (movement away from the midline).

>> **Circumduction:** The movement of a body part in circles, such as doing arm circles in gym class.

REMEMBER

It's helpful to think of the movements as pairs instead of individual terms. The plane, or type of motion, connects you to the pair. Moving your arms in the frontal plane (doing jumping jacks) would have to be either abduction or adduction; all the other choices are eliminated. From there, you only need to remember which is which.

No bones about it — that was a lot! Flip to the chapter quiz and check your understanding.

Practice Questions Answers and Explanations

1 **(a) Support of soft tissue, (c) Production of red blood cells, (e) Mineral storage** I wrote only "movement" for answer choice d on purpose to illustrate a common pitfall. It may seem like it's of no consequence, but saying movement is misleading. The skeletal system does not cause it because it can't generate forces by itself. The articulations dictate whether and how your body can move. Hemostasis, when broken down, would mean keeping blood *(hemo-)* the same *(-stasis)* and refers to clotting. Although the marrow does produce blood cells, including platelets, I wouldn't say that it's the marrow's job to make us stop bleeding. As odd as it seems to say, that's the blood's job.

2 **(c) Long bones**

3 **epiphysis**

4 **diaphysis**

5 **medullary cavity.** The bracket refers to the whole area; that's how you know that the answer isn't marrow.

6 **spongy/trabecular bone**

7 **compact/cortical bone**

8 **periosteum**

9 **(c) It creates a hollowed-out space inside the bone to house marrow.** Be careful not to confuse the periosteum and endosteum. *Endo-* sounds like *indoor,* so that's a good connection.

10 **(e) They differentiate into bone-building cells.** Progenitor cells do undergo mitosis (unlike osteoblasts), but they don't lay down new matrix. Because *matrix* is defined as the space between the cells within a tissue, that answer doesn't make sense anyway.

11 **Haversian and Volkmann's canals are both tunnels through hardened bone matrix, created to carry blood vessels and nerves. The Haversian, or central, canals form the center of an osteon because the matrix is filled with calcium compounds around the vessels inside. Volkmann's canals are formed the same way, but around branches. Rather than spanning the length of the bone, like Haversian canals, the Volkmann's canals are roughly perpendicular, allowing blood vessels to reach deeper into the bone.**

12 **Volkmann's canal: F**

13 **Canaliculi: G**

14 **Lacuna: D**

15 **Osteon: C**

16 **Periosteum: E**

17 **Trabecula: H**

18 **Osteoblasts bring calcium ions in from the interstitial fluid and build them into compounds such as hydroxyapatite (the most common). Then the cells release the compounds which**

then stick to the fibers in the area. **Many osteoblasts work simultaneously to completely cover the fibers, creating characteristic bony plates of spongy bone.**

19. **(b) Osteoblasts migrate to spongy bone, gather around blood vessels one ring at a time, and deposit matrix until they've walled themselves off.** Choice (a) is wrong because it implies that all of the space is filled at the same time instead of one layer at a time. Choice (c) is incorrect because a single osteoblast doesn't build an osteon. A common mistake is to think that the Haversian canal is a cell nucleus; then you get all mixed up. Choice (d) is wrong because the fibers aren't formed into rings; their distribution is random.

20. **These two processes have more commonalities than differences. When osteoblasts are triggered to deposit mineral compounds, the steps are the same: the process of ossification. The difference lies in how they create the space. Endochondral growth uses hyaline cartilage as a template, whereas intramembranous uses fibrous connective tissue.** In fetal development, all bones are formed via endochondral ossification, with the exception of the flat bones of the skull and the mandible (which use intramembranous growth). Throughout life, only long bones use endochondral growth (to increase their length) and only until you stop getting taller. All other bones use intramembranous growth to increase in size proportionally, one layer at a time. Long bones do this to increase their width.

21. **(d) GH.** Growth hormone (GH) triggers bone growth for all bones. Easy to remember, right? Calcitonin and PTH are hormones that trigger bone cells, but they affect matrix density, not bone size. Ossein is a structural protein — the specific type of collagen found in bones — not a hormone.

22. **Calcitonin is released by the thyroid when blood calcium levels are too high. This hormone triggers osteoblasts to build up bone matrix, storing the calcium in the bones for a rainy day. When calcium levels drop too low (that would be the rainy day), parathyroid hormone (PTH) is released from the parathyroid gland. This hormone triggers osteoclasts to break down matrix, which releases the calcium ions into blood flow.**

23. **Frontal**

24. **Sphenoid.** You can also see the sphenoid as the back of the eye socket.

25. **Nasal**

26. **Lacrimal.** Remember there is a portion of the maxilla that lies between the nasal and lacrimal. The lacrimal is often mistaken for the ethmoid for that reason, especially in a diagram like this one, in which you can't see the ethmoid.

27. **Zygomatic**

28. **Maxilla**

29. **Mandible**

30. **Parietal**

31. **Temporal**

32. **Occipital**

33. **Frontal sinus**

34. **Ethmoid sinus**

(35) Sphenoid sinus

(36) Maxillary sinus

(37) T4 or thoracic vertebra #4

(38) L1 or lumbar vertebra #1

(39) Intervertebral disc

(40) Axis or cervical vertebra #2 (C2)

(41) Spinous process

(42) Intervertebral foramen

(43) Coccyx

(44) heart/lungs

(45) lungs/heart

(46) true

(47) false

(48) floating

(49) hyoid

(50) H

(51) P

(52) B

(53) L

(54) E

(55) D

(56) Ilium

(57) Ischium

(58) Femur

(59) Patella

(60) Tibia

(61) Talus

(62) Fibula

(63) Calcaneus

64) **Cuboid**

65) **Lateral cuneiform**

66) **Navicular**

67) **(a) At the acromion process**

68) **(a) Olecranon**

69) **(a) Lower back**

70) **suture; synarthrotic**

71) **synovial – gliding; diarthrotic**

72) **synovial – hinge; diarthrotic**

73) **cartilaginous – amphiarthrotic**

74) **cartilaginous – synarthrotic**

75) **cartilaginous – amphiarthrotic**

76) **joint capsule: A.** Be careful to not mix this up with the periosteum (D), as it branches off from there.

77) **epiphyseal plate: C**

78) **synovial membrane: E.** This is continuous with the articular cartilage (B) and creates the synovial cavity (F).

79) **(b) Saddle.** This is the only joint between a metacarpal and wrist bone that isn't condyloid.

80) **(a) Between the radius and the ulna**

If you're ready to test your skills a bit more, take the following chapter quiz, which incorporates all the chapter topics.

Whaddya Know? Chapter 6 Quiz

Quiz time! Complete each problem to test your knowledge of the various topics covered in this chapter. You can find the solutions and explanations in the next section.

1. The constant breakdown and rebuilding of bone matrix is termed

   (a) endochondral growth

   (b) primary calcification

   (c) ossification

   (d) intramembranous growth

   (e) remodeling

2. Which is the largest and strongest tarsal bone?

   (a) The talus

   (b) The cuboid

   (c) The cuneiform

   (d) The calcaneus

   (e) The metatarsal

3. What is happening during hematopoiesis?

   (a) Bone marrow is converting from red type to yellow type.

   (b) Bone marrow is forming new blood cells.

   (c) Bone marrow is removing damaged bone matrix.

   (d) Bone marrow is converting back from yellow type to red type.

   (e) Bone marrow is releasing minerals into the bloodstream.

4. What does calcitonin do?

   (a) It maintains calcium homeostasis.

   (b) It controls the development of Haversian systems.

   (c) It influences the formation of bone marrow.

   (d) It breaks down calcium compounds in the matrix.

   (e) It encourages the action of osteoclasts.

5. What is the formal term for an immovable joint?

   (a) An amphiarthrosis

   (b) A synarthrosis

   (c) A syndesmosis

   (d) A diarthrosis

   (e) A synchondrosis

For questions 6–14, classify the following bones by shape. Each classification may be used more than once.

    (a) Flat bone(s)

    (b) Irregular bone(s)

    (c) Long bone(s)

    (d) Short bone(s)

6 _____ Vertebrae of the vertebral column

7 _____ Femur in thigh

8 _____ Sternum

9 _____ Tarsals in ankle

10 _____ Humerus in upper arm

11 _____ Phalanges in fingers and toes

12 _____ Scapulae of shoulder

13 _____ Vertebrae

14 _____ Carpals in wrist

15 Which of the following is not a function of the mineral compounds containing phosphorous, calcium, and magnesium?

    (a) To enhance the skeleton's structural integrity

    (b) To provide cushioning and nourishment to the joints

    (c) To ensure availability of these minerals

    (d) To form networks within the bones' structure

    (e) To support the weight of the body

16 Volkmann's canals (choose all that apply)

    (a) contain nerves

    (b) pass through the epiphysis

    (c) form the center of an osteon

    (d) supply blood to articular cartilage

    (e) join with Haversian canals

For questions 17–23, fill in the blanks in the paragraph about bone growth.

To grow in length, long bones use endochondral growth. This process occurs at the (17) _____, creating the new space to be ossified. Rather than osseous tissue, this area is made of (18) _____. In response to growth hormone (GH), the cells of this tissue will (19) _____ which pushes the ends of the bone farther apart. These cells, called (20) _____, will continue to grow until they effectively explode, creating even more space. Blood vessels will branch into the new area and (21) _____, our bone-building cells, will migrate. These cells will build (22) _____ and deposit them onto the fibers of the new tissue, creating the characteristic (23) _____ of spongy bone.

For questions 24–33, match the type of joint movement with its description.

(a) Upward or palm upward
(b) Decrease in the angle between two bones
(c) Turning the sole of the foot inward
(d) Downward or palm downward
(e) Increase in the angle between two bones
(f) Turning the sole of the foot outward
(g) Movement away from the midline of the body
(h) Forming a cone with the arm or leg
(i) Turning around an axis
(j) Movement toward the midline of the body

24 _____ Flexion

25 _____ Eversion

26 _____ Abduction

27 _____ Supination

28 _____ Rotation

29 _____ Pronation

30 _____ Adduction

31 _____ Extension

32 _____ Inversion

33 _____ Circumduction

34 What is inside the medullary cavity?

(a) Bone marrow

(b) Epiphyses

(c) Volkmann's canals

(d) Trabeculae

(e) Osteoclasts

35 What type of cartilage is found in a cartilaginous joint? Choose all that apply.

(a) Hyaline cartilage

(b) Elastic cartilage

(c) Fibrocartilage

(d) Articular cartilage

36 In what way does the skeleton make locomotion possible?

(a) It ensures that hematopoiesis occurs as needed.

(b) Muscles use the skeleton as a scaffold against which they can contract.

(c) It cushions the joints against erythrocytes.

(d) Bones generate the motion necessary for movement within the joints.

(e) It ensures that the thoracic cavity remains properly filled.

37 What is the formal term for the socket at the head of the femur?

(a) Obturator foramen

(b) Acetabulum

(c) Ischial tuberosity

(d) Symphysis pubis

(e) Greater sciatic notch

38 Which of the following cells would be found in yellow marrow? Choose all that apply.

(a) Leukocytes

(b) Adipocytes

(c) Thrombocytes

(d) Erythrocytes

(e) Osteocytes

For questions 39–52, give the proper term for each bone described.

39 _____ Backbone in your neck

40 _____ Shinbone

41 _____ Upper arm

42 _____ Middle hand bone

43 _____ Heel bone

44 _____ Ribs that are only in the back

45 _____ Forehead

46 _____ Jawbone

47 _____ Back of eye socket

48 _____ Bones of big toe

49 _____ Thigh bone

50 _____ Tailbone

51 _____ Collarbone

52 _____ Middle of nose

53 What does the presence of an epiphyseal plate indicate?

(a) Short and flat bones will no longer increase in length.

(b) A fracture is present in a bone.

(c) Long bones are still changing proportion.

(d) Irregular bones are still growing proportionally.

(e) The bone is healthy.

54 What are the four types of bone cells, and what does each one do?

_____

_____

_____

_____

55 Which of the following are unpaired facial bones? Choose all that apply.

(a) Vomer

(b) Zygomatic

(c) Lacrimal

(d) Maxilla

(e) Palatine

56 A male pelvis would have which of the following characteristics?

(a) A rounded pelvic inlet

(b) A straight and flexible coccyx

(c) A shallow cavity

(d) An acute subpubic angle

(e) A flared (outward) iliac crest

57 How are each of the following factors important for maintaining bone strength?

Calcium consumption:

_____

Exercise:

_____

Exposure to sunlight:

_____

Protein production:

_____

Answers to Chapter 6 Quiz

1. (e) remodeling

2. (d) The calcaneus. Your heel is the part of the foot that bears the most weight, so that bone needs to be the strongest.

3. (b) Bone marrow is forming new blood cells. The root *-hema* refers to blood cells, and *-esis* means *create*. So *hematopoiesis* means the creation of blood cells, which is the function of marrow.

4. (a) It maintains calcium homeostasis. When released from the thyroid, calcitonin triggers osteoblasts to uptake calcium ions from the blood and deposit them in the bone matrix.

5. (b) A synarthrosis

6. (b) irregular bone

7. (c) Long bone

8. (a) Flat bone

9. (d) Short bones

10. (c) Long bone

11. (c) Long bones

12. (a) Flat bone

13. (b) Irregular bone

14. (d) Short bones

15. (b) To provide cushioning and nourishment to the joints. The crystalline structure of these mineral compounds allows them to create a strong, rigid matrix. Because the minerals are also required nutrients for us, this makes a convenient storage system for them as well.

16. (a) contain nerves and (e) join with Haversian canals. Volkmann's canals are the ones that carry blood vessels and nerves deeper into the bones. They branch into the Haversian canals that carry the vessels along the length of the bone, forming the center of an osteon.

17. epiphyseal plate

18. hyaline cartilage

19. undergo mitosis

20. chondrocytes

21. osteoblasts

22. calcium compounds

23. trabeculae

(24) **(b) Decrease in angle between two bones**

(25) **(f) Turning the sole of the foot outward**

(26) **(g) Movement away from the midline of the body**

(27) **(a) Upward or palm upward**

(28) **(i) Turning around an axis**

(29) **(d) Downward or palm downward**

(30) **(j) Movement toward the midline of the body**

(31) **(e) Increase in the angle between two bones**

(32) **(c) Turning the sole of the foot inward**

(33) **(h) Forming a cone with the arm or leg**

(34) **(a) Bone marrow**

(35) **(a) Hyaline cartilage** and **(c) Fibrocartilage.** Elastic cartilage isn't found in joints, and artic- ular cartilage is the hyaline cartilage that lines the bones in synovial joints.

(36) **(b) Muscles use the skeleton as a scaffold against which they can contract.** Remember that the bones are moved only by the muscles.

(37) **(b) Acetabulum**

(38) **(b) adipocytes.** Yellow marrow is inactive; it doesn't make any blood cells. The adipocytes, or fat cells, of the marrow take in lipids to stay active so the marrow can be reactivated if needed.

(39) **Cervical vertebrae**

(40) **Tibia**

(41) **Humerus**

(42) **Third metacarpal**

(43) **Calcaneus**

(44) **Floating ribs**

(45) **Frontal**

(46) **Mandible**

(47) **Sphenoid**

(48) **Phalanges** (distal and proximal)

(49) **Femur**

(50) **Coccyx**

(51) **Clavicle**

(52) **Vomer**

(53) **(c) Long bones are still changing proportion.** Found only in long bones, the epiphyseal plate is where these bones increase in length, thus changing their proportion.

(54) **Osteoblasts: Deposit calcium compounds in the matrix. Osteoclasts: Secrete acid to break down matrix. Osteocytes: maintain bone. Progenitors: differentiate into osteoblasts.**

(55) **(a) Vomer.** All the rest are paired, including the maxilla. (That question always got me.)

(56) **(d) An acute subpubic angle.** The other choices are characteristics of a female pelvis.

(57) **Calcium consumption:** Because calcium is the mineral most often used to build bone, we need to make sure to consume enough of it. **Exercise:** Moving the bones stimulates remodeling, which strengthens bones in areas that absorb the most forces. **Exposure to sunlight:** It seems odd, but exposure to sunlight stimulates the production of vitamin D by skin cells. Without vitamin D, our bodies can't absorb calcium from the food we eat. **Protein production:** Calcium compounds can't be deposited in empty space; they attach to the fibers of connective tissue. The fibers are proteins.

Chapter 7

Getting in Gear: The Muscles

M uch of what we think of as the body centers on our muscles — what they can do, what we want them to do, and how tired we get trying to make them do it. With all that muscles do and are, it's hard to believe that the word *muscle* is rooted in the Latin word *mus*, which means *mouse*. The prefix *myo-* (which you'll see all over this chapter) is from the Greek work *mys*, which also means *mouse*. I'm not sure what they were trying to say back then, but the muscle is a mouse that roars! Muscles make up most of the fleshy parts of the body and account for 30 to 50 percent of the body's weight (depending on sex, with females being at the lower end). Layered over the skeleton, they largely determine the body's form. More than 500 muscles are large enough to be seen by the unaided eye, and thousands more are visible only through a microscope. So strap yourself in as we explore the roller coaster that is our muscular system.

Meet & Greet Your Mighty Muscles

Muscle tissues always have work to do. They pull things up and push things down and around. They move things inside you and outside you. And of course, they move you too. The function of this system is, in a word, movement: of food and air into and out of your body; of blood around your body; of different parts of your body relative to one another (such as when you shift position); and of your body through space, which you normally think of as movement.

All muscle tissue is strong. Most is enduring; some of it astoundingly so. Its cells are crowded with *mitochondria* — thousands of little factories that are constantly churning out molecules of adenosine triphosphate (ATP), the "energy molecule" our cells use to do their work. The muscle cells use this fuel to manufacture the strong, flexible proteins that will do the work of contraction, which also requires ATP. In fact, muscle activity alone is demanding 20 percent of your energy right now, and you're just sitting there!

There are three distinct types of muscle tissue, classified by structure and control, but all types have the following characteristics:

>> **Contractility:** Ability to shorten

>> **Extensibility:** Ability to stretch

>> **Elasticity:** Ability to return to resting length

>> **Excitability:** Ability to respond to stimulus

>> **Conductivity:** Ability to create an electrical signal

The three types of muscle tissue are smooth, cardiac, and skeletal. (For more on them, see the "Structure Matches Function" section later in this chapter.) They carry out the numerous functions of the muscular system:

>> **Body movement:** Skeletal muscles (those attached to bones) convert chemical energy to mechanical work, producing movement ranging from blinking to jumping by *contracting* — shortening, in other words. Reflex muscle reactions protect your fingers when you put them too close to a fire and startle you into watchfulness after an unexpected noise. Many purposeful movements require several groups of muscles to work in unison.

>> **Vital functions:** Without muscle activity, you die. Muscles are doing their job when your heart beats, when your blood vessels constrict, and when your intestines squeeze food along your digestive tract (a process known as *peristalsis*).

>> **Antigravity:** Perhaps that's overstating it, but muscles do make it possible for you to stand and move about in spite of gravity's ceaseless pull. Did your mother ever nag you to stand up straight? Just think how bad your posture would be without any muscles!

>> **Heat generation:** You involuntarily shiver when you're cold and may rub your arms and jog in place when you need to warm up. You do these things because chemical reactions in muscles create heat as a byproduct, helping maintain the body's temperature. If the heat generated by the muscles raises body temperature too high, other thermoregulatory processes, such as sweating, are activated.

>> **Joint stability:** Muscles and their tendons help the ligaments reinforce joints, keeping them properly aligned.

>> **Supporting and protecting soft tissues:** The body wall and floor of the pelvic cavity support the internal organs and protect soft tissues from injury.

>> **Controlling release:** Ring-shaped muscles called *sphincters* encircle openings to control actions such as swallowing, defecation, and urination.

Structure Matches Function

As you may remember from studying tissues (see Chapter 4 for details), muscle cells — called *fibers* — are some of the longest in the body. Some muscle fibers contract rapidly, whereas others move at a leisurely pace. Generally speaking, however, the smaller the structure to be moved, the faster the muscle action. The fibers (which, again, are the cells; you've got to make that connection really quick here so you don't get lost as the chapter goes on) will shorten to generate a force, allowing something somewhere to move.

On the move: The three tissue types

A muscle tissue type is not the same as a muscle. Your left bicep is a muscle, your quadriceps are muscles; you have hundreds of named muscles. You can see the largest and most superficial (on the outside) of these muscles in the color plates in the middle of this book. There are only three muscle tissue types, however: *smooth*, *cardiac*, and *skeletal*.

Smooth and cardiac muscle fibers are *involuntary*, meaning that their contraction is initiated and controlled by parts of the nervous system that are far from the conscious level of the brain (specifically, the *autonomic nervous system*, which I discuss in Chapter 8). You have no practical way to consciously control, or even become aware of, the smooth muscle contractions in your stomach that are grinding up this morning's muffin. The involuntary contractions that cause your heartbeat aren't even under the nervous system's constant control, as I discuss in Chapter 11.

Skeletal muscle is classified as *voluntary* because you can make a decision, at the conscious level, to move the muscle. You're walking into a building, so you reach out for the handle and give it a good pull. Skeletal muscle cells contract because you told them to, and the door swings open (that is, provided that it isn't a push door; then you just stand there looking silly). Now consider this scenario: The door handle is charged with static electricity. Instead of pulling open the door, your arm muscles pull your hand away before you're even consciously aware of being zapped. This *reflex* is still classified as voluntary movement, however, because it involves skeletal muscle, which is controlled by the somatic (voluntary) nervous system.

TIP Not everything in anatomy and physiology makes sense at first. Just keep in mind that skeletal muscle is classified as voluntary, but an involuntary override exists. Shivering, like the arm withdrawal reflex I just described, is another good example of this.

The differing shapes and structures of the muscle cells correspond to their functions at the tissue and organ levels. Figure 7-1 illustrates the differences, and the list below spells out what the differences mean, practically speaking

>> **Smooth muscle tissue:** So called because these spindle-shaped fibers don't have the striations typical of other kinds of muscle. The cells are randomly arranged but tightly packed into sheets and make up the walls of hollow organs such as the stomach, intestines, and bladder. The tissue's involuntary movements are relatively slow, so contractions last longer than those of other muscle tissues, and fatigue is rare. If the tissue is inside an organ and arranged in a circle, contraction constricts the cavity inside. If it is arranged lengthwise, contraction of the smooth muscle tissue shortens the organ.

» **Cardiac muscle tissue:** Found only in the heart. Cardiac muscle fibers are cylindrical, striated, and feature one central nucleus. These fibers align end to end, forming cellular junctions called *intercalated discs.* These connections are gap junctions that help move an electrical impulse throughout the heart. For more on cardiac muscle, flip to Chapter 11.

» **Skeletal muscle tissue:** The tissue that most people think of as muscle. In fact, this tissue is what we use when exploring how muscle physiology works. The long, striated, cylindrical fibers of skeletal muscle tissue contract quickly but tire just as fast. They are packed parallel to one another and span the length of the entire muscle that they compose. Some skeletal muscle fibers have *muscle spindles* wrapped around them. These specialty sensory cells function in *proprioception,* which means that they communicate with the brain to relay information on muscle actions that is translated into a sense of body position.

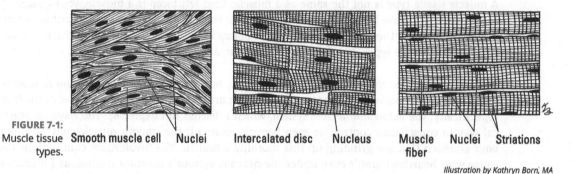

FIGURE 7-1:
Muscle tissue types.

Smooth muscle cell   Nuclei   Intercalated disc   Nucleus   Muscle fiber   Nuclei   Striations

Illustration by Kathryn Born, MA

No ordinary cells

While the cellular components of all three muscle types are very similar, the same cannot be said when comparing them to any other cell types in the body. They have unique structures that allow them to contract and relax repeatedly for your entire life. By "your entire life," I mean that the life span of a typical skeletal or cardiac muscle fiber is your life span; smooth muscle cells are the only ones that are able to divide into new cells.

TECHNICAL STUFF

Portions of skeletal muscle fibers can be replaced through stem cells that reproduce and fuse to a damaged area. This process is stimulated by physical injury in the location of damage (as opposed to just wear and tear). As far as researchers can determine, an entire cell in our larger muscles cannot be replaced. Replacing cardiac muscle cells is even more complicated, as those cells are connected in an intricate pattern that coordinates the contraction of the chambers. So if you try to look up the life span of either type of muscle fiber, you'll find reliable sources saying different things. Part of the difference is likely due to how the age of a cell is determined, and part of it could be due to the fact that we just haven't quite figured out the intricacies of muscle fiber repair and regeneration.

Most muscle cells have a typical nucleus, large and centrally located. Skeletal muscle fibers are the exception as these are formed during embryonic development by fusing together individual stem cells. In addition, each stem cell retains all its nuclei, so skeletal muscle cells end up being *multinucleate.* (This also helps explain why/how skeletal muscle cells are as long as the entire muscle). These nuclei are found underneath the cell membrane, which in muscle fibers is called the *sarcolemma*. The nuclei don't stay near the center of the cell because that area is packed full of proteins called *myofibrils*. These ropelike structures span the entire cell, occupying most of the space, so the nuclei and the numerous mitochondria get shoved to the edges.

The myofibrils are the stripes you see running end to end within the cells when observing muscle tissue under the microscope. Important note: These *are not* striations! (I'll get to what those are in the "Electric Slide" section.)

Wrapped around each myofibril is the *sarcoplasmic reticulum* (SR), which stores the calcium ions that are needed to begin the contraction process. The interconnected SRs form a giant network of tunnels connecting to the sarcolemma (cell membrane) through smaller tunnels called the *t-tubules*, or *transverse tubules*. Within each myofibril are *thin filaments* (called *actin*) and *thick filaments* (called *myosin*). The actin and myosin are arranged in a deliberate pattern that allows them to overlap, which is how the force of contraction is actually generated.

Cardiac muscle fibers have the same internal structure but have more space inside; they don't get filled to capacity with proteins because we don't need them to generate quite as much force. That means the nucleus doesn't get pushed around; it gets to remain in the center. In smooth muscle fibers, the myofibrils are not bundled together parallel within the cell (like they are in skeletal and cardiac fibers). And while smooth muscle fibers do have SRs, they're much smaller, and there are no t-tubules to connect them to the sarcolemma. The myofibrils here connect directly to the membrane in a random pattern that crisscrosses the interior of the cell. This causes the fiber as a whole to collapse from all directions when contracting rather than shorten from end to end as skeletal and cardiac fibers do.

Maybe you should take a myoment to process all that. Then try your hand at answering some practice questions.

 1 Name the cellular unit in muscle tissue.

   (a) Filament

   (b) Sphincter

   (c) Myofibril

   (d) Nucleus

   (e) Fiber

2 Identify the false statements. Choose all that apply.

   (a) Muscle fibers are some of the longest cells in the body.

   (b) Myofibrils within muscle cells contain filaments that slide during contraction.

   (c) Sphincters hold fibers together to form ring-shaped muscles.

   (d) Muscles create actions that both move the body and move things through it.

   (e) Posture is a result of muscle contraction.

3 Why do your muscles shiver when you're cold?

_____

_____

4 Intercalated discs

(a) anchor skeletal muscle fibers to one another.

(b) play a role in moving electrical impulses through the heart.

(c) are found only in the muscles of the back.

(d) overlap to cause contraction.

(e) contribute to tactile perception.

5 Answer true or false and then support your answer: It's possible to fully relax all the muscles in your body.

_____

_____

For questions 6–11, ID the structures of the skeletal muscle fiber illustrated in Figure 7-2.

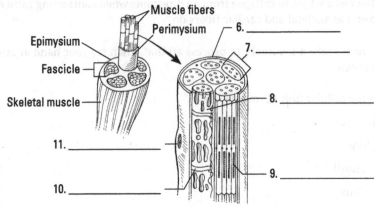

FIGURE 7-2:
Anatomy of a skeletal muscle fiber.

Illustration by Kathryn Born, MA

The Electric Slide

Now we're ready for the real fun: the physiology of muscle contraction, which is called the *sliding-filament theory*. The myofilaments — our actins and myosins — overlap to generate power.

REMEMBER This all happens at the level of individual molecules — molecules that make up a mere portion of a protein. It's like typing one character in the process of writing a term paper. When considered individually, it seems of no consequence. A single letter. A period. But when you put a bunch of them together, simultaneous and organized, you get a powerful result.

What this means is we have to keep looking at smaller and smaller structures. Fortunately, proteins rely on their 3D structure to function, so we get to stop there instead of identifying which atoms are bonding and separating. We tend to approach anatomy as the visual stuff and the physiology as the explanation of what happens; therefore, it's natural to separate the two topics when learning them. (Truly, it's a necessity when you don't have someone standing in front of you to model things.)

TIP

In this unit, though, it's critical that you continuously refer to the visuals when reading about the process. That's why I didn't just give you a diagram, talk about all the structures, and then go on to tell you what each one does. When I talk about myosin heads doing something, stop and look at the myosin heads in the diagram. Imagine them doing what I describe. Attaching these two types of information will go a long way toward helping you understand (and remember) the content.

I base this description of muscle on the muscle type that has generated the most study: skeletal muscle. These fibers are also the most neatly organized. Although most of the diagrams show skeletal muscle, the physiology of the actual contraction process is the same regardless of tissue type.

Use Figure 7-3 as a visual guide as you read the following sections.

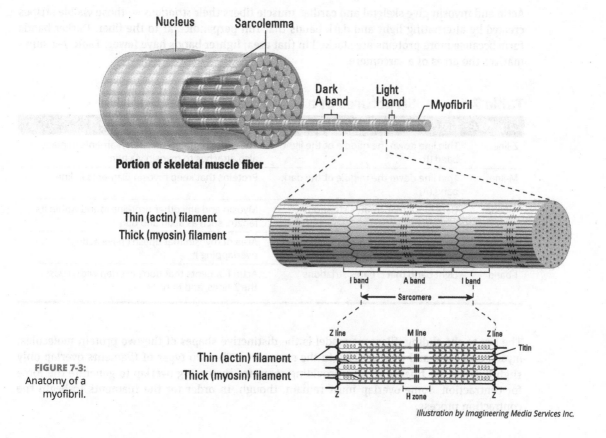

FIGURE 7-3:
Anatomy of a myofibril.

Illustration by Imagineering Media Services Inc.

What are filaments and how do they slide?

Each muscle fiber contains hundreds or even thousands of myofibril strands made up of myofilaments. Upon closer inspection (refer to Figure 7-3), you see that these myofilaments aren't actually long strands of proteins. Rather, they are made of even smaller filaments arranged in a repeating pattern called a *sarcomere*.

REMEMBER

A *fiber* is the functional unit of a muscle, and the *sarcomere* is the functional unit of the fiber (within the myofibril). Sarcomeres line up end to end along the myofibril to create the myofilaments. The key components of a sarcomere are myosin, actin, and the Z-line proteins.

The thick, or *myosin*, filaments are dense and rubbery. They're actually a bundle of individual pieces of myosin proteins, which resemble golf clubs. The myosin filaments stack on top of one another, though not touching, forming the center of a sarcomere, and are held in place by spiral-shaped proteins called *titin*. Each thin, or *actin*, filament is made of two twisted strands of the lighter (less dense) protein actin, which is springy. Each myosin block can interact with four actin filaments, one at each corner. The thin and thick filaments line up together in an orderly way to form a sarcomere. The actin of neighboring sarcomeres are held together by the Z-line proteins. Contraction happens by shortening the sarcomere or pulling the Z-lines closer together. (These structures are visible, in differing detail, in Figures 7-3 and 7-4.)

Actin and myosin give skeletal and cardiac muscle fibers their *striations* — those visible stripes created by alternating light and dark bands that run perpendicular to the fiber. Darker bands form because more proteins are stacked in that area; lighter bands have fewer. Table 7-1 summarizes the areas of a sarcomere.

Table 7-1   Structures of a Sarcomere

Structure	Description	Contents
Z-line	Thin line down the middle of the light band (I)	The protein that holds actin filaments in place; also attaches to myosin via titin
M-line	Thin line down the middle of the dark band (A)	Proteins that keep myosin filaments in line
A-band	Dark block that creates striations	Myosin and any other proteins found within its length, such as actin and titin
H-zone	Lighter block in the center of the dark band	Area of myosin that doesn't have actin overlapping it
I-band	Light band that creates striations	Actin filaments that don't overlap with myosin, the Z-lines, and titin

The key to the sliding-filament model is the distinctive shapes of the two protein molecules, *myosin* and *actin*, and their overlap in the sarcomere. The two types of filaments overlap only slightly when the fiber is at rest, providing room for increasing overlap to generate the force for contraction. Some overlap must remain, though, in order for the filaments to begin the contraction process.

The myosin proteins of the thick filaments have numerous club-shaped heads that point away from its center (toward the two Z-lines). At rest, these heads nearly touch the *binding sites* on the actin (the designated place of the myosin heads to attach, in other words). Figure 7-4a shows a sarcomere at rest. Note that the myosin heads are next to the actin but are not yet attached. (If they were, the fibers would be locked, which is exactly what happens in a cramp.)

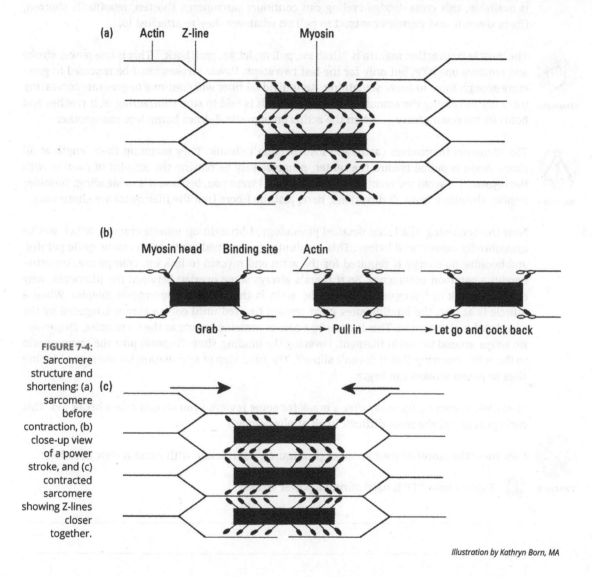

(a) Actin Z-line Myosin

(b) Myosin head Binding site Actin

Grab ⟶ Pull in ⟶ Let go and cock back

(c)

FIGURE 7-4:
Sarcomere
structure and
shortening: (a)
sarcomere
before
contraction, (b)
close-up view
of a power
stroke, and (c)
contracted
sarcomere
showing Z-lines
closer
together.

Illustration by Kathryn Born, MA

Contraction begins when the myosin heads grab onto the actin (at the binding sites), form *cross bridges*, and perform *power strokes*. (Refer to Figure 7-4b.) The first part of a power stroke requires no energy; cross bridges form, and the myosin heads reflexively pull in, shortening the sarcomere. You can see from the figure, though, that this likely is not going to generate enough force to move a body part; it's one tiny click. Clearly, we need to keep doing this, overlapping the filaments more with each power stroke. We've gotta finish the first one first, though!

To finish a power stroke, ATP comes in and binds to the myosin head, causing it to let go of the actin. Then ATP is catalyzed (broken down into ADP and P), and the energy released is used to cock the heads back into their original position. (Refer to Figure 7-4c.) Now the myosin heads are able to perform another power stroke, this time grabbing a binding site farther down on the actin filament. (Refer to Figure 7-4b.) As long as the muscle is being stimulated and ATP is available, this cross-bridge cycling can continue; sarcomeres shorten, myofibrils shorten, fibers shorten, and muscles contract to pull on whatever they're attached to.

REMEMBER

The muscle contraction mantra is "Grab on, pull in, let go, cock back." This is one power stroke and requires one ATP, but only for the last two steps. Power strokes must be repeated to generate enough force to move something. An individual fiber will continue to generate increasing force (by increasing the amount of overlap) until it is told to stop contracting or it reaches and holds its maximum force — when the actins on opposite Z-lines bump into one another.

WARNING

The filaments themselves (actin and myosin) don't shrink. They maintain their length at all times. Actin is pulled toward the center, progressively increasing the amount of overlap with the myosin. I'd avoid the term *shrink* altogether if I were you, because it's misleading. *Shrinking* implies shrinking from all directions; here, just the fibers (not the filaments) are shortening.

Near the beginning of all that detailed physiology, I brought up muscle cramps, which you've undoubtedly experienced before. This involuntary, sustained contraction can be quite painful, and because no energy is required for the actin and myosin to link up, cramps are (unfortunately) a common occurrence. So if there's always some overlap between the filaments, why *don't* they link up? Wrapping around the actin is the *troponin-tropomyosin complex*. When a muscle is at rest, the binding sites must remain covered until contraction is triggered by the associated motor neuron. This is done by a pair of proteins known as the *t-t complex*. *Tropomyosin* wraps around the actin filament, covering the binding sites. *Troponin* pins the tropomyosin to the actin, ensuring that it doesn't slip off. The final step of stimulation uncovers the binding sites so power strokes can begin.

That topic is coming up next, after a break for some review. (You should take a break too. This concept is one of the most difficult in all of physiology!)

Now show the power of your knowledge of muscle contraction with some review questions.

PRACTICE 12 Explain how ATP is used during a power stroke.

_____

_____

_____

For questions 13–19, label the sarcomere structures indicated in Figure 7-5.

a A-band   d I-band   f titin

b actin   e myosin   g Z-line

c H-zone

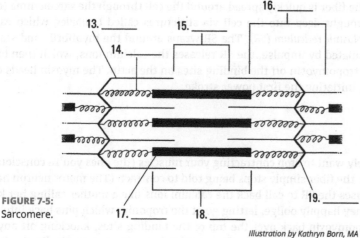

FIGURE 7-5:
Sarcomere.

13. _____
14. _____
15. _____
16. _____
17. _____
18. _____
19. _____

Illustration by Kathryn Born, MA

20 If the myosin heads rest right above the actin binding sites, why aren't they attached all the time?

_____

_____

_____

Get your motor runnin'

The first part of this chapter talks about how muscles generate force. That's all well and good, but we still need to figure out what stimulates them to do so. I cover the details of the nervous system in Chapter 8, but here, you can find out what's happening when an impulse (which is how the nervous system communicates) stimulates a skeletal muscle.

The *stimulus* makes its way from the central nervous system to the muscle, traveling through an *efferent* (outgoing) nerve until it reaches the *motor neuron* tasked with stimulating the fiber directly. Each individual muscle fiber must be stimulated by a motor neuron, and each fiber has a designated place along its sarcolemma for this task: the *motor end plate*. Fortunately, a single motor neuron has numerous branches that allow it to trigger multiple fibers at the same time, creating motor units. Our *large motor units* are able to generate great force, controlling our gross motor skills (such as walking). Our *small motor units*, on the other hand, trigger only a few fibers, so they get to control our fine motor skills (such as writing), sacrificing force to gain more-intricate control.

A motor neuron communicates with a fiber by using a neurotransmitter called *acetylcholine* (ACh). When it's released by the neuron, ACh moves across the space between the neuron and the muscle fiber — a space affectionately known as the *synaptic cleft*. Then ACh binds to a receptor on the motor end plate, triggering the fiber to generate an *impulse*. (It turns out that this process is the same one neurons use in nervous communication, which I describe in detail in Chapter 8. For now, picture a spark moving down a power line like in cartoons.)

REMEMBER The impulse created by the fiber is quickly spread around the cell through the sarcolemma (cell membrane). Then it's brought deep into the cell via structures called *t-tubules*, which carry the impulse to the *sarcoplasmic reticulum (SR)*. The SR wraps around the myofibrils and stores calcium ions. When stimulated by impulse, the SR releases the calcium ions, which then bind to troponin, shifting the tropomyosin off the binding sites on the actin. The myosin heads can then grab onto the actin, initiating the first power stroke.

Time to relax

At some point, you'll likely want to stop contracting your muscles (the ones you're consciously using, anyway). To relax, the fiber simply stops being told to contract. (The motor neuron halts its stimulation.) This causes the SR to call back the calcium ions like a mother calling her kids back home for dinner. They happily oblige, letting go of the troponin, which pins back into the actin. This pulls the tropomyosin back over the top of the binding sites, knocking off any of the myosin heads that were attached (breaking the cross bridges). Then the actin slowly slides back, restoring the length of the sarcomere to its original position.

That's the extent of nervous communication. Motor neurons are either releasing ACh or they aren't. The brain can't tell a muscle to contract a little or a lot. It's all or nothing; it's not even yes or no; it's "Do it" or the silent treatment.

Pile this on to the rest of the contraction process and do some more review.

PRACTICE

21 What does a sarcolemma do?

(a) It receives and conducts stimuli for skeletal muscle tissue.

(b) It provides a structure for moving electrical impulses through the heart.

(c) It forms a wall for hollow organs.

(d) It's made up of the filaments responsible for contraction.

(e) It prevents muscle fatigue.

22 Which statement best characterizes a large motor unit?

(a) Many neurons stimulate the same muscle, generating a large force.

(b) One neuron stimulates all the fibers of a muscle for precise control.

(c) It controls precise movements such as writing.

(d) Many neurons stimulate many fibers for a powerful contraction.

(e) One neuron stimulates many fibers for a powerful movement.

23 Outline the steps of contraction stimulus, ending with exposure of the binding sites.

24 Relaxation is considered to be a passive process because

(a) nothing; relaxation isn't a passive process.

(b) it decreases the force generated from contraction.

(c) the motor neuron releases different neurotransmitters that don't require ATP.

(d) ATP isn't required to halt the spread of impulse.

(e) when the Z–lines slide back to resting position, ADP is generated.

Push, Pull, & Hold

So here's a question: How do all of these microscopic clicks going on inside muscle tissue translate into big, powerful movements like running? Well, it's sort of like how one small spark can start a forest fire. For that image to make sense, though, we need to pull way back and look at the muscle as a whole. Follow along in Figure 7-6 as I walk you through the structures in the following sections.

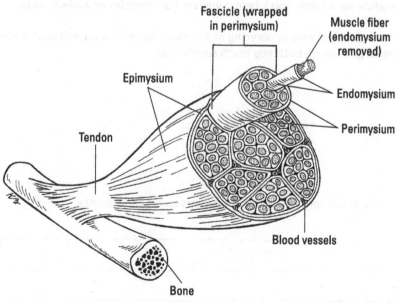

FIGURE 7-6: Organizational structure of a muscle.

Illustration by Kathryn Born, MA

Bundles of fun

The sarcolemma of each muscle fiber is surrounded by a layer of loose connective tissue called *endomysium* that keeps the fibers in perfect alignment with their neighbors. Another sheet of connective tissue called the *perimysium* wraps several fibers up into a bundle called a *fascicle*. All the fascicles together make up the *belly* of the muscle, which is surrounded by dense connective tissue called the *epimysium*. Blood vessels, lymphatic vessels, and nerves can be found wrapped in the epimysium as well as between the fascicles.

The collagen fibers of the epimysium weave together with those in the tendon; it's this weaving that attaches the tendon (quite firmly) to a bone. When you take a moment to remember that the job of skeletal muscle is to move bones, the fusion of muscle tissue to a tendon makes sense (especially if you keep in mind that the tendon is fused with the outer tissue of a bone as well; for more on that topic, see Chapter 6). You don't want to be exerting a large force with a muscle only to have it separate from the tendon or bone. If that were the case, the only movement that would happen would be you falling to the floor writhing in pain.

Bones also have ligaments to attach to one another, keeping them consistently aligned, pulling them along when being moved. It would seem logical, then, that we have something that connects muscles to other muscles. *Aponeuroses* are connective tissue sheets similar to tendons but broad and flat. We often have multiple muscles moving a bone, applying forces from different angles and power. Nearby muscles may not be involved, but we don't want to pull away from them, so an aponeurosis maintains that connection.

WARNING

How many ways can you say *fiber*? Anatomists need them all when they're talking about the muscular system. Make sure that you're thinking at the right level of organization (*subcellular, cellular,* or *tissue*) when you see the terms *filament, myofibril, fiber,* and *fascicle*. In everyday English, we'd call every one of these structures a fiber.

PRACTICE

Time to pause for a vocab check before you see how muscles as a whole work.

25) Starting with the outer covering of a muscle, list all the organizational structures you would go through until you reach sarcomeres.

26) Compare and contrast tendons, ligaments, and aponeuroses.

_____

_____

_____

_____

Leveraging power

Most of our body movements are not powered by a single muscle; it's a group effort. The muscle providing the most force for any action is called the *prime mover*. The muscles that help the prime mover create the movement are *synergists*. When you move your elbow joint, the biceps brachii is the prime mover, and the brachioradialis stabilizes the joint, thus aiding the motion.

Antagonistic muscles also act together to move a body part, but one group contracts while the other releases — a kind of push–pull. One example is flexing your arm. When you bend your forearm up toward your shoulder, your biceps muscle contracts (shortens), performing a *concentric contraction*. In the meantime, the triceps muscle in the back of your arm extends (is stretched), performing an *eccentric contraction*. The actions of the biceps and triceps muscles are opposite, but you need both actions to flex your arm, which is why both are confusingly referred to as contractions. Antagonistic actions lower your arm too, but this time the roles reverse; the biceps extend, and the triceps shortens.

As I mention in the earlier section "Time to relax," when it comes to contraction of a muscle fiber, it's an all-or-nothing affair. Bear in mind, though, that this applies to a fiber, not the muscle as a whole. While the actin begins to slide back once the myosin heads let go, they won't reach their resting position if the fiber is being continuously stimulated. When pulled in again, they overlap even more, increasing the force generated with each power stroke. This is referred to as *summation*. If the stimulus lasts long enough, there will be no relaxation between each contraction cycle, which is known as *tetanic contraction* (again, only of a single fiber, not an entire muscle or even a whole fascicle).

Multiple neurons forming multiple motor units allow us to contract only a portion of a muscle or the whole thing. Some motor units require greater stimulation than others, which allows us to control the force of contraction. Fibers that are easier to stimulate are triggered first. As the stimulus intensifies (with more ACh being released), more motor units are activated, which is referred to as *recruitment*. Recruitment ensures that muscle movements are smooth rather than jerky as the fibers *summate* to sustained contraction.

REMEMBER

In physiology, a muscle contraction is referred to as a *muscle twitch*. A twitch is the fundamental unit of recordable muscular activity. The twitch consists of a single stimulus-contraction-relaxation sequence in a muscle fiber. The *contraction* is the period when the cross bridges are cycling. The *relaxation* phase occurs when cross bridges detach and the muscle tension decreases. Complete *fatigue* occurs when no more twitches can be elicited, even with increasing intensity of stimulation.

Assuming the right tone

Even when a muscle is at rest, some of its fibers maintain a level of sustained contraction called *tone* or *tonus*. Without tone, we wouldn't be able to stand or even hold our heads up! A very close interaction outside your conscious control between some muscle cells and the nervous system keeps you not just upright, but also in balance. Nerve impulses throughout the muscular system cause muscles to contract or relax to oppose gravity. This is your *muscle tone*, and it's what is enabling you to hold your head up right now.

Muscle tone relies on the *muscle spindles* (the sensory receptors found around muscle fibers mentioned earlier in the "Structure Matches Function" section). They continuously send messages about your body position to the central nervous system (spinal cord and brain). To initiate the fine adjustments, signals are sent by the central nervous system to motor neurons, stimulating the appropriate muscle fibers to contract. (Remember, we don't tell cells to relax; we simply stop telling them to contract.)

The contraction process within the fibers remains the same whether or not your body is moving. To generate actual movement, muscles perform *isotonic contractions*. The actin and myosin overlap to generate a force that is greater than the resistance from the bone you're trying to move. If the muscle doesn't generate enough force to cause movement, but contraction is nevertheless occurring, the fibers are undergoing *isometric contractions*. (You do *isometrics* when you try to push over a large tree with your hands or do a wall-sit.) Sometimes, we can't generate enough force for movement. At other times, the point isn't generating movement at all; what we really want to do is develop muscle tone, which is the point of isometric exercises. The difference lies in whether the force generated matches or is less than the resistance. In the case of muscle tone, the force of resistance is our good friend gravity.

The individual fibers responsible for maintaining tone are different from those that generate movement. Your legs, for example, need both. We need big, powerful fibers that can generate a great deal of force quickly, such as when we're jumping up and down. These big, powerful fibers are called *fast-twitch fibers*. Something unique about these fibers is that they're able to perform *anaerobic respiration*. The fibers that maintain long-term contraction are *slow-twitch* and are solely *aerobic*. So we have a trade-off: power and quick response for endurance. The cost, though, is the anaerobic respiration of the fast-twitch fibers generates lactic acid. (Flip back to Chapter 2 if you need to review respiration types.)

WARNING

When lactic acid builds up, fibers can *fatigue* and are thus incapable of contracting. This is distinct from a *cramp*, which is a sustained, involuntary contraction. While cramps and fatigue can occur simultaneously within a muscle, cramps come about from an electrolyte imbalance in the fluid surrounding the cell, not the accumulation of lactic acid.

PRACTICE

All right, it's time to flex your knowledge of muscle tone and function by answering some practice questions.

For questions 27–31, use the following terms to fill in the blanks in the following paragraph. Not all the terms will be used.

Fascicles   Motor units   Muscle spindles   Recruitment
Relaxation   Summation   Tetanic   Tonus

Because we can't tell individual fibers how hard to contract, we must have a system in place to provide contractions that require a lot of force as opposed to ones that don't. If we want more power, we need more fibers contracting at the same time. We start with easy-to-stimulate 27 _____ that leads to numerous fibers starting to contract, all triggered by the same neurons. These fibers generate increasingly forceful contractions through 28 _____ and might even max out. While holding this 29 _____ contraction at maximum power, other fibers in the muscle are stimulated through the process of 30 _____. Some muscles have fibers that maintain a constant level of contraction but not at maximum force. This gives our muscles a quality called 31 _____ that allows us to stand upright.

32 A muscle contraction that doesn't generate enough force to move an object is referred to as

(a) isometric

(b) eccentric

(c) tetanic

(d) isotonic

(e) concentric

33 What do you call muscles that tend to counteract or slow an action?

(a) Antagonists

(b) Fixators

(c) Primary movers

(d) Agonists

(e) Synergists

34 Why are fast-twitch fibers more prone to fatigue than slow-twitch fibers?

_____

_____

_____

What's in a Name?

It may seem like a jumble of meaningless Latin at first, but muscles follow a convention that names them according to one or more characteristics, as shown in Table 7-2. Check out the "Muscular System" color plate in the center of the book as you go through this section and refer to Chapter 1 if necessary for the names of the body regions.

Table 7-2   Characteristics in Muscle Names

Characteristics	Examples
Muscle size	The largest muscle in the buttocks is the *gluteus maximus*. (*Maximus* means *large* in Latin.) A smaller muscle in the buttocks is the *gluteus minimus*. (*Minimus* means *small* in Latin.)
Muscle location	The *frontalis muscle* lies on top of the skull's frontal bone.
Muscle shape	The *deltoid muscle,* shaped like a triangle, comes from *delta* — the Greek letter that's shaped like a triangle.
Muscle action	The *extensor digitorum* is a muscle that extends the fingers or *digits.*
Number of muscle attachments	The *biceps brachii* attaches to bone in two locations, whereas the *triceps brachii* attaches to bone in three locations.
Muscle fiber direction	The *rectus abdominis muscle* runs vertically along your abdomen. (*Rectus* means *straight* in Latin.)

TIP

So that you may practice the labeling, figures have been left out of this section. Whenever you see a muscle you're unfamiliar with, see what clues are given in the name before finding it in a figure. The *serratus anterior*, for example, are the sawtooth-shaped muscles that start on the front side of the ribs. (*Serrated* means *with a jagged edge*, like a bread knife; *anterior* means *in front of*.) After reading about the location and function of each region of muscles, you'll see a set of labeling questions. It's important that you practice identifying anatomical structures in many ways and on many figures; otherwise, you might inadvertently memorize a specific image, and when presented with a different one, you can't label anything!

Heads up! Muscles of the head, neck, and shoulders

Your head contains muscles that perform three basic functions: chewing, making facial expressions, and moving your neck. Ear wiggling falls into this category too.

To chew, you use the muscles of *mastication* (a big fancy word that means *chewing*). The *masseter*, a muscle that runs from the *zygomatic bone* (your cheekbone) to the *mandible* (your lower jaw), is the prime mover for mastication, so its name is based on its action (*masseter, mastication*). The fan-shaped *temporalis muscle* works with the masseter to allow you to close your jaw. It lies on top of the skull's temporal bone, so its name is based on its location.

To smile, frown, or make a funny face, you use several muscles. The *frontalis muscle*, along with a tiny muscle called the *corrugator supercilii*, raises your eyebrows and gives you a worried or angry look, respectively, when you wrinkle your brow. (Think of the appearance of corrugated cardboard, and then feel the skin between your eyebrows when you wrinkle your brow. That, and you'll look 'super silly' doing it.) The *orbicularis oculi muscle* surrounds the eye. (The word *orbit*, as in *orbicularis*, means to encircle; *oculi* refers to the eye.) This muscle allows you to blink and to close your eyes; it also gives you those little crow's feet at the corners. The *orbicularis oris* surrounds the mouth. (*Or* refers to the mouth, as in *oral*.) You use this muscle to pucker up for a kiss.

If you've played a trumpet or another instrument that requires you to blow out, you're well aware of what your *buccinator muscle* does. This muscle is in your cheek. (*Bucc* means *cheek*, as in the word *buccal*, which refers to the cheek area.) It allows you to whistle and also helps keep food in contact with your teeth as you chew. Remember that your zygomatic is your cheekbone? Well, the *zygomaticus muscle* is a branched muscle that runs from your cheekbone to the corners of your mouth. This muscle pulls your mouth up into a smile when the mood strikes you.

When you want to nod, shake your head, or tilt it, your neck muscles come into play. You have two *sternocleidomastoid muscles,* one on each side of your neck. This name is annoyingly long, but it reflects the locations of its attachments: the sternum, clavicle, and mastoid process of the skull's temporal bone. When both sternocleidomastoid muscles contract, you can bring your head down toward your chest and flex your neck. When you turn your head to the side, one sternocleidomastoid muscle contracts — the one on the opposite side of the direction your head is turned. So if you turn your head to the left, your right sternocleidomastoid muscle contracts, and vice versa. If you lean your head back to look up at the sky or to shrug your shoulders, your *trapezius muscle* allows you to do so.

The trapezius is an antagonist to the sternocleidomastoid muscle. If you remember basic geometry, you can see that the trapezius is shaped like a trapezoid. It runs from the base of your skull to your *thoracic vertebrae* and connects to your scapula (shoulder blades). Therefore, the trapezius and sternocleidomastoid muscles connect your head to your torso and provide a nice segue to the next section . . . not before some labeling, though!

PRACTICE

For questions 35–44, label Figure 7-7.

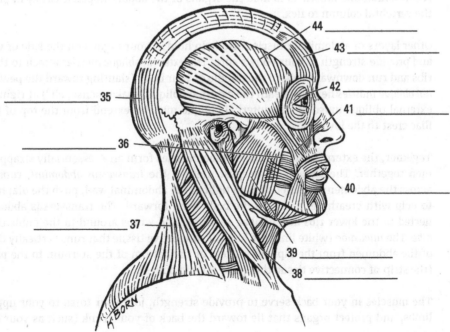

44 _____

43 _____

42 _____

41 _____

35 _____

36 _____

40 _____

37 _____

39 _____

38 _____

FIGURE 7-7:
The muscles
of the head
and neck.

Illustration by Kathryn Born, MA

Twisting the torso

REMEMBER

The torso muscles have important functions. They not only give support to your body, but also connect to your limbs to allow movement, allow you to inhale and exhale, and protect your internal organs. In this section, I cover the muscles that run along your front (called your *anterior* or *ventral* side) and then cover the muscles of your back (your *posterior* or *dorsal* side).

In your chest, your *pectoralis major muscles* connect your torso at the sternum and collarbones (clavicles) to your upper limbs at the humerus bone in the upper arm. Your pecs also help protect your ribs, heart, and lungs. You can feel your pectoralis major muscle working when you move your arm across your chest. Also in your chest are the muscles between and around the ribs. The *internal intercostal muscles* help raise and lower your rib cage as you breathe. The torso's largest muscles, however, are the abdominal muscles.

TIP

The *abdominal muscles* create the core of your body. If the abdominal muscles are weak, the back is weak because the abdominal muscles help flex the vertebral column. And if you can't flex your vertebral column easily, the muscles attached to it can become strained and weak. Now, as you probably know already, the muscles of the abdomen and back join to the upper and lower limbs. If the abdomen and back are weak, the limbs can have problems too.

The muscles of the abdomen are thin, but the fact that these muscle fibers run in different directions increases their strength. This woven effect makes the tissues much stronger than they would be if they all went in the same direction. Think about how a child connects building blocks and their likelihood of falling over. Laying a top layer of blocks perpendicular to the blocks underneath (like the game Jenga) helps the structure stay together, which is similar to how the abdominal muscles provide strength and stability.

The "six-pack" muscle of the abdomen, the *rectus abdominis*, forms the front layer of the abdominal muscles. It runs from the pubic bone up to the ribs and sternum. The function of the rectus abdominis muscle is to hold the organs of the abdominopelvic cavity in place and allow the vertebral column to flex.

Other layers of abdominal muscles also help hold in your organs on the side of your abdomen and provide strength to your body's core. The *external oblique muscles* attach to the eight lower ribs and run downward toward the middle of your body (slanting toward the pelvis). The *internal oblique muscles* lie underneath the external oblique (makes sense, eh?) at right angles to the external oblique muscles. The internal oblique muscles extend from the top of the hip at the iliac crest to the lower ribs.

Together, the external and internal oblique muscles form an X, essentially strapping the abdomen together. The abdomen's deepest muscle, the *transversus abdominis*, runs horizontally across the abdomen; its function is to tighten the abdominal wall, push the diaphragm upward to help with breathing, and help the body bend forward. The transversus abdominis is connected to the lower ribs and lumbar vertebrae and wraps around to the pubic crest and *linea alba*. The *linea alba* (white line) is a band of connective tissue that runs vertically down the front of the abdomen from the xiphoid process at the bottom of the sternum to the *pubic symphysis* (the strip of connective tissue that joins the hip bones).

The muscles in your back serve to provide strength, join your torso to your upper and lower limbs, and protect organs that lie toward the back of your trunk (such as your kidneys). The *deltoid muscle* joins the shoulder to the collarbone, scapula, and humerus. This muscle is shaped like a triangle. (Think of the Greek letter delta: Δ.) The deltoid muscle helps you raise your arm to the side (that is, laterally). The *latissimus dorsi muscle* is a wide muscle that's also shaped like a triangle. It originates at the lower part of the spine (thoracic and lumbar vertebrae) and runs upward on a slant to the humerus. Your lats allow you to move your arm down if you have it raised and also to pull when reaching, such as when you're climbing a tree or swimming.

PRACTICE

For questions 45–51, label Figure 7-8.

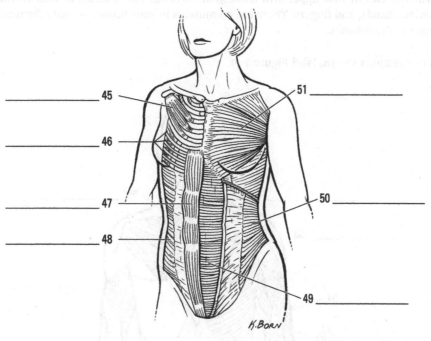

FIGURE 7-8: Anterior muscles of the chest and abdomen.

Illustration by Kathryn Born, MA

Spreading your wings

Your upper limbs have a wide range of motions. Obviously, your upper limbs are connected to your torso. One of the muscles that provide that connection, the *serratus anterior*, is below your armpit (the anatomic term for armpit is *axilla*) and on the side of your chest. The serratus anterior muscle connects to the scapula and the upper ribs. You use this muscle when you push something or raise your arm higher than horizontal. Its action pulls the scapula downward and forward.

Although the *biceps brachii* and *triceps brachii* are muscles located in the top (anterior) part of your upper arm, their actions allow your forearm (lower arm) to move. The name *biceps* refers to this muscle's two origins (points of attachment); it attaches to the scapula in two places. From there, it runs to the radius of the forearm (its point of insertion). The triceps brachii is the only muscle that runs along the back (posterior) side of the upper arm. The name *triceps* refers to the fact that it has three attachments: one on the scapula and two on the humerus. It runs to the ulna of the forearm. You can feel this muscle in motion when you push or punch. Other muscles of the arm include the *brachioradialis*, which helps you flex your arm at the elbow, and the *supinator*, which rotates your arm from a palm-down position to a palm-up position.

Your forearm contains muscles that control the fine movements of your fingers. When you type or play the piano, you're using your *extensor digitorum* and *flexor digitorum muscles* to raise and lower your fingers onto the keyboard and move them to the different rows of keys. As you lift your hands off the keyboard, the muscles of your wrist kick into gear. The *flexor carpi radialis* (attached to the radius bone) and *flexor carpi ulnaris* (attached to the ulna bone) allow your wrist to flex forward or downward. The *extensor carpi radialis longus* (which passes by the carpal bones), the *extensor carpi radialis brevis*, and the *extensor carpi ulnaris* allow your wrist to extend — that is, bend upward.

REMEMBER

The muscles in your upper arm move your forearm. The muscles of your forearm move your wrists, hands, and fingers. There are no muscles in your fingers — only the tendons that connect to those bones.

PRACTICE

For questions 52–58, label Figure 7-9.

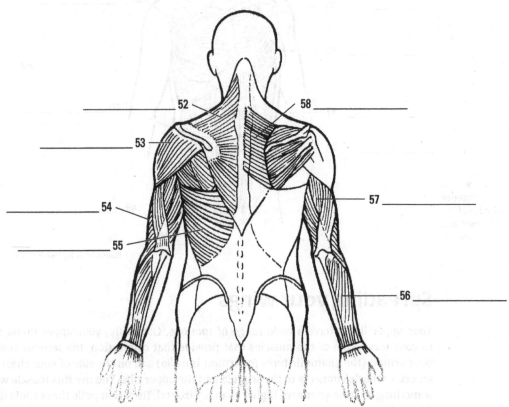

FIGURE 7-9:
Posterior view
of the torso
and arm
muscles.

Illustration by Kathryn Born, MA

Getting a leg up

Your lower limbs are connected to your buttocks, and your buttocks are connected to your hips. The *iliopsoas* connects your lower limb to your torso and consists of two smaller muscles: the *psoas major*, which joins the thigh to the vertebral column, and the *iliacus*, which joins the hipbone's ilium to the thigh's femur bone. Originating on the iliac spine of the hip and joining to the inside surface of the tibia (a bone in your shin), the *sartorius muscle* is a long, thin muscle that runs from the hip to the inside of the knee. These muscles stabilize your lower limbs and provide strength for them to support your body's weight and to balance your body against the pressure of gravity.

Some muscles in the lower limb allow the thigh to move in a variety of positions. The buttocks muscles allow you to straighten your lower limb at the hip and extend your thigh when you walk, climb, or jump. The *gluteus maximus* — the largest muscle in the buttocks — is the largest muscle in the body. The gluteus maximus is antagonistic to the stabilizing *iliopsoas muscle,* which flexes your thigh. The *gluteus medius muscle,* which lies behind the gluteus maximus, allows you to raise your leg to the side so that you can form a 90-degree angle with your legs. (This action is *abduction* of the thigh.) Several muscles serve as *adductors* — that is, they move an abducted thigh back toward the midline. These muscles include the *pectineus* and *adductor longus,* which become injured when you pull a groin muscle, as well as the *adductor magnus* and *gracilis,* which run along the inside of your thigh.

Other muscles in the thigh serve to move the lower leg. Along your thigh's front and lateral side, four muscles work together to allow you to kick. These four muscles — the *rectus femoris, vastus lateralis, vastus medialis,* and *vastus intermedius* — are better known as the *quadriceps (quadriceps femoris). Quad* means *four,* as in *quadrilateral* or *quadrant.* Moving across these at an angle from the outside hip toward the inside of the knee is the *sartorius,* which is both a hip and knee flexor.

The *hamstrings* are a group of muscles that are antagonistic to the quadriceps. The hamstrings — the *biceps femoris, semimembranosus,* and *semitendinosus* — run down the back of the thigh and allow you to flex your lower leg and extend your hip. They originate on the ischium of the hipbone and join (insert at) the tibia of the lower leg. You can feel the tendons of your hamstring muscles behind your knee.

Your lower leg's shin and calf muscles move your ankle and foot. The *gastrocnemius,* better known as the calf muscle, begins (originates) at the femur (thigh bone) and joins (inserts at) the *Achilles tendon* that runs behind your heel. You can feel your gastrocnemius muscle contracting when you stand on your toes. The antagonist of the gastrocnemius, the *tibialis anterior,* starts on the surface of the tibia (shinbone), runs along the shin, and connects to your ankle's metatarsal bones. You can feel this muscle contract when you raise your toes and keep your heel on the floor. The *fibularis longus* and *fibularis brevis (brevis* means *short,* as in *brevity)* run along the outside of the lower leg and join the fibula to the ankle bones. In doing so, the fibularis muscles help move the foot. The *extensor digitorum longus* and the *flexor digitorum longus* muscles join the tibia to the feet and allow you to extend and flex your toes, respectively.

For questions 59–64, label Figure 7-10.

PRACTICE For questions 65–71, label Figure 7-11.

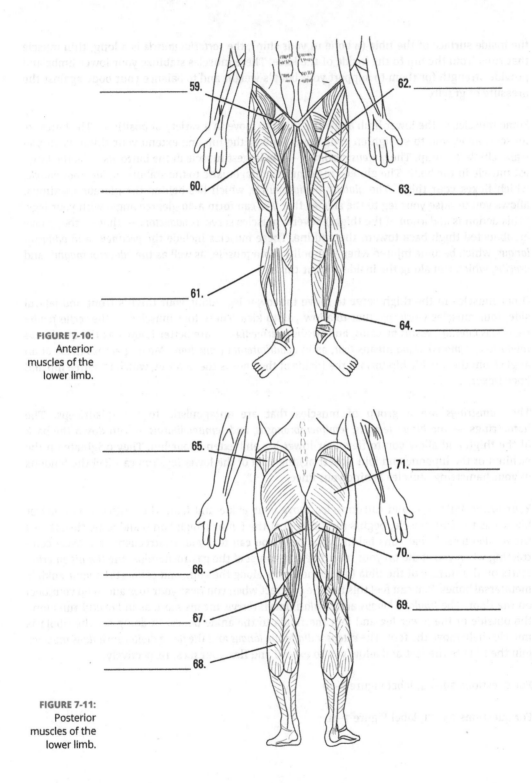

59. _____

62. _____

60. _____

63. _____

61. _____

64. _____

FIGURE 7-10:
Anterior
muscles of the
lower limb.

65. _____

71. _____

70. _____

66. _____

69. _____

67. _____

68. _____

FIGURE 7-11:
Posterior
muscles of the
lower limb.

Practice Questions Answers and Explanations

1. **(e) Fiber.** Filaments are the overlapping proteins, and they assemble to form myofibrils, but the term *fiber* means *cell* in the context of the muscular system.

2. **(c) Sphincters hold fibers together to form ring-shaped muscles.** Sphincters *are* ring-shaped muscles. All the other statements are accurate.

3. **When ATP is used, heat is created as a byproduct.** When your body temperature drops, your skeletal muscles are triggered (without your conscious approval) to perform quick, repeated contractions, which you know as shivering.

4. **(b) play a role in moving electrical impulses through the heart.** Unlike skeletal muscles, cardiac muscle fibers can pass a signal to a neighboring cell through a gap junction, the intercalated disc.

5. **False.** Each individual fiber is able to relax, but we can't relax all of them at the same time. And since smooth and cardiac muscles are involuntary, you wouldn't be able to relax them by choice anyway.

6. **Sarcolemma** (the proper term for a muscle cell membrane)

7. **Myofibril**

8. **Sarcoplasmic reticulum** (SR)

9. **Myosin**

10. **T-tubule**

11. **Nucleus**

12. Energy is needed only for the latter half of a power stroke. If the binding sites are exposed, no ATP is required for the myosin heads to grab on. As a result of the attachment, the heads will pull in, conveniently pulling the opposite Z-lines closer together. We, however, want to keep doing power strokes for as long as the fiber is being told to contract, which is where energy comes in. **The mere presence of ATP causes the myosin heads to let go. Then it is broken down into ADP to provide the power for the myosin heads to shift back to their initial position, where they can grab on to the actin again, this time farther down.**

13. **(b) Actin**

14. **(f) Titin**

15. **(c) H-zone**

16. **(d) I-band**

17. **(e) Myosin**

18. **(a) A-band**

19. **(g) Z-line**

20 Wrapping around the actin is the t-t complex, a pair of proteins that covers the binding sites when the muscle is at rest. The tropomyosin is the spiral-shaped protein that physically blocks off the binding sites. The troponin is a much smaller protein that effectively pins the tropomyosin to the actin, keeping the tropomyosin in place. (I like to call it tropo-*pin*, because that's exactly what it does!)

21 **(a) It receives and conducts stimuli for skeletal muscle tissue.** Keep in mind that stimuli are just triggering events. So receptors on the sarcolemma will receive the ACh and then the membrane conducts (spreads) the impulse to begin the actual shortening of the fiber.

22 **(e) One neuron stimulates many fibers for a powerful movement.**

23 The steps should go something like this:

    Neuron releases ACh.

    ACh spreads across the synaptic cleft.

    ACh binds to the motor end plate.

    Muscle fiber generates an impulse.

    The impulse is spread through sarcolemma and then through the t-tubules.

    The impulse stimulates the SR which releases calcium ions.

    Calcium ions bind to troponin.

    Tropomyosin (or t-t complex) slides around, uncovering the binding sites.

24 **(d) ATP isn't required to halt the spread of impulse.** The motor neurons stop releasing ACh, and when there's no more stimulus, there's no more contraction.

25 **Epimysium ⇨ belly (body of muscle) ⇨ perimysium ⇨ fascicle ⇨ endomysium ⇨ sarcolemma ⇨ fiber ⇨ myofibril ⇨ sarcomere**

26 Tendons, ligaments, and aponeurosis are all pieces of fibrous connective tissue, full of collagen proteins (which in any other chapter we could call fibers). These link up with the collagen in either the epimysium or the *periostem* (outer membrane of bone). Tendons and ligaments are ropes that connect bone to muscle and bone to bone, respectively. An aponeurosis is more of a broad sheet that connects two muscles.

27 **Motor units**

28 **Summation**

29 **Tetanic**

30 **Recruitment**

31 **Tonus** or tone

32 **(a) isometric.** Isotonic contractions pair eccentric with concentric and cause movement, so all three of those are out. And although a tetanic contraction can be isometric, the amount of force generated by it is irrelevant.

33. (a) Antagonists

34. **Fast-twitch fibers are the ones that can do anaerobic respiration. In doing so, instead of producing CO_2 as a byproduct waste, they make lactic acid. Even in small amounts, a drop in pH (from the acid) can prevent an enzyme from functioning or disrupt ion concentrations, keeping the fibers from being stimulated.** No stimulus, no contraction, thus fatigue.

35. Occipitalis

36. Buccinator

37. Trapezius

38. Sternocleidomastoid

39. Masseter

40. Orbicularis oris

41. Zygomaticus

42. Orbicularis oculi

43. Frontalis

44. Temporalis

45. Pectoralis minor

46. Serratus anterior

47. Rectus abdominus

48. Internal oblique

49. Transversus abdominus

50. External oblique

51. Pectoralis major

52. Trapezius

53. Deltoid

54. Brachialis

55. Latissimus dorsi

56. Brachioradialis

57. Triceps brachii

58. Rhomboids

59. Pectineus

60. Vastus lateralis

(61) **Fibularis longus**

(62) **Sartorius**

(63) **Adductor longus**

(64) **Soleus**

(65) **Gluteus medius**

(66) **Gracilis**

(67) **Semimembranosus**

(68) **Gastrocnemius**

(69) **Biceps femoris**

(70) **Adductor magnus**

(71) **Gluteus maximus**

If you're ready to test your skills a bit more, take the following chapter quiz, which incorporates all the chapter topics.

Whaddya Know? Chapter 7 Quiz

Time to muscle up and flex that knowledge on the chapter quiz! You can find the solutions and explanations in the next section.

1 During contraction, what does *not* occur?

(a) Actin and myosin filaments shorten.

(b) Thick and thin filaments slide past each other.

(c) Muscle fibers shorten.

(d) Myosin heads pull on actin filaments.

(e) Troponin binds with calcium ions.

2 Which of the following is/are *not* part of the muscle group referred to as hamstrings? Choose all that apply.

(a) Biceps femoris

(b) Gracilis

(c) Rectus femoris

(d) Semimembranosus

(e) Semitendinosus

For questions 3–7, identify the statement as true or false. When the statement is false, identify the error.

3 A weak stimulus causes a muscle fiber to contract with less force.

_____

4 Exercise forms new muscle fibers.

_____

5 Although they're structurally different, all three muscle fiber types contract their myofibrils by using the sliding-filament model.

_____

6 Motor neurons communicate with the motor end plates on the fascicles.

_____

7 Skeletal muscle isn't considered to be an involuntary muscle even though it can be stimulated involuntarily.

_____

8 When looking at a tissue sample under the microscope, you notice cells with long stripes inside running from end to end, but there are also lines that seem to be at right angles to these stripes. Unfortunately, you can't see any nuclei. The cells look slightly rounded and are a variety of sizes; some are long, others are short and join to other cells at their ends. What types of muscle tissue could you have?

(a) Cardiac

(b) Skeletal

(c) Smooth

9 Explain how the force of a muscle contraction is generated (after it has been stimulated).

_____

_____

_____

10 A muscle's name is most often based on one or more of its characteristics. Which characteristic is used in all three of these muscles: the latissimus dorsi, the rectus abdominis, and the serratus anterior?

(a) Shape

(b) Size

(c) Location

(d) Fiber direction

(e) Action

11 What muscle is found in both the arms and the legs? (The x refers to the second part of the muscle name, which differs based on location.)

(a) Quadriceps x

(b) Triceps x

(c) Biceps x

(d) Rectus x

(e) Adductor x

12 This muscle divides the thoracic cavity from the abdominal cavity.

(a) Diaphragm

(b) Oblique

(c) Transversus abdominis

(d) Rectus abdominus

(e) Intercostal

13 Compare and contract small and large motor units.

_____

_____

_____

For questions 14–24, match the muscles with their locations.

    (a) Head or neck

    (b) Abdomen

    (c) Back

    (d) Arm

    (e) Thigh

    (f) Lower leg

14 _____ Latissimus dorsi

15 _____ Internal oblique

16 _____ Soleus

17 _____ Masseter

18 _____ Gastrocnemius

19 _____ Occipitalis

20 _____ Orbicularis oris

21 _____ Rhomboids

22 _____ Brachialis

23 _____ Gracilis

24 _____ Vastus medialis

25 Which of these terms _doesn't_ belong in the following list?

    (a) Anisotropic

    (b) Actin

    (c) Myosin

    (d) Sarcolemma

    (e) Troponin

For questions 26–30, match the muscles with their actions.

(a) Semitendinosus

(b) Temporalis

(c) Biceps brachii

(d) Latissimus dorsi

(e) Trapezius

26 _____ Rotation of scapula

27 _____ Flexion of leg at knee joint

28 _____ Extension at shoulder joint

29 _____ Mastication

30 _____ Flexion of forearm

For questions 31–35, match the muscle structures with their descriptions.

(a) Perimysium

(b) Aponeurosis

(c) Epimysium

(d) Fascicles

(e) Endomysium

31 _____ Areolar tissue that surrounds fibers

32 _____ Bundles of muscle fibers

33 _____ Connective tissue that wraps muscle fibers into bundles

34 _____ Connective tissue that creates a functional muscle

35 _____ Flat, sheetlike tendon that serves as insertion for a large, flat muscle

36 Which of the following steps of muscle contraction require ATP? Choose all that apply.

(a) Cross bridges break off from the binding sites.

(b) Myosin heads cock back to their original positions.

(c) Myosin heads pull actin closer.

(d) Calcium ions are released.

(e) Binding sites are uncovered.

For questions 37–40, match the term to its definition.

(a) Summation

(b) Tone

(c) Recruitment

(d) Tetanus

37 _____ As stimulation increases, more motor units are triggered to contract.

38 _____ Twitches occur before the previous one ends, generating more force.

39 _____ Stimulation occurs so frequently that no relaxation occurs.

40 _____ Constant, sustained contraction of some fibers occurs within a muscle.

41 Which of the following tissues lack striations? Choose all that apply.

(a) Aponeuroses

(b) Cardiac

(c) Skeletal

(d) Smooth

(e) Tendons

For questions 42–46, label the structures of the neuromuscular junction illustrated in Figure 7-12.

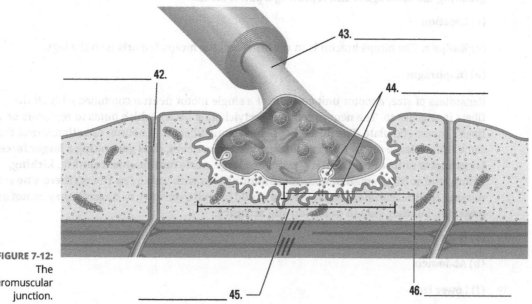

FIGURE 7-12:
The
neuromuscular
junction.

43. _____

42. _____

44. _____

45. _____

46. _____

Photo credit: aldonagriskeviciene

Answers to Chapter 7 Quiz

1. **(a) Actin and myosin filaments shorten.** Everything shortens except the filaments, which overlap. That's why the theory is called the sliding-filament theory, not the shrinking-filament one.

2. **(b) Gracilis** and **(c) Rectus femoris.** If you think of what the biceps brachii in your arm does when you bend your elbow, you'll know which femoris is which; the biceps femoris is going to bend your knee, so it has to be in the back. The gracilis run straight down your inner thigh.

3. **False.** A fiber is either contracting or it isn't; there's no weak versus strong stimulus.

4. **False.** Exercise will not trigger the creation of new muscle cells. It does lead to protein production, though. The increase in myofibril number inside a fiber leads to the bulking-up.

5. **True.** They're just assembled in a different pattern in smooth muscle.

6. **False.** The motor end plate is on the sarcolemma of the fiber (cell membrane of the cell).

7. **True.**

8. **(a) Cardiac.** Seeing two sets of lines inside the cell indicates striations. And though you can't see that it would have a single, large nucleus, the fact that it meets up with other cells confirms it. Skeletal fibers are stacked parallel, never connecting to each other.

9. Once the binding sites are uncovered, the myosin heads link to the binding sites on the actin. Then they immediately shift position, pulling inward, which shortens the sarcomere by pulling the Z-lines closer together. ATP comes in and causes the myosin heads to release. Then the ATP is broken down to provide energy for the myosin heads to cock back, grabbing the actin again and repeating a power stroke.

10. **(c) Location**

11. **(c) Biceps x.** The biceps brachii is in the arms and the biceps femoris is in the legs.

12. **(a) Diaphragm**

13. Regardless of size, a motor unit consists of a single motor neuron combined with all the fibers it attaches to. The neuron releases acetylcholine (ACh), which binds to receptors on a fiber's motor end plate. The difference is how many fibers that neuron can stimulate at the same time. **A large motor unit stimulates numerous fibers, thus generating a larger force. Large motor units are particularly useful for large-scale movements: running, kicking, lifting, and so on. A small motor unit communicates with only a few fibers (there's no set number to differentiate, but a large one would be well over 100); as a result, they're not as powerful, but they gain in precision and accuracy.**

14. **(c) Back**

15. **(b) Abdomen**

16. **(f) Lower leg**

17. **(a) Head or neck** (head/facial)

18. **(f) Lower leg**

19. **(a) Head or neck** (top of neck/bottom of head)

(20) **(a) Head or neck** (head/facial)

(21) **(c) Back**

(22) **(d) Arm**

(23) **(e) Thigh** (inner)

(24) **(e) Thigh** (front)

(25) **(d) Sarcolemma** All the other terms are parts of a sarcomere; the sarcolemma is the cell membrane.

(26) **(e) Trapezius**

(27) **(a) Semitendinosus**

(28) **(d) Latissimus dorsi**

(29) **(b) Temporalis**

(30) **(c) Biceps brachii**

(31) **(e) Endomysium.** Keep in mind that the sarcolemma is the normal cell membrane, and the endomysium is a layer of connective tissue that wraps the outside of it.

(32) **(d) Fascicles**

(33) **(a) Perimysium**

(34) **(c) Epimysium**

(35) **(b) Aponeurosis**

(36) **(a) Cross bridges break off from the binding sites and (b) Myosin heads cock back to their original positions.** This question is tricky due to how I worded it — saying "requires ATP" as opposed to "uses ATP." Besides remembering that the initial force generation doesn't require it, you should know that both steps *require* ATP, but only the cocking back *uses* it.

(37) **(c) Recruitment**

(38) **(a) Summation**

(39) **(d) Tetanus**

(40) **(b) Tone**

(41) **(a) Aponeuroses, (d) Smooth, and (e) Tendons.** Smooth muscle is so named because it lacks striations. Aponeuroses and tendons are connective tissues, so you'd see structures resembling myofibrils; they lack the perpendicular (up-and-down) stripes that are the striations.

(42) **T-tubules** (or transverse tubules)

(43) **Motor neuron**

(44) **Acetylcholine (ACh)**

(45) **Motor end plate**

(46) **Synaptic cleft**

3
Talking to Yourself

In This Unit . . .

Chapter **8**

Your Electric Personality

n organism's awareness of itself and of its environment depends on communication between one part of its body and another. This internal messaging is accomplished by several mechanisms that fall into two categories: those that use chemicals and those that use electricity. I cover the chemical messaging system (hormones) in Chapter 9. This chapter is devoted to the body's electrical messaging system, which, to keep things interesting (or perhaps complicated), also uses chemicals.

Our nervous system is the single most distinctive feature of the human species. Its structures reach into every organ and participate one way or another in nearly every physiological reaction. Perceiving the beauty and aroma of a flower in bloom and digesting your healthy, balanced breakfast can happen simultaneously, and each action is dependent on the nervous system.

The human brain in particular functions differently from the brain of any other species. Although just how this system contributes to human nature (consciousness, for example) is being actively researched, our goal here is to understand the mechanics.

In this chapter, you get a feel for how the nervous system is put together. You practice identifying the parts and functions of nerves and the brain itself, as well as the structure and activities of the central, peripheral, and autonomic systems. In addition, the chapter touches on the sensory organs that bring information into the human body.

Meet & Greet the Far-Reaching Nervous System

The nervous system is the communications network that goes into nearly every part of the body, *innervating* your muscles (supplying them with nerves), pricking your pain sensors, and letting you reach beyond yourself into the larger world. The core of communication is the brain and spinal cord, comprising what we call the *central nervous system* (CNS). The information coming in and instructions going out follow *nerves*, the roads for communication in the *peripheral nervous system* (PNS).

More than 80 major nerves make up this intricate network, communicating with billions of *neurons* (individual nervous cells). There are three categories of neurons: sensory, motor, and interneurons. These cells are highly specialized for the initiation and transmission of electrical signals, which we call *impulses.* In an instant, the neuron is able to receive stimuli from many other cells; process this incoming information; and decide whether to generate its own signal to be passed on to other neurons, muscles, or gland cells.

Supporting our neurons are the other guys — the *glial cells*, or *neuroglia.* Ignored for more than a century because they were thought to merely provide nutrients and physical support, the glial cells actually outnumber neurons (though exactly by how much is quite the debate among neuroscientists). They do indeed provide physical support for neurons, as well as manufacture biomolecules so that neurons can devote their resources to their role as communicators, but these little guys can do much more. We're still discovering all the ways that glial cells contribute to the function of nervous tissues.

It's through the complex network between the PNS and CNS that you respond to external and internal stimuli, demonstrating a characteristic called *irritability*, which is the capacity to respond to stimuli, not the tendency to yell at annoying people.

Put another way, there are three functions of the human nervous system as a whole:

>> **Sensation:** The ability to generate nerve impulses in response to changes in the external and internal environments. Specialized neurons called *sensory receptors* collect information from the entire body, create an impulse, and transmit the impulse to the CNS.

>> **Coordination:** The ability to receive, sort, and direct those signals to channels for response, also referred to as *integration.* The CNS makes sense of the input it receives from all around the body and decides whether a response is necessary.

>> **Reaction:** The ability to translate orders to the *effectors*, which carry out the body's reaction to the stimulus. In response to the integration of the sensory input, impulses are sent back through the PNS to muscles and glands. Sometimes, as with *reflexes*, the brain isn't even involved; the spinal cord coordinates the response.

Does this three-step process ring any bells for you? It should sound familiar, because it's the model of *homeostasis*: input, integration, and output. In the nervous system, though, you'll find it helpful to think of it as a five-step process:

Stimulus ⇨ Sensory input ⇨ Integration ⇨ Decision-making ⇨ Motor output

Nervous communication starts with a *stimulus* — a change that can be noticed by sensory receptors. These receptors don't necessarily need to send a message, however. Consider this: You're presumably wearing clothes right now, but you don't feel them. Well, now you might because you're thinking about them, but in a moment, you'll go back to not noticing them on your body. Touch receptors are constantly stimulated by your shirt, but they stop sharing that information with your brain. Imagine that you were constantly being made aware that you are in fact wearing clothes. While that awareness might help prevent the "showing up to work or school naked" nightmare, the brain has far too much to do to process such irrelevant information, even if you weren't being made consciously aware of it. Some receptors that are being constantly stimulated can adapt so they stop sending the same message over and over. This is called *neural accommodation*. Otherwise, these cells generate an impulse to pass the sensory input along *afferent nerves* (*a* for *approach*) to the CNS. The brain (and/or spinal cord) receives and routes the input to the appropriate area, combining it with other relevant sensory input, which is the integration part. From here, a decision is made: Is this worthy of a response? If the answer is no, that's it; the communication is stopped dead in its tracks (or tracts, which is a great pun that you'll appreciate more after you get through more of this chapter). If some sort of action is necessary, an impulse is sent along an *efferent nerve* (*e* for *exit*) to a motor neuron. The last motor neuron in the line stimulates a cell outside the nervous system to carry out the desired effect, which is why it is dubbed an effector.

WARNING We call all the divisions of the nervous system . . . nervous systems, which can get pretty disorienting. You've already seen that the brain and spinal cord form the CNS and the nerves are the PNS. The afferent nerves create the sensory nervous system, and the efferent ones create the motor nervous system. Because orders come from the brain either involuntarily or consciously, the motor system is also divided. Nerves that carry your consciously made orders to contract skeletal muscles form the *somatic nervous system*. Those that are under involuntary control form the *autonomic nervous system*. (That one's easy to remember because *autonomic* looks and sounds like *automatic*, which is what it is.) All your automatic processes can be stimulated (started or speeded) or inhibited (slowed or stopped), depending on the needs of the body as a whole. As such, the autonomic nervous system is further divided into the *sympathetic* and *parasympathetic nervous systems*, which I elaborate on later in this chapter.

Integrating the Input with the Output

The nervous system is comprised mostly of neural tissue, along with connective tissues for providing structure as well as for housing blood and lymphatic vessels. There's a lot of fluid-filled space between the two categories of cells: *neurons* and *glial cells (neuroglia)*. The nervous system also contains *nerves* — a commonly misused term. Nerves are not cells; they're an organizational structure that bundles together parts of the neurons, wrapping them in connective tissue. The structure of a nerve mirrors muscle structure (which you can see illustrated in Chapter 7's Figure 7-6), except here the term *fiber* annoyingly refers to yet another thing.

Neurons

The nervous cell responsible for the communication at the core of the nervous system's functions is the neuron. Its properties include that irritability that I mention in the introduction of this chapter, as well as *conductivity*, otherwise known as the ability to transmit a *nerve impulse*.

The central part of a neuron is the *cell body*, or *soma*. It contains a large nucleus with one or more nucleoli, the organelles common to all other cells, and *Nissl bodies*. (See Chapter 3 for an overview of a cell's primary parts.) The Nissl bodies are also found in the dendrites and are specialized pieces of rough ER (endoplasmic reticulum) thought to make proteins for the cell itself. The ER found in the cell body manufactures *neurotransmitters* that the neuron uses to communicate with other cells. Extending from the cell body are numerous extensions forming *dendrites* and *axons*. (See Figure 8-1.) Dendrites transmit a stimulus to the cell body, and *axons* conduct impulses away from it. (*Axon* and *away* both start with *a*.) While a neuron can have numerous dendrites, it has only one axon. But each axon can have many branches, called *axon collaterals*, that enable communication with many target cells. The point of attachment on the cell body is called the *axon hillock* or *trigger zone*. The opposite end of the axon enlarges into a *synaptic knob* or *bouton*.

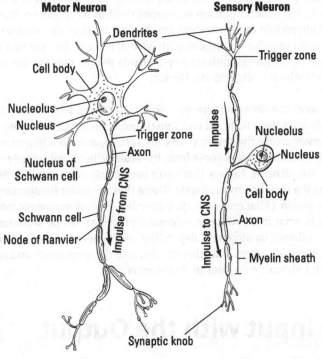

Motor Neuron   **Sensory Neuron**

Dendrites

Trigger zone

Cell body

Nucleolus

Nucleus

Trigger zone

Axon

Nucleus of Schwann cell

Impulse

Nucleolus

Nucleus

Impulse from CNS

Schwann cell

Cell body

FIGURE 8-1:
The motor neuron on the left and the sensory neuron on the right show the cell structures and the paths of impulses.

Node of Ranvier

Impulse to CNS

Axon

Myelin sheath

Synaptic knob

Illustration by Kathryn Born, MA

Neurons in different parts of the nervous system perform diverse functions and therefore vary in shape, size, and electrochemical properties. There are four general types of neurons:

>> **Multipolar:** Multiple dendrites are separated from the axon by the cell body. Most motor, or efferent, neurons are of this type, connecting the CNS to its effectors — the structures that carry out the effect. (Get it?)

>> **Unipolar:** Both the axon and dendrites come from the same cytoplasmic extension, meaning the dendrites connect directly to the axon rather than through the cell body. Most sensory, or afferent, neurons are unipolar, carrying information about stimuli to the CNS.

>> **Bipolar:** A single dendrite and single axon (though they both have branches) meet the cell body on opposite sides. Bipolar neurons are found almost exclusively in the CNS, the only exception being specialized sensory neurons from the eyes, nose, and ears.

>> **Anaxonic:** Axons can't be distinguished from dendrites. These neurons are found in the brain, but their function isn't yet well understood. They're thought to regulate motor neuron networks and may play a role in learning.

Many neurons have beadlike structures around their axons called *myelin sheaths.* These sheaths are like blankets of lipids that wrap around the axons to speed up the impulse. Made by glial cells, they're also thought to provide a pathway for regenerating axons. Unfortunately, neurons don't divide and are not readily replaced when lost. They can regrow an axon, however, like a starfish that has lost an arm. It's important, though, that the new axon ends up in the same place as the previous one. (When they don't, this is a reason for memory loss with brain injury.)

Glial cells

The numerous cells of nervous tissue that aren't neurons are collectively called *glial cells* (or *neuroglia).* There's still much to learn about these cells. They come in a wide variety of shapes that are tailored to their specialized functions. They can communicate with one another, but they lack axons and dendrites and do not generate impulses. Table 8-1 outlines the functions of glial cells.

Table 8-1   Glial Cells

Glial Cell	Description	Function
Astrocyte	Star-shaped, found between neurons and blood vessels in CNS	Form blood-brain barrier, take up excess neurotransmitters and ions, repair damage, and create new neuron connections
Ependyma	Often ciliated, forms simple cuboidal epithelium in CNS	Create a filter controlling which molecules can move in and out of the membranes surrounding the CNS and the nervous tissue itself
Microglia	Small cell with tiny branches in CNS	Phagocytosis of debris, waste, and pathogens
Oligodendrocyte	Rounded cell with extensions that wrap around axons in the CNS	Form myelin sheaths, produce neurotrophic factors (growth and development of neurons)
Satellite cell	Flat cell surrounding cell bodies in ganglia of PNS	Protection, regulate nutrients
Schwann cell	Small cell with extensions that wrap around axons in the PNS	Form myelin sheaths

Nerves

A *nerve* is a bundle of axons from peripheral neurons. An individual axon plus its myelin sheath is called a *nerve fiber.*

Whereas neurons are the basic unit of the nervous system, nerves are the cablelike bundles of axons that organize them.

There are three types of nerves:

>> **Afferent nerves** are composed of sensory nerve fibers (axons) grouped to carry impulses from receptors to the central nervous system.

>> **Efferent nerves** are composed of motor nerve fibers carrying impulses from the central nervous system to effector organs, such as muscles and glands.

>> **Mixed nerves** are composed of both afferent and efferent nerve fibers.

From the inside out, nerves are composed of the following:

>> **Axon:** The impulse-conducting process (extension of the cell) of a neuron, also known as a fiber.

>> **Myelin sheath:** An insulating envelope that protects the axon and speeds impulse transmission. (This process is discussed in detail in the "Feeling Impulsive?" section later in this chapter.)

>> **Neurolemma (or neurilemma):** A thin membrane present in many peripheral nerves that surrounds the myelin sheath and its enclosed axon.

>> **Endoneurium:** Loose, or areolar, connective tissue surrounding the length of individual fibers.

>> **Fascicle:** Bundles of fibers within a nerve.

>> **Perineurium:** The same kind of connective tissue as endoneurium; creates the fascicles.

>> **Epineurium:** The same kind of connective tissue as endoneurium and perineurium; bundles the fascicles together, creating the nerve.

Nerves organize the transport of signals both to and from the brain. Think about traveling across the country by car. Which is more efficient: taking an interstate or city streets? Nerves are the interstates of our nervous system's roads.

Ganglia and plexuses

Because nerves contain only the axons of peripheral neurons, a different structure organizes the cell bodies (and their dendrites). This structure, called a *ganglion*, bundles the cell bodies of related neurons. Ganglia serve as relay points among the body's neurological structures, especially at the spinal cord, where they act as the junction between the CNS and the PNS.

There are 12 pairs of cranial nerves that relay information from the head and neck to the brain. There are 31 pairs of spinal nerves that do the same for the rest of the body. Each nerve has its own ganglion as it meets the CNS.

Plexus is a general term for a network of anatomical structures, such as lymphatic vessels, nerves, or veins. (The term comes from the Latin *plectere*, meaning *to braid*.) A *neural plexus* is a network of intersecting nerves. The *solar plexus* serves the internal organs. The *cervical plexus* serves the head, neck, and shoulders. The *brachial plexus* serves the chest, shoulders, arms, and hands. The *lumbar*, *sacral*, and *coccygeal plexuses* serve the lower body.

PRACTICE

Before you dig in to neurophysiology, see whether you can answer the following questions about the cells and structures of the nervous system:

For questions 1–10, fill in the boxes of the graphic organizer in Figure 8-2.

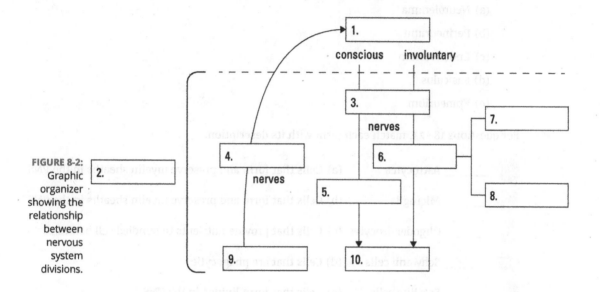

FIGURE 8-2: Graphic organizer showing the relationship between nervous system divisions.

Use the following terms:

Afferent	Autonomic	Central	Efferent
Motor	Parasympathetic	Peripheral	
Sensory	Somatic	Sympathetic	

For questions 11–15, fill in the blanks.

The communicating cells of the nervous system, or (11) _____, have two essential properties. First is (12) _____, the tendency to respond to a stimulus. Second is the ability to generate electricity, called (13) _____. This flow of electricity is known as (14) _____ but it can't be used to communicate with other neurons. For this purpose, (15) _____ are used.

16 How many dendrites and axons are on a multipolar neuron?

(a) Many dendrites and one axon

(b) Three dendrites and three axons

(c) Four or more dendrites and two axons

(d) A single dendrite and a single axon

(e) Numerous dendrites and numerous axons

17 The connective tissue that surrounds a bundle of fibers in a nerve is known as the

(a) Neurolemma

(b) Perineurium

(c) Endoneurium

(d) Fasiculus

(e) Epineurium

For questions 18–23, match each term with its description.

18 _____ Astrocytes   (a) Cells that form and preserve myelin sheaths in the PNS

19 _____ Microglia   (b) Cells that form and preserve myelin sheaths in the CNS

20 _____ Oligodendrocytes   (c) Cells that provide nutrients to bundled cell bodies

21 _____ Schwann cells   (d) Cells that are phagocytic

22 _____ Satellite cells   (e) Cells that form linings in the CNS

23 _____ Ependyma   (f) Cells that contribute to the repair process of the CNS

Feeling Impulsive?

All right. Here we are. This topic is arguably the most difficult to understand in the entire book. If you've been sitting for a while, take a break! Stand up, and get that blood pumping. Have a snack. Do whatever you feel like doing so you can come back with a fresh head on your shoulders.

You ready? Great, because you got this!

Neurons communicate by using both electrical and chemical signals. The electrical signal comes from the movement of ions, much like the flow of electrons we know as electricity. The chemical signal comes in the form of *neurotransmitters*. The entire communication process occurs in three phases: stimulus of the dendrites, impulse through the axon, and synaptic transmission from the synaptic knob. Figure 8-3 summarizes impulse transmission — how the spark is made and how it spreads down the power line. You'll want to refer to the figure often as you read this section.

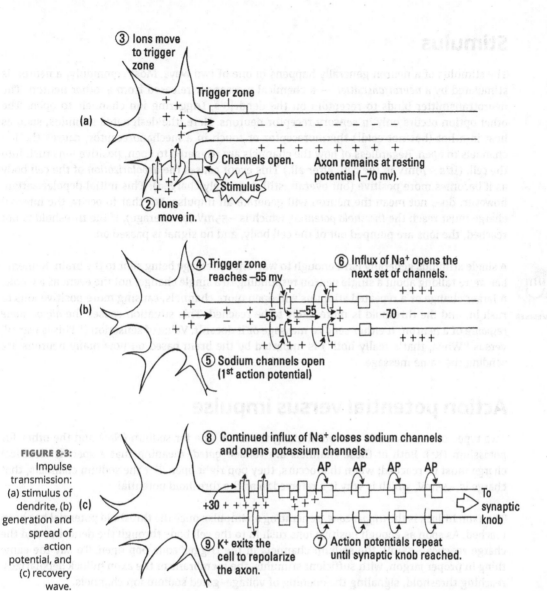

③ Ions move to trigger zone

Trigger zone

(a)

① Channels open.

Axon remains at resting potential (−70 mv).

Stimulus

② Ions move in.

④ Trigger zone reaches −55 mv

⑥ Influx of Na⁺ opens the next set of channels.

++++

(b)

−55   −55

−70 − − − −

++++

⑤ Sodium channels open (1ˢᵗ action potential)

⑧ Continued influx of Na⁺ closes sodium channels and opens potassium channels.

FIGURE 8-3: Impulse transmission: (a) stimulus of dendrite, (b) generation and spread of action potential, and (c) recovery wave.

(c)

AP   AP   AP   AP

+30 + + + + +

To synaptic knob

⑨ K⁺ exits the cell to repolarize the axon.

⑦ Action potentials repeat until synaptic knob reached.

Illustration by Kathryn Born, MA

Before diving in to the communication process, you must first understand that a neuron is a *polarized* cell. Compared with its surroundings, the inside of the cell is more negatively charged — to the tune of −70mV. This negative charge is due to the high concentration of *anions* (negatively charged molecules) held there and that the neuron maintains when at rest — hence the term *resting potential*. Additionally, the axon has a high concentration of potassium ions (K⁺) inside and a high concentration of sodium ions (Na⁺) outside.

Don't let the positive charge of potassium ions throw you off. Numerous anions counteract the K⁺ inside the cell, creating the resting potential of −70mV.

TIP

Stimulus

The stimulus of a neuron generally happens in one of two ways. Most commonly, a neuron is stimulated by a *neurotransmitter* — a chemical messenger released from another neuron. The neurotransmitter binds to receptors on the dendrites, triggering ion channels to open. The other option occurs only in sensory receptor neurons. Here, the designated stimulus, such as heat in a hot (but not cold!) thermoreceptor or touch in a mechanoreceptor, causes the ion channels to open. Regardless of how the channels are triggered to open, positive ions rush into the cell. (It's −70mV in the cell, after all.) This event causes the *depolarization* of the cell body as it becomes more positive (but overall, still negatively charged). This initial depolarization, however, does not mean the neuron will generate an impulse. For that to occur, the internal charge must reach the *threshold potential*, which is −55mV (on average). If the threshold is not reached, the ions are pumped out of the cell body, and no signal is passed on.

REMEMBER

A single stimulus generally isn't enough to warrant a message being sent to the brain. Remember we're talking about a single neuron responding to a single change, not the event as a whole. A larger change or a repeated stimulus will open more channels, causing more positive ions to rush in, and the threshold is more likely to be reached. This situation causes the *all-or-none response* of a neuron: It either sends a message or it doesn't. Value of sensation ("This is warm" versus "Whoa, that's really hot!") is imparted by the brain based on how many neurons are sending the same message.

Action potential versus impulse

Two types of ion channels are found along the axon: one for sodium (Na^+) and the other for potassium (K^+). Both of these channels are *voltage-gated*, meaning that a specific electrical charge must be reached; when that occurs, they pop right open. For the sodium channels, that charge is −55mV, which is why it's referred to as the threshold potential.

The axon hillock is the *trigger zone*, generating an impulse once the threshold potential has been reached. As soon as enough positive ions rush in to the cell body through the dendrites and the charge reaches −55mV, the sodium channels in the trigger zone pop open. To say the same thing in proper jargon, with sufficient stimulus, the membrane of the axon hillock depolarizes, reaching threshold, signaling the opening of voltage-gated sodium ion channels.

Either way you say it, the generation of an impulse is actually the beginning of the first *action potential*. A single action potential occurs in the following steps:

1. Sodium gates open.

2. Na^+ rushes in, depolarizing that section of the axon.

3. This process continues until enough Na^+ comes in to trigger the voltage-gated channels to close, which is +35mV on average.

4. Reaching 35mV also triggers potassium gates to open, and K^+ flows out.

5. Now the axon segment is repolarized to the resting potential (−70mV).

6. Too many K^+ leave, causing *hyperpolarization* of the area.

7. The hyperpolarization kick-starts the *sodium-potassium pumps* (so to speak) to move the ions against their concentration gradients to their original locations.

The Na/K pumps are the only step of action potential that requires ATP. Until all the ions have been replaced, the axon segment is in a *refractory period*, meaning that it's unable to respond to another stimulus.

An action potential carries us down only one small segment of the axon, nowhere near its entire length. This is where *impulse*, the sequential generation of new action potentials, comes in. (Refer to Figure 8-3.) When Na+ first rushes in, some of the ions flow down the axon, increasing the charge within. When enough slide over to hit −55mV, a new set of sodium gates open, generating the next action potential. This process repeats all the way to the synaptic knob while the individual segments continue to proceed through the process of action potential.

TIP

Remember learning how to sing in a round (or singing rounds) in elementary school? Students were split into groups to sing "Row, row, row your boat," and each group started at a different time. This is how impulse works. If you follow the impulse, all you hear is "row, row, row" over and over again until the synaptic knob is reached. But if you listen to only the first group, you hear the entire song — the action potential.

TECHNICAL
STUFF

Many neurons are *myelinated*, where axons are wrapped in insulating lipids created by glial cells — *oligodendrocytes* in the CNS and *Schwann cells* in the PNS. (Both of the neurons illustrated in Figure 8-1 are myelinated.) The myelin is said to speed up impulse transmission through a process called *saltatory conduction*. The myelin does not cover the entire axon; small sections called the *nodes of Ranvier*, are left exposed. When Na+ enters the axon, the ions continue to flow down the axon, under the insulation of the myelin. A new action potential can be initiated only at the nodes (because any Na+ gates would be covered). Thus, a myelinated neuron needs far fewer action potentials to span the length of the axon, saving energy (less ion pumping for recovery) and increasing signal speed by up to 100 times. This is called saltatory conduction because *saltus* is Latin for *jump*; it's like the action potentials jump from node to node.

Across the synapse

Unfortunately, most neurons don't touch, so they can't pass the spark directly to the next cell in line. A gap called a *synapse* or *synaptic cleft* separates the axon of one neuron from the dendrite of the next. To bridge this gap, a neuron releases a chemical called a *neurotransmitter* that may (or may not) cause the next neuron to generate an electrical impulse. This process, called *synaptic transmission*, is shown in Figure 8-4.

When the impulse reaches the synaptic knob, it again triggers ion channels to open. This time, though, they're calcium gates, and calcium ions (Ca2+) rush in. The Ca2+ that enter effectively push the packets (vesicles) of neurotransmitters toward the membrane. Then the neurotransmitters are released into the synapse, where they can bind with dendrites in the area that have the matching receptor — or any other cell, for that matter, because this process is how ACh (acetylcholine) stimulates muscle contraction.

Each type of neurotransmitter has its own type of receptor. Whether the *postsynaptic* (receiving) *neuron* is excited or inhibited depends on the neurotransmitter and its effect. If the neurotransmitter is *excitatory*, the Na+ channels open, the neuron membrane becomes depolarized, and the impulse travels through that neuron. If the neurotransmitter is *inhibitory*, the K+ channels open, the neuron membrane becomes hyperpolarized as the ions exit, and any ongoing impulse is stopped.

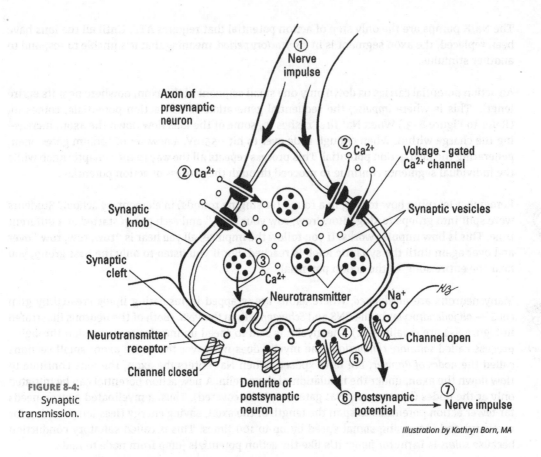

Nerve
impulse

Axon of
presynaptic
neuron

② Ca²⁺

Voltage - gated
Ca²⁺ channel

② Ca²⁺

Synaptic
knob

Synaptic vesicles

Synaptic
cleft

③
Ca²⁺

Neurotransmitter

Na⁺

Neurotransmitter
receptor

Channel open

Channel closed

Dendrite of
postsynaptic
neuron

④

⑤

⑥ Postsynaptic
potential

⑦

→ Nerve impulse

FIGURE 8-4:
Synaptic
transmission.

Illustration by Kathryn Born, MA

After the neurotransmitter produces its effect (excitation or inhibition), the receptor releases it, and the neurotransmitter goes back into the synapse. From here, one of three things happens to the neurotransmitter:

>> It's broken down by an enzyme.

>> It's taken back up by the cell that released it (a form of recycling).

>> It simply floats away into the surrounding fluid.

PRACTICE

You made it to the end of this detailed-packed section. Now get your neurons firing by answering some questions for a crucial memory-building repetition.

For questions 24–28, match each term with its description.

24 _____ All-or-none response

(a) A differential in charge from outside the cell to inside

25 _____ Hyperpolarization

(b) Negatively charged molecule concentrated inside the cell

26 _____ Anion

(c) Threshold of excitation determines ability to respond

27 _____ Polarization

(d) A drop in charge inside the cell past resting potential

28 _____ Depolarization

(e) An increase in charge inside the cell

29 Place the steps of action potential in order from 1 to 11:

_____ Potassium gates close.

_____ Pumps restore polarization.

_____ Cell depolarizes.

_____ Sodium rushes in.

_____ Potassium flows out.

_____ Sodium gates open.

_____ Sodium gates close.

_____ Potassium gates open.

_____ Ions return to initial concentrations.

_____ Cell repolarizes.

_____ Cell is hyperpolarized.

30 What is the role of calciumions (Ca^{2+}) in nervous communication?

(a) They repolarize the cell during stimulation.

(b) They exit the cell to depolarize it during impulse transmission.

(c) They enter the cell to depolarize it during action potential.

(d) They repolarize the cell to activate the synaptic knob.

(e) They enter the cell to trigger the release of neurotransmitters.

31 Explain the difference between impulse and action potential.

_____

_____

_____

Minding the Central Nervous System

The central and peripheral nervous systems are physically separate but functionally integrated networks of nervous tissue. Working together, these networks perceive and respond to internal and external stimuli to maintain homeostasis and simultaneously move the genetic development program forward. It's the job of the brain and spinal cord to make sense of all the input and then send out marching orders.

Show some spine

The spinal cord is the vice president to the brain's chief executive officer (CEO). It organizes and reorganizes all of the input, bundling relevant information and putting it on the path to the appropriate department within the brain. It's also empowered to make some decisions on its own without bothering the higher-ups. In other words, the spinal cord has two primary functions: It conducts nerve impulses to and from the brain, and it processes sensory information to produce a spinal reflex without initiating input from the higher brain centers.

The spinal cord begins at the brainstem and extends down into the lumbar region of the vertebral column. In an adult, its bundled, organized structure ends between the first and second lumbar vertebrae, roughly where the last ribs attach. From there, it branches into separate strands that taper off, forming what's called the *cauda equina* (horsetail).

The vertebral column encases the spinal cord as it feeds through the cylindrical opening created by the vertebrae (see Figure 8-5). Creating a protective layer around the spinal cord are three tough layers of connective tissue called the *meninges*. Within these layers is the *cerebrospinal fluid* (CSF), which circulates around the spinal cord and brain to transport material through the CNS.

Two types of solid material make up the inside of the cord: *gray matter* (which is indeed grayish in color), containing unmyelinated neurons, dendrites, cell bodies, and neuroglia; and *white matter*, so-called because of the whitish tint of its myelinated nerve fibers. At the cord's midsection is a small *central canal* filled with CSF and surrounded first by gray matter in the shape of the letter *H* and then by white matter, which fills in the surrounding area. The top, skinny legs of the *H* are the *posterior horns*, and the bottom, thicker sections are the *anterior horns*. The crossbar is the *gray commissure*, which houses the central canal.

The white matter consists of thousands of myelinated nerve fibers arranged in three *funiculi* (columns) on each side that convey information up and down the spinal cord to the brain. The funiculi contain *nerve tracts* that further organize the flow of information. Akin to nerves, these tracts can be ascending (transmitting sensory information to the brain), descending (transmitting motor information from the brain), or bidirectional.

Finding your way around Grand Central Station

The brain is one of the largest organs in the human body; it makes up 3 percent of our body weight but uses about 20 percent of our energy. As the taskmaster, the brain manages its workload by compartmentalizing its functions. The parts of the brain are in constant contact and certainly influence one another, but each part is responsible for carrying out different functions.

The brain's major parts are the *cerebrum, cerebellum, diencephalon,* and *brainstem.* The brain's four connecting, fluid-filled cavities are called *ventricles.* Let's see if we can explore the brain without getting you lost!

TIP

As in Chapter 7, you won't see any labeled figures of brain anatomy in this section. All the structures are addressed descriptively. Immediately after this section, you'll see a brain to label. See what structures you can identify based solely on how they were described, or at least make your best guess. Then refer to the answers, and make corrections. You'll have a labeled diagram for future reference and also a solid foundation for building the details into your memory.

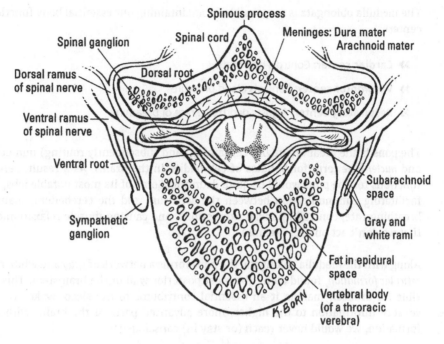

Spinous process

Spinal ganglion

Spinal cord

Meninges: Dura mater
Arachnoid mater

Dorsal ramus
of spinal nerve

Dorsal root

Ventral ramus
of spinal nerve

Ventral root

Subarachnoid
space

Sympathetic
ganglion

Gray and
white rami

Fat in epidural
space

Vertebral body
(of a throracic
verebra)

K. BORN

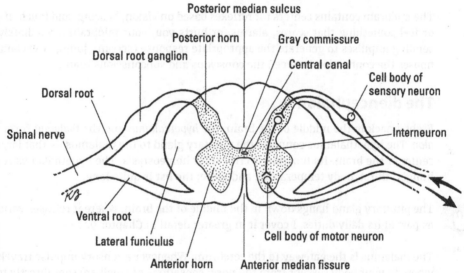

Posterior median sulcus

Posterior horn

Gray commissure

Dorsal root ganglion

Central canal

Dorsal root

Cell body of
sensory neuron

Spinal nerve

Interneuron

Ventral root

Lateral funiculus

Cell body of motor neuron

Anterior horn

Anterior median fissure

FIGURE 8-5:
Cross section
of the spinal
cord.

Illustration by Kathryn Born, MA

The brainstem

The *brainstem* consists of the *medulla oblongata*, *pons*, and *midbrain*. The medulla oblongata is continuous with the spinal cord, beginning where the cord passes through the *foramen magnum*, the hole at the bottom of the skull. As the brainstem widens, the oval-shaped *pons* juts out. Posterior to (behind) it is the *midbrain*, which continues upward.

The medulla oblongata is responsible for maintaining our essential body functions. It has three centers:

>> **Cardiac center:** Controls heart rate

>> **Vasomotor center:** Controls blood pressure

>> **Respiratory center:** Adjusts depth and rate of breathing

The pons is the great organizer, containing (and subsequently routing) numerous nerve tracts and *nuclei* (the term for bundles of cell bodies in the brain). As a result, generating a list of functions for the pons would be quite daunting. Some of its most notable jobs, though, include facilitating communication between the cerebrum and the cerebellum, maintaining regular breathing rate, controlling facial expressions, and causing muscle relaxation during sleep (so that we don't act out our dreams).

Along with the medulla oblongata, the pons forms a network of gray and white matter called the *reticular formation,* found along the most posterior wall of the brainstem. This area is responsible for arousal, making it an essential contributor to the sleep–wake cycle; it also routes sensory information to the higher, more advanced parts of the brain. Without the reticular formation, we would never reach (or stay in) consciousness.

The midbrain contains centers for reflexes based on vision, hearing, and touch. If you see, hear, or feel something that scares, alarms, or hurts you, your midbrain immediately responds by sending impulses to generate the appropriate response (scream, jump, or exclamation). It also houses the connection between the conscious and subconscious brain.

The diencephalon

Right smack in the middle of the brain, the *hypothalamus* and the *thalamus* form the *diencephalon.* The hypothalamus connects the pituitary gland to the thalamus — that large circle in the center of the brain. Its function is critical for homeostasis. The hypothalamus regulates sleep, hunger, thirst, body temperature, and more; the list is quite long.

REMEMBER

The pituitary gland hangs down in the middle of the brain, where it releases various hormones as part of its daily duties. I cover it in greater detail in Chapter 9.

The thalamus is the gateway to the cerebrum. Whenever a sensory impulse travels from somewhere in your body (except from the nose; sensations of smell are sent directly to the brain by the olfactory nerve), it passes through the thalamus. Then the thalamus relays the impulse to the proper location in the cerebrum, which interprets the message.

Also in the diencephalon are the following:

>> **Optic chiasma:** Here's where the optic nerves carrying input from the eyes cross over to ensure that both halves of the brain get the same information.

>> **Mammillary bodies:** These structures play a role in memory recall.

>> **Pineal gland:** This gland controls your circadian rhythm. That is, it interprets the environmental dark/light cycle, releasing *melatonin* to signal that it's bedtime.

The cerebellum

The *cerebellum* lies just below the back half of the cerebrum. A narrow stalk called the *vermis* connects the cerebellum's left and right hemispheres. The inner white matter resembles a tree and is called the *arbor vitae*. (*Arbor* means *tree*, and *vitae* means *of life*.) The remainder is gray matter.

The cerebellum coordinates your skeletal–muscle movements, making them smooth and graceful instead of stiff and jerky. The cerebellum also maintains normal muscle tone and posture, using sensory information from the eyes, inner ear, and muscles.

The cerebrum

If you're conscious, you're using your cerebrum, which is the brain's largest part and controls our more advanced functions.

The cerebrum is divided into left and right halves, called the *left* and *right cerebral hemispheres*, and each half has four lobes: *frontal, parietal, temporal,* and *occipital.* The names of the lobes come from the skull bones that overlay them. (See Chapter 6 for more on the old skull and bones.)

The *cerebral cortex* is the brain's outer layer of gray matter. It covers the entire surface of the cerebrum and overlays the deeper white matter. The brain's elevations are called *gyri* (singular *gyrus*). The shallow grooves that separate the elevations are called *sulci* (singular *sulcus*). Deep grooves in the brain are called *fissures*.

When you look at the top of a brain, you notice a deep groove running down the middle of the cerebrum. This groove is the *longitudinal fissure*, which incompletely divides the cerebrum into the left and right hemispheres. The *corpus callosum*, located within the brain at the bottom of the longitudinal fissure, contains myelinated fibers that connect the left and right hemispheres.

Table 8-2 gives you an overview of what each lobe of the cerebrum controls.

Table 8-2   Functions of Lobes within Cerebral Hemispheres

Lobe	Functions
Occipital	The primary vision center. It processes all visual information and combines images with other sensory information.
Temporal	The primary auditory center. It houses situational memories such as visual scenes and music. It also plays a role in understanding and producing speech.
Parietal	The primary site for integration of sensory information. This area is responsible for our conscious perception of touch, temperature, and pain. The *somatosensory cortex* is located here, as is *Wernicke's area*, which is responsible for language comprehension. The parietal lobe is also responsible for spatial awareness and maintaining attention.
Frontal	The most advanced part of the brain. It's responsible for our highest intellectual functions, such as planning, logical reasoning, morality, and social behaviors. It contains the *primary motor cortex*, which initiates voluntary movements. The frontal lobe also contains *Broca's area*, which is responsible for the generation of speech.

The meninges and ventricles

Like the spinal cord, the brain is surrounded by protective membranes called *meninges.* In fact, the tissue is continuous. The outermost layer is the *dura mater,* which is followed by the *arachnoid mater.* Between these two layers flows a fluid much like interstitial fluid. The innermost membrane, the *pia mater,* contacts the nervous tissue. Between the arachnoid and pia flows a fluid unique to the CNS called *cerebrospinal fluid* (CSF). Cavities are created in the *subarachnoid space* (between the arachnoid and pia layers), forming ventricles, which are basically pools of CSF.

Each cerebral hemisphere contains a *lateral ventricle* (the first and second ventricles). The other two ventricles are, believe it or not, the third and fourth ventricles. (A *ventricle* is a connecting, fluid-filled cavity.) The *third ventricle* lies just about in the center of your brain, and the *fourth ventricle* lies at the top of the brainstem. The *cerebral aqueduct* connects the third and fourth ventricles. From the inferior portion of the fourth ventricle, a narrow channel called the *central canal* continues down through the spinal cord. See Figure 8-6 for a view of the ventricles.

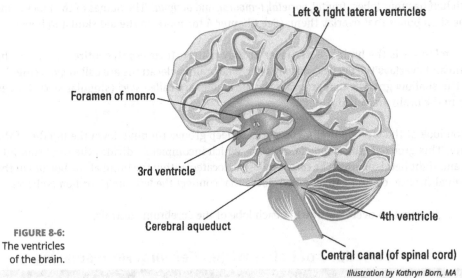

Left & right lateral ventricles

Foramen of monro

3rd ventricle

Cerebral aqueduct

4th ventricle

Central canal (of spinal cord)

FIGURE 8-6:
The ventricles
of the brain.

Illustration by Kathryn Born, MA

TIP

When you think of an aqueduct, you may think of Rome. The Romans built aqueducts as a system for distributing water. Well, in your central nervous system, the ventricles and aqueducts serve as a system to circulate CSF.

The CSF picks up waste products from the CNS cells and delivers them to the bloodstream for disposal. The CSF also cushions the CNS, creating a protective layer in and around your brain and spinal cord. The brain's capillaries are less permeable than elsewhere in the body, so there's much greater control of what molecules can actually exit the blood flow there. The tightly bound epithelial cells along the pia mater control what is allowed out of the CSF and into the brain tissue itself. This collective control is referred to as the *blood–brain barrier.*

PRACTICE

For questions 32–40, label the parts of the brain shown in Figure 8-7.

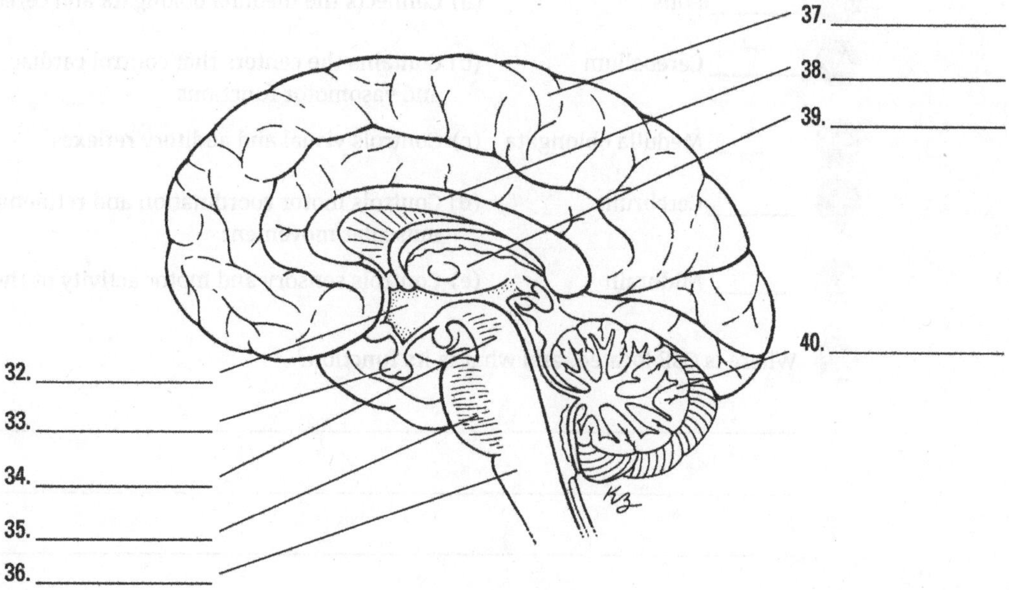

37. _____

38. _____

39. _____

40. _____

32. _____

33. _____

34. _____

FIGURE 8-7:
Sagittal view of
the brain.

35. _____

36. _____

Illustration by Kathryn Born, MA

For questions 41–45, match each term with its description.

41 _____ White matter

42 _____ Reticular formation

43 _____ Ganglion

44 _____ Corpus callosum

45 _____ Gray matter

(a) Has the capacity to arouse the brain to wakefulness

(b) Myelinated fibers

(c) Coordinates communication between right and left brain

(d) Bundle of neuron cell bodies

(e) Unmyelinated fibers, neuron cell bodies, and neuroglia

46 Where within the cerebrum does visual activity take place?

(a) Parietal lobe

(b) Temporal lobe

(c) Occipital lobe

(d) Frontal lobe

47 What does the cerebellum do? Choose all that apply.

(a) Controls visual activity

(b) Assesses balance and body position

(c) Organizes memories

(d) Interprets auditory stimulation

(e) Coordinates fine motor control

For questions 48–52, match the brain structure with its function.

48 _____ Pons   (a) Connects the medulla oblongata and cerebellum

49 _____ Cerebellum   (b) Contains the centers that control cardiac, respiratory, and vasomotor functions

50 _____ Medulla oblongata   (c) Controls visual and auditory reflexes

51 _____ Cerebrum   (d) Controls motor coordination and refinement of muscular movement

52 _____ Midbrain   (e) Controls sensory and motor activity of the body

53 Where is CSF located, and what is its function?

_____

_____

_____

Taking the Side Streets: The Peripheral Nervous System

The *peripheral nervous system* (PNS) consists of the nerves and ganglia outside the brain and spinal cord. Unlike the CNS, the PNS isn't protected by bone or by the blood–brain barrier, leaving it exposed to toxins and mechanical injuries. Structures of the PNS include

>> **Cranial nerves:** Twelve pairs of nerves that emerge directly from the brain and brainstem. Each pair is dedicated to particular functions. Some pairs bring information from the sense organs to the brain, and others control muscles, but most have both sensory and motor functions. Some pairs are connected to glands or internal organs, such as the heart and lungs. The *vagus nerves,* for example — the longest pair among the cranial nerves — pass through the neck and chest into the abdomen. They relay sensory impulses from part of the ear, tongue, larynx, and pharynx; motor impulses to the vocal cords; and motor and secretory impulses to some abdominal and thoracic organs.

>> **Spinal nerves:** Thirty-one pairs of nerves that emerge from the spinal cord. Each pair contains thousands of *afferent* (sensory) and *efferent* (motor) fibers.

>> **Sensory nerve fibers:** Nerve fibers all over the body that send impulses to the CNS via the cranial nerves and spinal nerves.

>> **Motor nerve fibers:** Nerve fibers that connect to muscles and glands and send impulses from the CNS via the cranial and spinal nerves.

Getting on your nerves

Spinal nerves connect with the spinal cord via two bundles of nerve fibers, or *roots*. The *dorsal root* contains afferent fibers that carry sensory information to the CNS. Cell bodies of sensory neurons lie outside the spinal cord in a bulging area called the *dorsal root ganglion*, allowing the spinal cord to organize incoming information and route it to the appropriate ascending nerve tract. A second bundle, the *ventral root*, contains efferent motor fibers; these cell bodies lie inside the spinal cord. In each spinal nerve, the two roots join outside the spinal cord to form what's called a *mixed spinal nerve*.

Catlike reflexes

Spinal reflexes, or *reflex arcs*, occur when a sensory neuron transmits a signal — such as a sensation of burning heat — through the dorsal root ganglion. An interneuron in the spinal cord receives the stimulus and passes the signal directly to a motor neuron (or efferent fiber) that stimulates a muscle, which immediately pulls the burning body part away from heat. Other reflexes, like the knee-jerk, or *patellar*, reflex, forgo the interneuron, and the sensory neuron stimulates the motor neuron directly.

After a spinal nerve leaves the spinal column, it divides into two small branches. The posterior, or *dorsal ramus*, goes along the back of the body to supply a specific segment of the skin, bones, joints, and longitudinal muscles of the back. The ventral, or *anterior ramus*, is larger than the dorsal ramus; it supplies the anterior and lateral regions of the trunk and limbs.

Have some sympathy

The efferent nerves are also grouped by their control; whether or not they're somatic or autonomic. *Somatic* nerves are under voluntary control; pick any one of those to follow out from the spinal cord, and you'll wind up at a skeletal muscle. The rest are *autonomic* — involuntarily controlled, in other words. Our maintenance processes, such as digestion, breathing, and blood circulation, happen automatically and are stimulated by autonomic nerves. But these processes aren't as simple as one nerve saying "Turn on" to everything while the other says "Turn off." We need to prioritize which processes get the energy based on the situation. When we're exercising, for example, we'd rather not expend valuable energy digesting our breakfast smoothie when we need that energy for muscle contractions, not to mention deeper breathing and increased heart rate to deliver that critical oxygen. On the other hand, when we're lying on the couch watching a movie and regretting our choice to eat that extra slice of pizza, we want our digestive system to do its thing so that our discomfort fades quickly. As a result, the autonomic nervous system is split into the sympathetic and parasympathetic nervous systems (see Figure 8-8):

>> **Sympathetic nervous system:** Responds to stress and is responsible for the increase in your heart rate, blood pressure, respiration, and numerous other physiological changes. This pathway inhibits functions such as digestion, urination, and immunity. This system is responsible for your "fight or flight" response.

>> **Parasympathetic nervous system:** Affectionately termed "rest and digest," this pathway is used most when you're relaxed. It opposes all the actions of the sympathetic pathway, such as constricting the pupils, dilating blood vessels, and slowing breathing. It stimulates our digestive and urinary functions, kicking them into high gear.

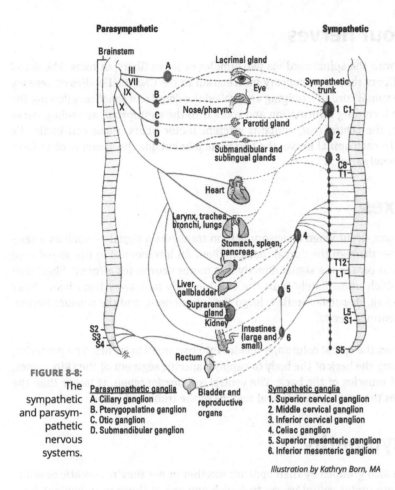

FIGURE 8-8:
The sympathetic and parasympathetic nervous systems.

Parasympathetic

Brainstem

III
VII
IX
X

A Lacrimal gland
B Eye
C Nose/pharynx
D Parotid gland
   Submandibular and sublingual glands

Heart

Larynx, trachea, bronchi, lungs

Stomach, spleen, pancreas

Liver, gallbladder

Suprarenal gland
Kidney

Intestines (large and small)

Rectum

S2
S3
S4

Bladder and reproductive organs

Sympathetic

Sympathetic trunk

1 C1
2
3
   C8
   T1

4

   T12
   L1

5

   L5
   S1

6

   S5

Parasympathetic ganglia
A. Ciliary ganglion
B. Pterygopalatine ganglion
C. Otic ganglion
D. Submandibular ganglion

Sympathetic ganglia
1. Superior cervical ganglion
2. Middle cervical ganglion
3. Inferior cervical ganglion
4. Celiac ganglion
5. Superior mesenteric ganglion
6. Inferior mesenteric ganglion

Illustration by Kathryn Born, MA

WARNING

Connecting the sympathetic nervous system to your fight-or-flight (or freeze) fear response is quite helpful, because it serves to remind you of all its effects (the things you want your body to do if you're running for your life or about to fight for it). But be careful that you don't fall into the trap of thinking that it's an all-or-nothing thing. When you catch a matinee at your favorite movie theater, you could be exiting the dark theater into the bright, squint-inducing afternoon sun. The dramatic change in light requires your pupils to constrict — an order that would come in on a sympathetic nerve. You don't need all the other sympathetic responses to join the action, however. In fact, your parasympathetic nerves are still orchestrating the digestion of the popcorn you ate. The actions oppose each other, but you don't have to fire impulses down every single sympathetic (or parasympathetic) nerve at the same time.

Sensory Overload

Some anatomists consider the sensory system to be part of the PNS; others regard it as a separate system. Either way, there are as many as 21 senses (depending on how they're grouped). All the senses use varieties of four categories of receptors. See Table 8-3 for their description.

Table 8-3   Sensory Receptor Types

Receptor	Stimulus	Examples
Chemoreceptor	Chemical change	Smell and taste, blood glucose, pH
Thermoreceptor	Temperature change	Found in skin; separate receptors for warm and cold
Photoreceptor	Light	Rods and cones in retina of eye
Mechanoreceptor	Pressure/mechanical forces	Touch receptors, blood pressure, stretch (in bladder and lungs)

You may be saying to yourself, "Wait, I thought we had five senses! Have I been taught wrong my whole life?" The answer is not really. The five senses you're familiar with — touch, hearing, sight, smell, and taste — are your *perceived senses*. That is, you're aware that your brain has interpreted the sensory input. You say "Ouch, that's hot!" or "Wow, that smells nice!" The other sensory inputs are processed, of course, but that processing occurs outside of your conscious thought. Otherwise, you'd be constantly bombarded with thoughts about pH and blood pressure monitoring.

Out of touch

The sensory receptors all over the skin perceive at least five types of sensation: pain, heat, cold, touch, and pressure. The five are often grouped together as the single sense of touch.

Receptors vary in terms of their overall abundance (pain receptors are far more numerous than cold receptors) and their distribution over the body's surface (the fingertips have far more touch receptors than the skin of the back). Areas where the touch sensors are packed in are especially sensitive.

The structure of the touch-related receptors varies with their function. *Free nerve endings* (dendrites) are responsible for itching and pain. Both the cold and warm thermoreceptors are free nerve endings as well. Other receptors are modified; the dendrites are wrapped in connective tissue fibers forming mechanoreceptors. Two varieties of these receptors are in your skin. *Meissner's* (or *tactile*) *corpuscles* respond to light touch, and *Pacinian* (or *lamellated*) *corpuscles* respond to a heavier, more forceful touch (pressure).

Seeing is believing

Although there are many romantic notions about eyes, the truth is that an eyeball is simply a hollow sphere bounded by a trilayer wall and filled with a gelatinous fluid called, oddly enough, *vitreous humor*. The outer fibrous coat is made up of the *sclera* (the white of the eye) and the *cornea* in front. The sclera provides mechanical support, protection, and a base for attachment of eye muscles, which assist in the focusing process. The cornea covers the anterior with a clear window.

The *palpebrae* (eyelids) extend from the edges of the eye orbit, into which roughly five sixths of the eyeball is recessed. A mucous membrane called the *conjunctiva* covers the inner surface of each eyelid and the anterior surface of the eye, but not the cornea. (The infamous pink eye is the inflammation of this membrane.) Up top and to the side of the orbital cavities are *lacrimal glands,* which secrete tears that are carried through a series of lacrimal ducts to the conjunctiva of the upper eyelid. Ultimately, these secretions flow across the eyes and drain into the *naso-lacrimal ducts* in the inside corners.

An intermediate layer called the *uvea* provides blood and lymphatic fluids to the eye; regulates the amount of light that enters; and secretes and reabsorbs *aqueous humor*, the thin, watery liquid that fills the anterior chamber of the eyeball in front of the iris and the posterior chamber between the iris and the lens. The uvea has three components:

>> **Iris:** Contains blood vessels and smooth muscle fibers to control the pupil's diameter. The pigments that provide our eye color prevent light from entering anywhere other than the pupil.

>> **Choroid coat:** A thin, dark brown, vascular layer lining most of the sclera on the back and sides of the eye. The choroid contains arteries, veins, and capillaries that supply nutrients to the *retina* and sclera, and it contains pigment cells to absorb light and prevent reflection and blurring. (We have the choroid coat to thank for the dreaded red eyes that appear in photos.)

>> **Ciliary body:** An extension of the choroid coat that leads to the iris. It also attaches to the lens via *suspensory ligaments* and controls the lens' shape to aid in focusing.

The crystalline lens consists of concentric layers of protein. This lens is *biconvex* in shape, bulging outward and, with help from the cornea, bends light, projecting it into the inner tunic. When the muscles of the ciliary body contract, the shape of the lens changes, altering the visual focus. This process — *accommodation* — allows the eye to see objects both at a distance and close-up.

The *inner tunic* is the internal lining of the eye. Here, the *retina* lies atop the choroid coat, containing the photoreceptors that will communicate with the occipital lobe of the brain. The space created, called the *posterior cavity*, is filled with vitreous humor, a jelly-like fluid that maintains the shape of the eye. The photoreceptors of the retina come in two varieties: *rods* and *cones,* whose axons are bundled into the optic nerve that exits out the back. The rods are dim light receptors that provide the general shape of the image. The cones detect bright light and provide sharpness and color to the image. The retina has an *optic disc*, which is essentially a blind spot incapable of producing an image. The eye's blind spot is a result of the absence of photoreceptors in the area of the retina where the optic nerve leaves the eye.

The *macula* is an oval, highly pigmented (yellow) area near the center of the retina. A small dimple near the center of the macula is called the *fovea centralis*. (*Fovea* is Latin for *pit*.) This dimple is where the eye's vision is sharpest and where most of the eye's color perception occurs. Only cones can be found in this area, smaller and more closely packed than anywhere else in the retina.

Have you heard?

We hear because the ear changes sound waves to nerve impulses that pass through the eighth cranial nerves to both sides of the brainstem and up to the temporal lobes. Sound moves through air in waves of pressure. Your outer ear, the *auricle*, acts as a funnel to channel sound waves through the *external auditory canal*, causing the *tympanic membrane* (eardrum) to vibrate. (The outer ear isn't necessary for hearing, but it certainly helps.) The ear bones, called *ossicles*, receive and amplify the vibration and transmit it to the *cochlea* of the inner ear. It moves first through the *malleus*, the *incus*, and the *stapes*; then the vibrations create ripples through the fluid within the cochlea. The mechanoreceptors that line it, called *hair cells*, release neurotransmitters when stimulated, generating the impulse that we perceive as sound.

Starting just below the ossicles and connecting to the nasopharynx (back of the nasal cavity) are the *eustachian tubes.* Besides draining any excess secretions from the middle ear, these tubes regulate the air pressure behind the eardrum. When the air pressure within the ear doesn't match the pressure of the air in the environment, hearing is difficult. At rest, the eustachian tubes are closed, but they open to equalize the pressure when you swallow or yawn.

Also housed in the inner ear are the *semicircular canals,* which transmit information to your brain regarding what position your head is in; that is, whether you're horizontal or vertical, spinning or still, or moving forward or backward. Therefore, your ears are the key organ of balance. The process of transmitting information to the brain about your body position is basically the same as that of hearing. When you're moving, the inner ear's fluid moves and causes hair cells in the semicircular canals to bend, sending impulses to the cerebellum. Our sense of balance and the awareness of our limbs is termed *proprioception.*

You have good taste

Your sense of taste *(gustation)* has a simple goal: to help you decide whether to swallow or spit out whatever is in your mouth. This extremely important decision can be made based on a few taste qualities. The tongue, the sense organ for taste, has chemoreceptors to detect certain aspects of the chemistry of food and some toxins, especially poisons made by plants to deter predation by animals.

The human tongue has about 10,000 *taste buds,* each with 50 to 150 chemoreceptor cells. A taste bud's *gustatory cells* generate nerve impulses that are transmitted to the sensory interpretation areas of the brain via the seventh, ninth, and tenth cranial nerves. The sense of taste is closely tied to the endocrine and digestive systems. The brain interprets the yummy stimulus and triggers the release of the digestive enzymes needed to break down that food. That is, if we don't reflexively spit it out because the input signaled a potentially harmful chemical.

Taste buds have receptors for sweet, sour, bitter, and salty sensations, as well as a fifth sensation called *umami* (savoriness, the flavor associated with meat, mushrooms, and many other protein-rich foods). Salty and sour detection is needed to control salt and acid balance. Bitter detection warns of foods that may contain poisons; many of the poisonous compounds that plants produce for defense are bitter. Sweet detection provides a guide to calorie-rich foods and umami detection to protein-rich foods.

Each receptor cell in a taste bud responds best to one of the five basic tastes. A receptor can respond to the other tastes, but it responds strongest to a particular taste. Also, the taste buds detect only the rather unsubtle aspects of flavor. The nose is responsible for our enjoyment of more-complex, subtler flavors.

TECHNICAL STUFF

The structure and function of taste buds is an area of active research and controversy, especially in the food industry. Even the exact number of basic tastes and their characterizations are controversial, with new basic tastes being proposed from time to time. Regardless, the majority of what we perceive as flavor (good or bad) comes from olfactory rather than gustatory input.

You smell well

Embedded in the epithelium of your nasal cavity are the *olfactory receptor cells*. As you inhale, and the air swirls around in your nose, *odorant* molecules stimulate the receptors. The axons of these highly specialized receptor neurons are bundled into the *olfactory bulbs*. These extensions of brain tissue lie atop the *cribriform plate* of the *ethmoid bone*, that forms the superior (top) border of your nasal cavity.

Smell is the only sense that isn't routed through the thalamus before reaching the cerebral cortex. But as with all other sensory input, the fact that a message reaches the brain doesn't mean that you'll become consciously aware of it. This direct pathway, though, is why smell is the sense with the strongest connection to memory.

 Do you have a good sense of your senses? It's time to find out.

PRACTICE For questions 54–67, identify the structures of the eye illustrated in Figure 8-9.

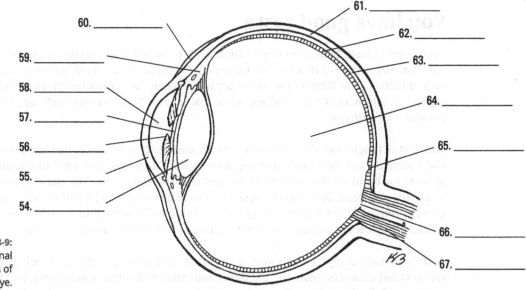

60. _____

61. _____

62. _____

63. _____

64. _____

65. _____

66. _____

67. _____

59. _____

58. _____

57. _____

56. _____

55. _____

54. _____

FIGURE 8-9:
The internal
structures of
the eye.

Illustration by Kathryn Born, MA

(a) Anterior chamber

(b) Choroid coat

(c) Ciliary body

(d) Conjunctiva

(e) Cornea

(f) Fovea centralis

(g) Iris

(h) Lens

(i) Optic disc

(j) Optic nerve

(k) Posterior cavity

(l) Pupil

(m)Retina

(n) Sclera

For questions 68–77, identify the structures of the ear illustrated in Figure 8-10.

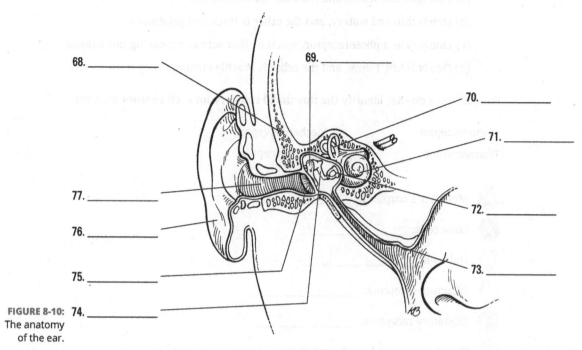

68. _____

69. _____

70. _____

71. _____

72. _____

73. _____

77. _____

76. _____

75. _____

FIGURE 8-10: 74. _____
The anatomy
of the ear.

Illustration by Kathryn Born, MA

(a) Auricle

(b) Cochlea

(c) Eustachian tube

(d) External auditory canal

(e) Incus

(f) Malleus

(g) Semicircular canal

(h) Stapes

(i) Temporal bone

(j) Tympanic membrane

78 What is proprioception? What sensory input does the brain use for this function?

_____

_____

_____

79 What is the difference between aqueous humor and vitreous humor?

(a) One lines the sclera, and the other bathes the iris.

(b) One is thin and watery, and the other is thick and gelatinous.

(c) One acts as a photoreceptor, and the other acts as a focusing mechanism.

(d) One is subtly funny, and the other is brashly obvious.

For questions 80–84, identify the functional category for each sensory receptor.

Chemoreceptor   Mechanoreceptor

Photoreceptor   Thermoreceptor

80 Meissner's corpuscle: _____

81 Cone cells: _____

82 Hair cells: _____

83 Olfactory receptors: _____

84 Gustatory receptors: _____

85 Which statement best describes the function of ossicles?

(a) They bind chemicals to communicate a savory flavor.

(b) They transmit vibrations from sound waves.

(c) They control the amount of light passing through the lens.

(d) They send a message of pain when tissue has been damaged.

(e) They combine shape and color input for vision.

Practice Questions Answers and Explanations

1. **Central**
2. **Peripheral**
3. **Efferent**
4. **Afferent**
5. **Somatic**
6. **Autonomic**
7. **Sympathetic/Parasympathetic** (either one)
8. **Parasympathetic/Sympathetic** (either one)
9. **Sensory**
10. **Motor**

> When you have a large of group of closely related terms, it's far more beneficial to review them like this — finding commonalities and the identifying differences. Thinking of each term as an independent vocabulary term may mean that you can recite a definition, but you won't necessarily be able to apply that knowledge.

TIP

11. **neurons**
12. **irritability.** It's helpful to connect this characteristic with the everyday English-language irritability: The more irritable you are, the less patience you have and the easier it is for you to "go off." The term means the same thing here, but we're referring to generating an impulse.
13. **conductivity**
14. **impulse.** We use this word to refer to nervous communication in general. Keep in mind, though, that impulse is only the electrical part — the spread of action potentials down an axon.
15. **neurotransmitters**
16. **(a) Many dendrites and one axon.** A neuron may have one, many, or no dendrites, but it always has a single axon.
17. **(b) Perineurium**
18. **(f) Cells that contribute to the repair process of the CNS**
19. **(d) Cells that are phagocytic**
20. **(b) Cells that form and preserve myelin sheaths in the CNS**
21. **(a) Cells that form and preserve myelin sheaths in the PNS**
22. **(c) Cells that provide nutrients to bundled cell bodies**

(23) (e) Cells that form linings in the CNS

(24) (c) Threshold of excitation determines ability to respond

(25) (d) A drop in charge inside the cell past resting potential

(26) (b) Negatively charged molecule concentrated inside the cell

(27) (a) A differential in charge from outside the cell to inside

(28) (e) An increase in charge inside the cell

(29) **8. Potassium gates close. 10. Pumps restore polarization. 3. Cell depolarizes. 2. Sodium rushes in. 6. Potassium flows out. 1. Sodium gates open. 4. Sodium gates close. 5. Potassium gates open. 11. Ions return to initial concentrations. 7. Cell repolarizes. 9. Cell is hyperpolarized.** Though steps 4 and 5 happen simultaneously (Na$^+$ gates open and K$^+$ gates close), it makes more sense to address the open sodium gates first.

(30) (e) They enter the cell to trigger the release of neurotransmitters.

(31) **Impulse is the stepwise process that occurs in a single segment of axon. It begins with voltage-gated sodium channels opening, leading to local depolarization, and ends with the sodium/potassium pumps returning the ions to their original positions. As the sodium ions rush in and depolarize that segment of axon, however, the threshold is reached in the next segment, opening the next set of sodium channels.** So the steps early in each action potential trigger the start of the next one. This progression of action potentials constitutes the impulse.

(32) **Hypothalamus**

(33) **Pituitary gland**

(34) **Midbrain**

(35) **Pons**

(36) **Medulla oblongata**

(37) **Cerebrum**

(38) **Corpus callosum**

(39) **Thalamus**

(40) **Cerebellum**

(41) (b) Myelinated fibers

(42) (a) Has the capacity to arouse the brain to wakefulness

(43) (d) Bundle of neuron cell bodies. The neurons have to have their bodies to gang up on you.

(44) (c) Coordinates communication between right and left brain

(45) (e) Unmyelinated fibers, neuron cell bodies, and neuroglia

(46) (c) Occipital lobe. To remember, use the word *occipital* to bring to mind the word *optic*, which of course is related to visual activity.

(47) **(b) Assesses balance and body position** and **(e) Coordinates fine motor control.** Anything related to moving or body position is going to involve the cerebellum, even if the signal initiated as a conscious thought.

(48) **(a) Connects the medulla oblongata and cerebellum**

(49) **(d) Controls motor coordination and refinement of muscular movement**

(50) **(b) Contains the centers that control cardiac, respiratory, and vasomotor functions**

(51) **(e) Controls sensory and motor activity of the body**

(52) **(c) Controls visual and auditory reflexes**

(53) **The CSF circulates around the central nervous system. It's found between the arachnoid and pia layers surrounding the brain and spinal cord, and flowing through the ventricles and central canal. The structure of the meninges allows for an added layer of filtration, controlling what is allowed out of the blood and into the CSF and then from the CSF into the interstitial fluid of the actual brain and spinal cord tissue.** Thus, its job is to provide nutrients and transport wastes to and from the nervous tissue.

(54) **(h) Lens**

(55) **(e) Cornea**

(56) **(g) Iris**

(57) **(l) Pupil**

(58) **(a) Anterior chamber**

(59) **(c) Ciliary body**

(60) **(d) Conjunctiva**

(61) **(n) Sclera**

(62) **(m) Retina**

(63) **(b) Choroid coat**

(64) **(k) Posterior cavity**

(65) **(f) Fovea centralis**

(66) **(i) Optic disc**

(67) **(j) Optic nerve**

(68) **(i) Temporal bone**

(69) **(f) Malleus**

(70) **(g) Semicircular canal**

(71) **(b) Cochlea**

(72) **(h) Stapes**

(73) (c) Eustachian tube

(74) (e) Incus

(75) (j) Tympanic membrane

(76) (a) Auricle

(77) (d) External auditory canal

(78) Proprioception is your sense of balance; it's how your brain knows whether you're right-side up or upside down, and where your arms are in relation to the glass you're about to pick up or the itch you're about to scratch. Visual input is useful for this purpose, but you can also feel that you're upside down even with your eyes closed. The semicircular canals of the ears provide the essential information regarding balance.

(79) **(b) One is thin and watery, and the other is thick and gelatinous.** Vitreous humor (which is thick and gelatinous) fills the hollow sphere of the eyeball, but aqueous humor (which is thin and watery) fills two chambers toward the front of the eyeball.

(80) Mechanoreceptor

(81) Photoreceptor

(82) Mechanoreceptor

(83) Chemoreceptor

(84) Chemoreceptor

(85) **(b) They transmit vibrations from sound waves.** The ossicles are the tiniest bones of your body, found in your middle ear.

If you're ready to test your skills a bit more, take the following chapter quiz, which incorporates all the chapter topics.

Whaddya Know? Chapter 8 Quiz

Are you ready to think about thinking? See whether you made some strong neuron connections to help you through this quiz. Answers and explanations are in the following section.

1 What is a mixed spinal nerve?

(a) The point outside the spinal cord where the dorsal root and the ventral root join

(b) The bulging area inside the spinal cord where afferent fibers deliver sensory information

(c) A neuromere that interacts with a plexus

(d) The point along the spinal cord where the dorsal ramus and the anterior ramus join

(e) A nerve that runs in and then directly out of the spinal cord, creating a reflex arc

2 A synapse between neurons is best described as the transmission of

(a) a continuous impulse.

(b) an impulse through chemical and physical changes.

(c) an impulse through a physical change.

(d) an impulse through a chemical change.

3 The most sensitive region of the retina, producing the greatest visual acuity, is the

(a) optic disc.

(b) cornea.

(c) fovea centralis.

(d) choroid coat.

(e) ciliary body.

4 The cerebrum is divided into two major halves, called _____, by the _____.

(a) cerebellar hemispheres, transverse fissure

(b) cerebral spheres, central sulcus

(c) cerebellar spheres, lateral sulcus

(d) cerebral hemispheres, longitudinal fissure

5 The accommodation, or focusing, of the eye involves which of these?

(a) Sphincter of the pupil

(b) Contraction of the iris

(c) Action of the ciliary muscles

(d) Adjustment of the cornea

(e) Contraction of the pupil

6 Which of the following statements best describes the resting potential of a neuron?

(a) The inside of the cell has a higher concentration of positively charged ions.

(b) The inside of the cell has a higher concentration of negatively charged potassium ions.

(c) The outside of the cell has an electrical difference that's more negative than inside.

(d) The inside of the cell maintains a charge that's more negative than outside.

(e) The outside of the cell has more negatively charged sodium ions than inside.

7 Which part of a neuron is responsible for conducting electrical signals toward the cell body?

(a) Axon

(b) Soma

(c) Bouton

(d) Dendrite

(e) Hillock

For questions 8–16, use the word bank to identify the parts of the pain reflex arc shown in Figure 8-11.

(a) Afferent nerve

(b) Cell body of motor neuron

(c) Cell body of sensory neuron

(d) Dorsal root

(e) Effector

(f) Efferent nerve

(g) Interneuron

(h) Nociceptor

(i) Ventral root

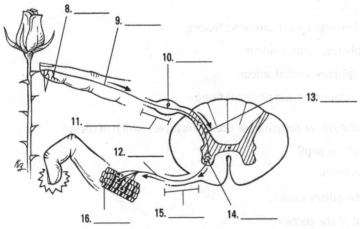

FIGURE 8-11:
A reflex arc
responding to
pain.

8. _____
9. _____
10. _____
11. _____
12. _____
13. _____
14. _____
15. _____
16. _____

Illustration by Kathryn Born, MA

17 What is the primary function of the autonomic nervous system?

(a) It activates in response to emotional stress.

(b) It controls involuntary body functions.

(c) It's the primary source of sensory input to the body.

(d) It controls the contraction of skeletal muscles.

(e) It works to block signals that may overload the overall nervous system.

18 True or false: Neurotransmitters bind to receptors on the dendrite, opening channels for positive ions to rush in and depolarize the cell.

19 Which statement best describes how neurons communicate?

(a) Axons generate an impulse that is passed to dendrites through the cell body.

(b) Axons receive a stimulus that is passed to the cell body, triggering the dendrite to release a chemical.

(c) Cell bodies respond to a stimulus by generating an electrical signal to pass on to dendrites and axons; the former will pass the signal on chemically, and the latter will do so electrically.

(d) In response to stimulus, the dendrites generate an electrical signal that is passed to the cell body. Then the axon spreads the impulse down its length, releasing a chemical at the end.

20 You've been in an accident and injured the left side of your cerebrum. Which areas of your body would you most likely have trouble moving?

(a) Right side

(b) Lower body

(c) Upper extremities

(d) Left side

21 Embedded in your skin in varying amounts, depending on the location, _____ are receptors that give you the sensation of being gently touched (as opposed to grabbed).

(a) Meissner's corpuscles

(b) nociceptors

(c) Pacinian corpuscles

(d) proprioceptors

(e) Ruffini's corpuscles

22 This type of glial cell plays a critical role in immunity should a pathogen make it through the blood–brain barrier.

(a) Ependyma

(b) Schwann

(c) Microglia

(d) Astrocyte

(e) Oligodendrocyte

23 When the interior of a neuron reaches +35mV, it responds by _____ the sodium channels and _____ the potassium channels to balance the charges back to resting potential.

(a) closing, closing

(b) closing, opening

(c) opening, closing

(d) opening, opening

24 What does the choroid plexus do?

(a) Filters cerebrospinal fluid

(b) Absorbs cerebrospinal fluid

(c) Produces cerebrospinal fluid

(d) Eliminates exhausted cerebrospinal fluid

25 Outline the path of light through the eye.

26 Which of the following best describes a signal that is sent to your pancreas to release its digestive enzymes into the small intestine?

(a) Efferent, autonomic

(b) Afferent, autonomic

(c) Efferent, somatic

(d) Afferent, somatic

27 What is the role of the semicircular canals?

(a) They provide structure and support to the orbit.

(b) They contain rods, cones, and hair cells to receive stimulus.

(c) They translate vibration of the ossicles to sound.

(d) They equalize pressure in the ear's chambers.

(e) They transmit information about the body's position.

For questions 28–33, label the following structures, shown in Figure 8-12.

(a) Cerebral aqueduct

(b) Fourth ventricle

(c) Lateral ventricles

(d) Optic chiasma

(e) Pineal gland

(f) Third ventricle

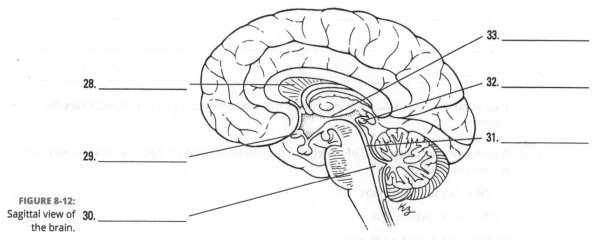

33. _____

28. _____

32. _____

31. _____

29. _____

FIGURE 8-12:
Sagittal view of
the brain.

30. _____

Illustration by Kathryn Born, MA

 34 Which structure is found at the end of an axon?

(a) Soma

(b) Dendrite

(c) Nissl body

(d) Ganglion

(e) Synaptic knob

 35 Which of the following statements is true of the autonomic nervous system?

(a) It has two parts: the parasympathetic, which carries out the stress response, and the sympathetic, which controls all normal functions.

(b) It's the nervous system that controls all reflexes.

(c) It doesn't function when the body's under stress.

(d) It has two divisions that are antagonistic, meaning that one counteracts the effects of the other.

(e) It controls the voluntary override for the body's autonomic actions.

36 What enters the nervous system through a dorsal root ganglion?

(a) A dorsal ramus

(b) An anterior ramus

(c) An efferent nerve

(d) An afferent nerve

(e) The lumbosacral plexus

37 Why are taste buds so much more receptive when you're able to produce adequate amounts of saliva?

_____

_____

_____

38 True or false: Nervous tissue is essentially neurons connected end to end, with very little matrix.

39 What triggers the migration of vesicles to the cell membrane so that neurotransmitters are released into the synapse?

(a) the influx of sodium ions

(b) the exit of sodium ions

(c) the influx of potassium ions

(d) the exit of potassium ions

(e) the influx of calcium ions

(f) the exit of calcium ions

For questions 40–47, use the choices to complete the paragraph.

(a) auditory canal

(b) cochlea

(c) hair cells

(d) incus

(e) malleus

(f) occipital lobe

(g) ossicles

(h) semicircular canals

(i) stapes

(j) temporal lobe

(k) tympanic membrane

(l) vestibule

Sound waves travel through the ear canal and cause the **40** _____, or eardrum, to vibrate. Those vibrations then are picked up by three small bones, called **41** _____. The **42** _____ is the first to receive the vibrations. Then it passes them to the **43** _____ and the **44** _____. These vibrations are transferred to the **45** _____, where fluid is jostled, stimulating the **46** _____, which communicate the sounds to the auditory cortex in the **47** _____ of the brain.

For questions 48–54, identify the portion of the brain (see Figure 8-13) responsible for the functions listed.

48 Controls body position: _____

49 Makes sense of auditory input: _____

50 Area where processing and planning occurs: _____

51 General interpretation area, speech generation: _____

52 Processes visual input: _____

53 Area of pain perception: _____

54 Center of creativity and personality: _____

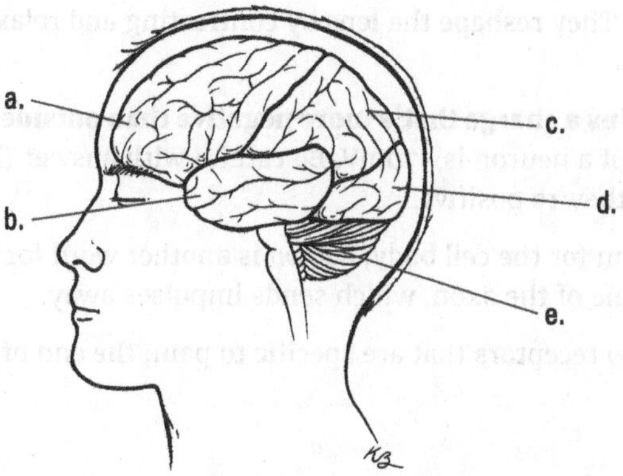

FIGURE 8-13:
Basic brain
anatomy.

Illustration by Kathryn Born, MA

55 Explain the role of the axon hillock in the stimulus and impulse generation of neurons.

_____

_____

_____

_____

Answers to Chapter 8 Quiz

1. **(a) The point outside the spinal cord where the dorsal root and the ventral root join.** That juncture occurs in each of the 31 pairs of spinal nerves; the afferent (incoming) and efferent (outgoing) nerves are wrapped together for organization's sake; they don't communicate with each other.

2. **(d) an impulse through a chemical change.** The term *impulse* is often used to refer to nervous communication in general. Pay attention to context; if the question is using (or asking for) the proper definition, it's the flow of electricity (ions) through the axon of one neuron.

3. **(c) fovea centralis.** It's the central location of your vision.

4. **(d) cerebral hemispheres, longitudinal fissure.** Cerebrum = cerebral, and two halves = hemispheres. You can remember the fissure's name by equating it to Earth's prime meridian, which separates the Eastern and Western hemispheres, and longitudinal is the most likely position for an equal division.

5. **(c) Action of the ciliary muscles.** They reshape the lens by contracting and relaxing as needed to bring things into focus.

6. **(d) The inside of the cell maintains a charge that's more negative than outside.** Compared with the outside area, the inside of a neuron is -70mV. Be careful with answer (b); there are more potassium ions inside, but they're positive.

7. **(d) Dendrite.** *Soma* is another term for the cell body; *bouton* is another word for the synaptic knob; and *hillock* is the trigger zone of the axon, which sends impulses away.

8. **(h) Nociceptor** This term refers to receptors that are specific to pain; the end of the word should help clue you in.

9. **(a) Afferent nerve**

10. **(c) Cell body of sensory neuron.** The bulge is the dorsal root ganglion, which is a bundle of cell bodies.

11. **(d) Dorsal root**

12. **(f) Efferent nerve**

13. **(g) Interneuron**

14. **(b) Cell body of motor neuron**

15. **(i) Ventral root**

16. **(e) Effector.** It carries out the effect.

17. **(b) It controls involuntary body functions.** That's the only answer with a sense of automation.

18. **False.** This question is a bit of a trick question, because that's exactly what excitatory neurotransmitters do. But there are also inhibitory neurotransmitters, which can lead to hyperpolarization instead.

(19) **(d) In response to stimulus, the dendrites generate an electrical signal that is passed to the cell body. Then the axon spreads the impulse down its length, releasing a chemical at the end.** Note that the electrical signal in a neuron isn't called an impulse until it's in the axon.

(20) **(a) Right side.** In the cerebrum, right = left and up = down. Clear as mud?

(21) **(a) Meissner's corpuscles**

(22) **(c) Microglia**

(23) **(b) closing, opening.** +35mV is the trigger to reset so that the axon segment can perform another action potential.

(24) **(c) Produces cerebrospinal fluid.** This one requires rote memorization. Sorry!

(25) **Cornea, anterior chamber, pupil, posterior chamber, lens, posterior cavity, retina**

(26) **(a) Efferent, autonomic**

(27) **(e) They transmit information about the body's position.**

(28) **(c) Lateral ventricles**

(29) **(d) Optic chiasma**

(30) **(b) Fourth ventricle**

(31) **(a) Cerebral aqueduct**

(32) **(e) Pineal gland**

(33) **(f) Third ventricle**

(34) **(e) Synaptic knob.** Though you could argue that (a) is true as well, it's not the end of the axon; it's the beginning.

(35) **(d) It has two divisions that are antagonistic, meaning that one counteracts the effects of the other.** As a result, the body achieves homeostasis.

(36) **(d) An afferent nerve.** The dorsal root brings in sensory information. Besides, all the other terms are efferent structures.

(37) **Taste buds contain chemoreceptors that, when bound to a matching chemical, send a signal to the brain. When you have plenty of saliva, the chemicals mix in with the water and are swished around in the mouth (so to speak) as you chew. When there isn't much liquid for the chemicals to spread around in, they don't make it down into the pits of the taste buds, so they don't trigger as many receptors.** Plus, not having enough saliva makes the food harder to chew and swallow.

(38) **False.** Abundant space is visible between neurons, dotted with glial cells and numerous "fibers," which are actually axons of other neurons.

(39) **(e) the influx of calcium ions**

(40) **(k) tympanic membrane**

(41) **(g) ossicles**

(42) **(e) malleus**

43 (d) incus

44 (i) stapes

45 (b) cochlea

46 (c) hair cells

47 (j) temporal lobe

48 (e) Cerebellum

49 (b) Temporal lobe

50 (a) Frontal lobe

51 (c) Parietal lobe

52 (d) Occipital lobe

53 (c) Parietal lobe

54 (a) Frontal lobe

55 **The axon hillock is the trigger zone, where the axon meets the cell body. When the stimulus triggers ion channels on the dendrites to open, positive ions rush in and move toward the cell body. This process depolarizes the area (because the charge is slowly increasing from -70mV). Because the axon hillock is connected, it also depolarizes. If enough ions rush in to make that area increase to -55mV — the threshold potential, in other words — the sodium channels along that first section of axon open. The ions rush in, depolarizing the axon section by section until it reaches the end (which is impulse).** That's why we call the axon hillock the trigger zone. If the stimulus was very brief or didn't cover much area, not as many dendrite channels will have been opened, so not as many positive ions will have entered. The trigger zone does increase in charge but maybe reaches only -60mV, which isn't high enough to start the first action potential, so the neuron won't send an impulse.

IN THIS CHAPTER

» Absorbing what endocrine and exocrine glands do

» Honing your hormone knowledge

» Checking in with the coordinators: The pituitary and hypothalamus glands

» Surveying the supporting glands

Chapter **9**

Raging Hormones: The Endocrine System

T he human body has two command and control systems that work in harmony most of the time but also work in very different ways. Designed for instant response, the nervous system cracks its cellular whip by using electrical signals that make entire systems hop to their tasks with no delay. (You can get a feel for the nervous system in Chapter 8.) By contrast, the endocrine system's glands use chemical signals called *hormones* that behave like the steering mechanism on a fully loaded ocean tanker; small changes can have big effects, but it takes some time for any evidence of the change to make itself known. Parts of the nervous system stimulate or inhibit the secretion of hormones, and some hormones are capable of stimulating or inhibiting the flow of nerve impulses.

The word *hormone* originates from the Greek word *hormao*, which literally translates as *I excite*. That's exactly what hormones do. Each chemical signal stimulates specific parts of the body, known as *target tissues* or *target cells*. The body needs a constant supply of hormonal signals to grow, maintain homeostasis, reproduce, and conduct myriad processes.

Meet & Greet Your Body's Chemical Messengers

The glands of the endocrine system and the hormones they release influence almost every cell, organ, and function of your body. The endocrine system is instrumental in regulating mood, tissue function, metabolism, growth and development, as well as sexual function and reproductive processes. Our hormones are designed to make the journey to their target cells via the bloodstream without errantly triggering an unrelated event along the way. This is due to their structural specificity; that is, each different hormone has a unique shape that is tailor-made for the receptors of its target cells.

Technically, there are ten or so primary endocrine glands, made up of epithelial cells embedded in connective tissue. (There is no consensus among anatomists about which ones count and which ones don't.) Flip to the color plates in this book and scope them out to see whether any of them sound familiar. Among glands, the pituitary, thyroid, and adrenal glands are most well-known. These organs have no significant function other than to produce hormones. Then there are the pancreas, ovaries, and testes, which secrete hormones in addition to performing their other familiar functions. Several other glands and hormones, though less well-known, are just as important in controlling vital bodily functions. In fact, much of the tissue in your body is in some way endocrine tissue.

"Wait," you say. "You're telling me my skin makes hormones?" Indeed, I am (and one of those hormones is growth hormone!) Not all endocrine tissue is assembled into stand-alone organs. You're no doubt quite familiar with your sweat glands. and while sweat isn't a hormone, it's a good example of the numerous glands sprinkled around our bodies. In fact, a *gland* is simply a group of cells that secretes a substance that carries out a particular function. If the substance is released through a duct, destined for a body surface, the gland is an *exocrine gland.* (Keep in mind your body has both external and internal surfaces.) If the chemical is secreted into blood flow, it's an *endocrine gland.* If the chemical released is a signal to cells elsewhere in the body, you have a *hormone.*

Hormones play a big part in homeostasis. When the blood passes certain checkpoints, hormone levels are measured. If the level of a certain hormone is too low, the gland that produces that hormone is stimulated to produce more of it. If a hormone level is too high, the gland that produces the hormone either doesn't receive any further hormonal stimulation or is instructed to stop or slow production. The hormones themselves communicate the needs of the body. To keep our bodies in balance, our nervous and endocrine systems must work together.

REMEMBER

Hormone functions include controlling the body's internal environment by regulating its chemical composition and volume, activating responses to changes in environmental conditions to help the body cope, influencing growth and development, enabling several key steps in reproduction, regulating components of the immune system, and regulating metabolism.

Honing In on Hormones

A hormone is an *endogenous substance* (one that's produced within the body) that has its effects on specific target cells. Hormones are many and varied in their sources, chemical natures, target tissues, and effects. But they're characterized by the facts that they're synthesized in one place (gland or cell) and that they travel via the blood until they reach their target cell. Hormones are bound by specific receptors on (or in) their target cells. The binding of the hormone on the receptor triggers a response by the cell.

We have good chemistry

Hormones can be classified as either *steroid* (derived from cholesterol) or *nonsteroid* (derived from amino acids and other proteins). Nonsteroid hormones work as just described; they grab on to a receptor to stimulate a response inside the cell. Because they are lipid-soluble, steroid hormones can enter the cell. To trigger the desired response, however, a receptor molecule inside the cell must bind the hormone and carry it to the nucleus. Because all your cell membranes are made of lipids (refer to Chapter 3), steroid hormones can enter any type of cell. This is why the steroid hormones — which include testosterone, estrogen, and cortisol — are the ones most closely associated with emotional outbursts and mood swings; they can stimulate cells other than their intended source, including nervous cells.

REMEMBER

The presence of a specific hormone receptor makes that cell a target for the hormone. Because hormones have very specific shapes, they'll bind only to their matching receptor. Without the target receptor, the hormone has no effect.

The nonsteroid hormones are divided among four classifications:

>> *Modified amino acids* include such hormones as epinephrine and norepinephrine, as well as melatonin.

>> *Peptide-based hormones* (at least three amino acids) include antidiuretic hormone (ADH) and oxytocin (OT).

>> *Glycoprotein-based hormones* include follicle-stimulating hormone (FSH), and luteinizing hormone (LH), both of which are closely associated with the reproductive system (covered in Chapter 16).

>> *Protein-based nonsteroid hormones* include such crucial substances as insulin and growth hormone (GH), as well as prolactin and parathyroid hormone.

Because these hormones are water-soluble, they are unable to enter the target cell. Instead, they bind to receptors on the cell's surface, triggering a cascade of chemical reactions by using a *second messenger* inside the cell, ultimately leading to the desired effect.

How function dictates structure

Not so long ago, by definition a hormone was produced in an endocrine gland, and an endocrine gland was a structure that produced one or more hormones. But as more hormone substances and forms have been discovered and described, the definition has expanded to include similar, sometimes identical substances that have a similar mechanism of action wherever they're produced. (Hence the title of this section being the familiar phrase "structure dictates function," except backwards.) Check out all the sources of hormones:

>> **Endocrine glands:** An *endocrine gland* is an organ that synthesizes a hormone within a specialized cell type. The anterior pituitary gland, for example, has cells that specialize in the production of growth hormone. Specialized cells within the thymus synthesize hormones that control the maturation of immune cells.

>> **Various organs:** Some organs that anatomists don't usually include in the endocrine system have cells and tissues specialized for the production of hormones. Here are a few examples:

- While part of the pancreas is busy secreting enzymes for the digestion of food, some specialized cells produce insulin, and others produce glucagon.

- The stomach and intestines synthesize and release hormones that control both the physical and chemical aspects of digestion.

- Specialized cells in the ovaries and testes transform cholesterol molecules into molecules of estrogen and testosterone, respectively.

- Even the heart produces hormones, the secretion of which has an immediate effect on blood pressure.

>> **Neurons:** Neurons make hormones that function as neurotransmitters. This seems a bit surprising, but if you think of hormones as molecules that deliver messages with considerable subtlety, it makes sense. The transmission of nerve impulses across a synapse is exactly that. (Flip to Chapter 8 for more on the nervous system.) The only difference between epinephrine synthesized in the adrenal glands and epinephrine synthesized in nerve cells is the distance the molecules travel to their target site.

Table 9-1 lists the body's major hormones, their sources, and their functions. I provide more information about these hormones throughout this chapter.

Table 9-1   Important Hormones: Sources and Primary Functions

Hormone	Source	Function(s)
Adrenocorticotropic hormone (ACTH)	Pituitary gland (anterior part)	Stimulates secretion of steroid hormones by the cortex of the adrenal gland
Aldosterone	Cortex of adrenal gland	Stimulates kidneys to reabsorb water and release excess potassium ions
Antidiuretic hormone (ADH)	Pituitary gland (posterior part)	Creates thirst sensation and stimulates the kidneys to reabsorb water
Calcitonin	Thyroid gland	Targets the kidneys and intestines to reduce the level of calcium in the blood; stimulates bone cells to deposit calcium into bone matrix
Cortisol	Cortex of adrenal gland	Regulates metabolism, slows inflammation response, and mediates stress response
Epinephrine/ Norepinephrine	Medulla of adrenal gland	Stimulates the heart and other muscles during the fight-or-flight response; increases the amount of glucose in the blood.
Estrogen	Ovaries	Stimulates the maturation and release of ova (eggs); targets muscles, bones, and skin to develop female secondary sex characteristics
Glucagon	Pancreas	Causes liver, muscles, and adipose tissue to release glucose into the bloodstream
Growth hormone (GH)	Pituitary gland (anterior part)	Targets the bones and soft tissues to promote cell division and synthesis of proteins
Insulin	Pancreas	Allows glucose into cells; stimulates liver, muscles, and adipose tissue to store it to lower blood glucose level
Melatonin	Pineal gland	Targets a variety of tissues to mediate control of biorhythms (the body's daily routine)
Oxytocin	Pituitary gland (posterior part)	Promotes social bonding; stimulates uterine contractions during childbirth and mammary glands to release milk
Parathyroid hormone	Parathyroid glands	Stimulates the cells in bones, kidneys, and intestines to release calcium so that blood calcium level increases
Progesterone	Ovaries	Prepares the uterus for implantation of an embryo and maintains the pregnancy
Prolactin (PRL)	Pituitary gland (anterior part)	Targets the mammary gland to stimulate production of milk
Releasing hormones	Hypothalamus	Stimulates anterior pituitary to make and release the corresponding hormone
Testosterone	Testes	Stimulates the production of sperm in testes; causes skin, muscles, and bones to develop male sex characteristics
Thyroid-stimulating hormone (TSH)	Pituitary gland (anterior part)	Stimulates the thyroid gland to produce and release its thyroxine
Thyroxine (T4)	Thyroid gland	Distributed to all tissues to increase metabolic rate; involved in regulation of development and growth

Let's see whether all this hormone-speak is sinking in.

PRACTICE For questions 1–4, identify the statement as true or false. When the statement is false, identify the error.

1 The endocrine system brings about changes in the metabolic activities of the body tissue.

_____

2 The amount of hormone released is determined by the body's need for that hormone at the time.

_____

3 Endocrine glands aren't functional in reproductive processes; those glands are separate.

_____

4 Hormones are synthesized from either amino acids or cholesterol.

_____

5 _____ glands secrete their product through ducts, whereas _____ glands secrete their product into the interstitial fluid, which flows into the blood.

(a) Endocrine; exocrine

(b) Endocrine; paracrine

(c) Paracrine, exocrine

(d) Exocrine; endocrine

(e) Exocrine; paracrine

6 Explain how, due to their structural differences, steroid and nonsteroid hormones elicit responses from their target cells in different ways.

_____

_____

_____

_____

The Ringmasters

The ups and downs of hormones can leave some people feeling a bit like a traveling circus has taken up residence in their bodies. There's an overall sense of order, but sometimes, the monkeys refuse to move, and the clowns go on strike. It's up to the ringmaster to carry out the expectations of the owner and keep the circus going (and bringing in the dollars). In us, the ringmaster is the *hypothalamus-pituitary axis*, the bridge between the nervous and endocrine systems. The pituitary gland speaks the language of all the performers (the other glands), but it doesn't really understand the language of the decision-makers that sit back and collect all the money (the brain). That's where the hypothalamus comes in. It serves as a translator of sorts (like an intern, perhaps). Together, they keep the circus profitable night after night, as illustrated in Figure 9-1.

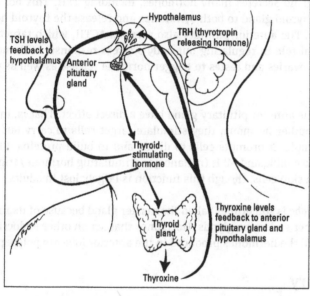

FIGURE 9-1: The working relationship of the hypothalamus and the pituitary gland.

Illustration by Kathryn Born, MA

The hypothalamus

The hypothalamus sits just above the pituitary gland, which is nestled in the middle of the cranium in a depression of the skull's sphenoid bone called the *sella turcica*. As part of the brain, the hypothalamus receives communication from both sensory neurons and other parts of the brain. Additionally, it contains special cells that can analyze the composition of the blood as it circulates through, measuring the amount of certain hormones (such as TSH, the example in Figure 9-1). It contains other specialized cells that generate chemical messengers, called *releasing hormones*, in response to the body's needs. Alternately, the hypothalamus can release inhibiting hormones if an ongoing response needs to be stopped.

The hypothalamus connects to the anterior portion of the pituitary gland via a narrow stalk called the *infundibulum*. Inside is a specialized network of blood vessels that carries the releasing hormones directly to the anterior pituitary, triggering its cells to release their corresponding hormone.

The pituitary gland

The pituitary gland has two parts — the *anterior pituitary* and *posterior pituitary* — that have different roles related to the hypothalamus. The cells of the anterior pituitary are the target of the releasing (and inhibiting) hormones. (I provide more detail about all these hormones in the relevant chapters and also list them in Table 9-1 earlier in this chapter.)

Anterior pituitary

The anterior pituitary gland secretes many hormones, including TSH. This hormone directly stimulates cells in the thyroid gland to both synthesize and release the thyroid hormones that control our metabolism. The anterior pituitary also secretes ACTH, which targets the adrenal glands, playing a critical role in regulating our body's stress response. FSH and LH are also made here, traveling to ovaries and testes to trigger hormone release as well as production of gametes (sex cells).

Other hormones from the anterior pituitary gland have a direct effect. That is, instead of triggering the release of another hormone, they stimulate target cells to carry out their specific function. GH is an example; it prompts cells to divide and to build proteins. PRL stimulates mammary cells to produce milk, and MSH (melanocyte stimulating hormone) triggers pigment (melanin) production in skin cells, though this function is largely lost in adults.

TECHNICAL STUFF

The pituitary's *anterior lobe* is sometimes called the master gland because of its role in regulating and maintaining other endocrine glands. Hormones that act on other endocrine glands are called *tropic hormones*; all the hormones produced in the anterior lobe are polypeptides.

Posterior pituitary

The *posterior pituitary gland* is largely connective tissue; it doesn't actually produce any hormones. It is responsible for storing and releasing hormones, however, which is why it's still considered to be the source for the two hormones that enter the bloodstream from there.

In Chapter 8, I talk about how neurotransmitters are released from neurons to stimulate the next cell in line. They do this by binding to receptors on the receiving cell's surface, just as most hormones do. The difference lies in what happens within the cell after that. It shouldn't come as much of a surprise, then, that some neurotransmitters could function as hormones if they were allowed into the bloodstream. This is where the posterior pituitary comes in. Two neurotransmitters, *vasopressin* and *oxytocin*, have important effects in the rest of the body.

Vasopressin, as it's called when we're talking about the neurotransmitter, is the *antidiuretic hormone* (ADH). Neurons in the hypothalamus produce ADH, which travels down the axons to be stored in the synaptic knobs. (Flip back to Chapter 8, Figure 8-4 to see this). When blood pressure falls below the ideal range, a nervous impulse causes the release of ADH into the posterior pituitary gland, where it's pulled into the blood. The ADH travels to its target kidney cells and binds to the receptors on their surfaces. The kidney cells remove more water from the urine and add it back into the blood, increasing the blood volume, which then increases blood pressure. (The process isn't as simple as just removing water from the urine. I provide a closer look in Chapter 15.)

The posterior pituitary also releases *oxytocin*, known as the love hormone. Its release in the brain leads to social bonding, particularly between parent and child. As a hormone, it causes smooth-muscle contraction. It plays a major role in the progression of uterine contractions during labor, as well as *milk letdown* (release of milk from the mammary glands). In males, it has been shown to cause contraction of the spermatic duct during ejaculation.

Are you storing any of this information? Take a brief concept check before you dive into an overview of the rest of the hormones.

PRACTICE

7 How do the anterior and posterior pituitary glands differ in structure and function?

_____

_____

_____

8 Why is the hypothalamus considered to be just as important as the pituitary gland?

(a) It contracts and relaxes to regulate the so-called master gland.

(b) It brings in substances from the thyroid.

(c) It tells the pituitary gland what to release and when.

(d) It controls how large the other glands can grow.

(e) It influences the other glands more directly than the pituitary does.

The Cast of Characters

Although the pituitary gland orchestrates the show at center stage, the circus doesn't go on without all its performers. They entertain on command and stop when the curtain drops (though I suppose there's not a curtain at a circus; maybe when the spotlight shifts?). A continuous checks-and-balance system keeps the show running smoothly, the body in tune.

Take a peek at the "Glands of the Endocrine System" color plate to refresh your memory on the shape and location before we take a look at the performers.

Topping off the kidneys: The adrenal glands

The adrenal glands lie atop the kidneys like little hats. The central area of each gland is called the *adrenal medulla,* and the outer layer is the *adrenal cortex.* (That seems backward, I know. Just think *m* for *middle.*) Each glandular area secretes different hormones. The cells of the cortex produce more than 30 steroids, including *aldosterone* and *cortisol,* as well as the sex hormones collectively called *androgens*. The medullary cells secrete *epinephrine* (you may know it as *adrenaline*) and *norepinephrine*.

The adrenal cortex

The cortex is made up of closely packed epithelial cells and is loaded with blood vessels. Layers form outer, middle, and inner zones, each composed of a different cellular arrangement that secretes different steroid hormones. Following are some of the hormones produced by the cortex:

>> *Aldosterone* regulates electrolytes (particularly sodium and potassium ions) in the body. It promotes the conservation of water and reduces urine output in the kidneys. (This hormone, along with ADH, is discussed in more detail in Chapter 15.)

>> *Cortisol* is released in response to stress and helps trigger the fight-or-flight response. It acts as an anti-inflammatory agent and can increase blood pressure. It also acts as an antagonist to insulin, increasing blood sugar.

>> *Androgens* and *estrogens* are cortical sex hormones. If you thought that as a woman, you don't have any testosterone or that as a man, you don't have any estrogen or progesterone, you're wrong. Admittedly, so-called male hormones are secreted in small amounts in women and have little influence on the development of their reproductive systems, just as the limited female hormones in men have little effect. Apparently, the primary responsibility of these hormones is to heighten sex drive. But excess production of their usually low amounts can lead to *feminization* in the male and *masculinization* in the female.

TECHNICAL STUFF

Cortisol and other corticosteroids affect the immune system by decreasing the number of circulating immune cells as well as decreasing the size of the lymphatic tissues (anti-inflammatory action). Under severe stress and with large amounts of glucocorticoids circulating in the blood, lymphocytes are unable to produce antibodies. (See Chapter 12 for more on the lymphatic system and immunity.) The role of corticosteroids in susceptibility to infectious disease is an active area of medical research.

The adrenal medulla

The *adrenal medulla* is made of irregularly shaped cells arranged in groups around blood vessels. The sympathetic division of the autonomic nervous system controls these cells as they secrete *epinephrine* and *norepinephrine* (in a ratio of about 4:1). Both hormones act quickly and have similar molecular structure and physiological functions:

>> *Epinephrine* accelerates the heartbeat, stimulates respiration, slows digestion, increases muscle efficiency, and helps muscles resist fatigue.

>> *Norepinephrine* does similar things but also raises blood pressure by stimulating constriction of muscular arteries.

TIP

The terms *adrenaline* and *noradrenaline* are interchangeable with the terms *epinephrine* and *norepinephrine*. You're likely to encounter both in textbooks and exams.

Skin (Cross Section)

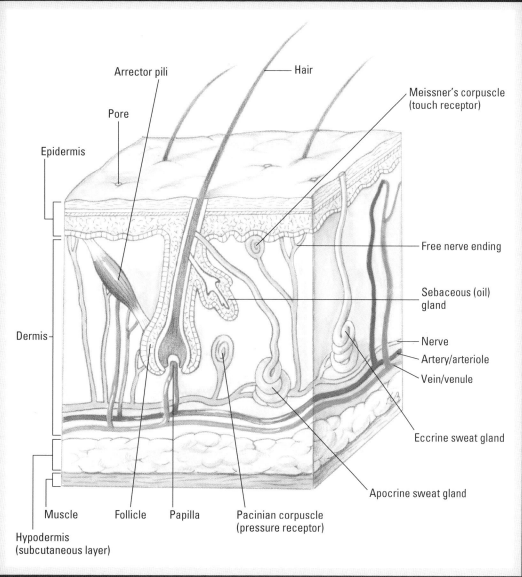

The skin's many layers protect the body from the environment. See Chapter 5.

Major Bones of the Skeleton

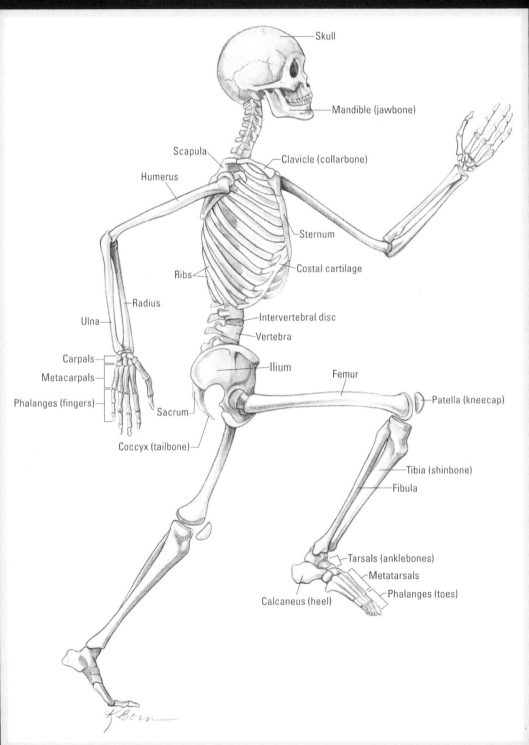

ILLUSTRATION BY KATHRYN BORN, MA

Skull

Mandible (jawbone)

Scapula

Clavicle (collarbone)

Humerus

Sternum

Costal cartilage

Ribs

Radius

Intervertebral disc

Ulna

Vertebra

Carpals

Ilium

Metacarpals

Femur

Phalanges (fingers)

Patella (kneecap)

Sacrum

Coccyx (tailbone)

Tibia (shinbone)

Fibula

Tarsals (anklebones)

Metatarsals

Phalanges (toes)

Calcaneus (heel)

The skeleton comprises the bones and the joints that connect them. See Chapter 6.

Major Skeletal Muscles

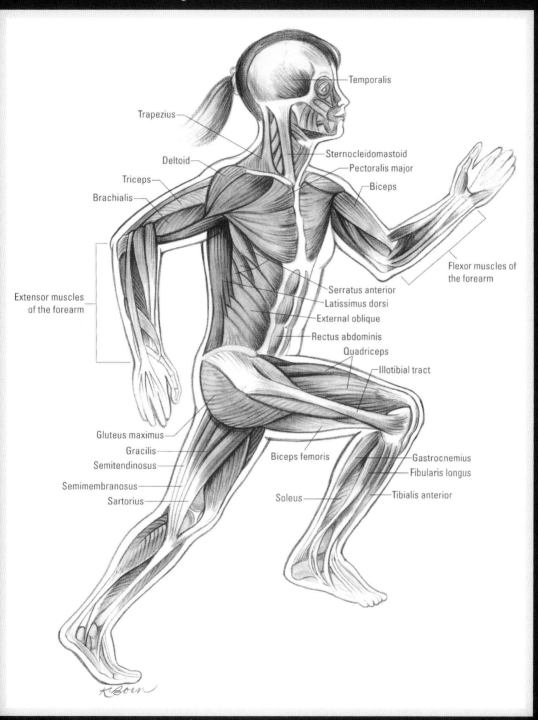

ILLUSTRATION BY KATHRYN BORN, MA

- Temporalis
- Trapezius
- Sternocleidomastoid
- Deltoid
- Pectoralis major
- Triceps
- Biceps
- Brachialis
- Flexor muscles of the forearm
- Extensor muscles of the forearm
- Serratus anterior
- Latissimus dorsi
- External oblique
- Rectus abdominis
- Quadriceps
- Illotibial tract
- Gluteus maximus
- Gracilis
- Semitendinosus
- Biceps femoris
- Gastrocnemius
- Fibularis longus
- Semimembranosus
- Sartorius
- Soleus
- Tibialis anterior

The major skeletal muscles generate the forces to move our skeleton.
See Chapter 7.

Nervous System

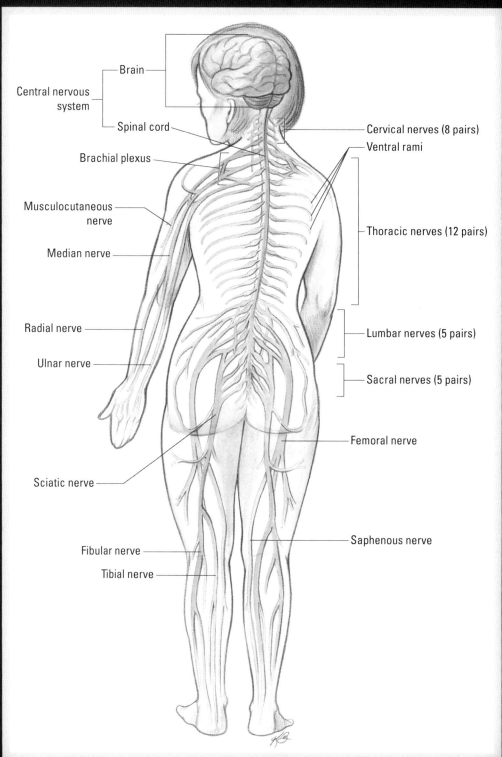

Central nervous system
- Brain
- Spinal cord

Brachial plexus

Musculocutaneous nerve

Median nerve

Radial nerve

Ulnar nerve

Sciatic nerve

Fibular nerve

Tibial nerve

Cervical nerves (8 pairs)

Ventral rami

Thoracic nerves (12 pairs)

Lumbar nerves (5 pairs)

Sacral nerves (5 pairs)

Femoral nerve

Saphenous nerve

The nervous system comprises the central nervous system and the peripheral nervous system. See Chapter 8.

Glands of the Endocrine System

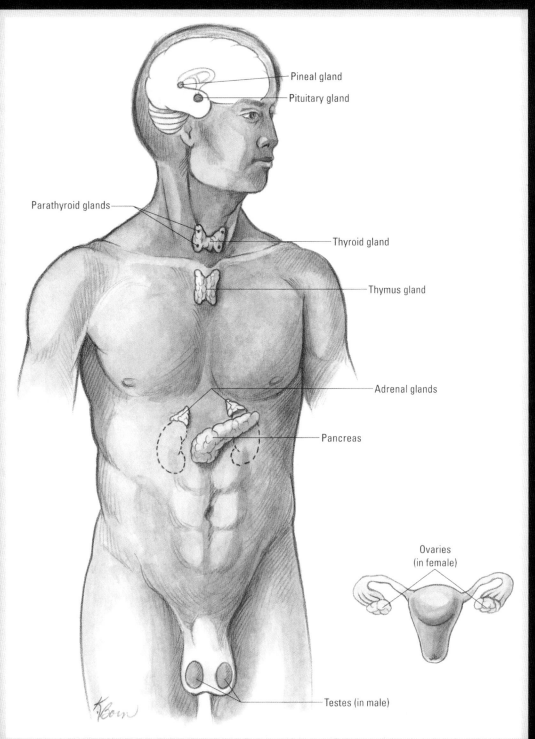

Pineal gland

Pituitary gland

Parathyroid glands

Thyroid gland

Thymus gland

Adrenal glands

Pancreas

Ovaries
(in female)

Testes (in male)

Endocrine glands produce hormones that are distributed throughout the body in the blood. See Chapter 9.

Heart

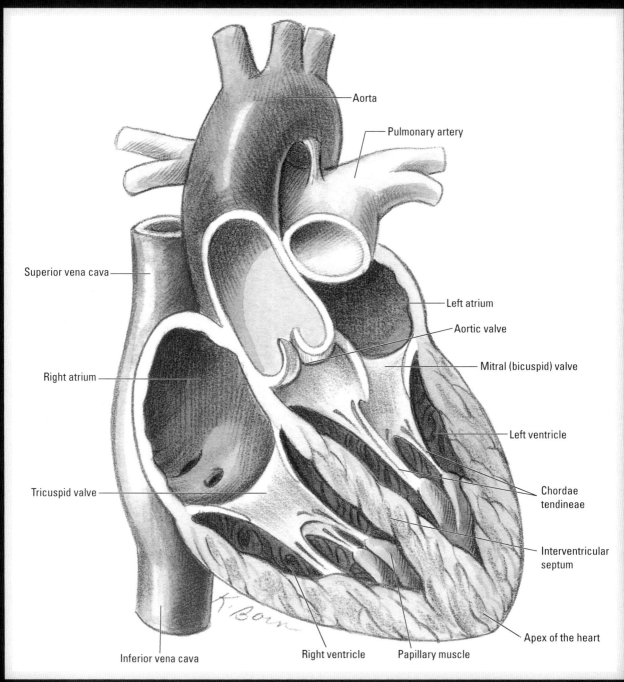

Aorta

Pulmonary artery

Superior vena cava

Left atrium

Aortic valve

Mitral (bicuspid) valve

Right atrium

Left ventricle

Chordae tendineae

Tricuspid valve

Interventricular septum

Apex of the heart

Inferior vena cava

Right ventricle

Papillary muscle

Effective blood circulation depends on the coordinated effort of the heart's muscular wall and the actions of its internal components. See Chapter 11.

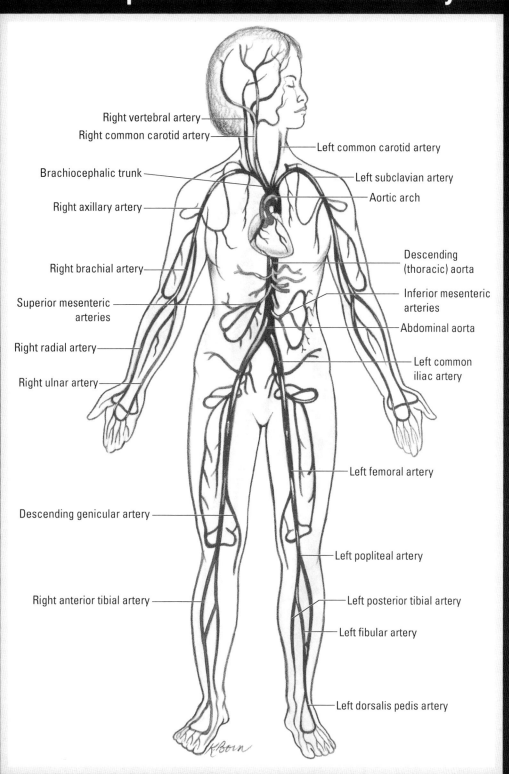

Right vertebral artery

Right common carotid artery

Left common carotid artery

Brachiocephalic trunk

Left subclavian artery

Aortic arch

Right axillary artery

Right brachial artery

Descending (thoracic) aorta

Superior mesenteric arteries

Inferior mesenteric arteries

Abdominal aorta

Right radial artery

Left common iliac artery

Right ulnar artery

Left femoral artery

Descending genicular artery

Left popliteal artery

Right anterior tibial artery

Left posterior tibial artery

Left fibular artery

Left dorsalis pedis artery

K Born

ILLUSTRATION BY KATHRYN BORN, MA

The arteries carry oxygenated blood from the heart to all parts of the body. Deoxygenated blood returns to the heart via the corresponding veins (not shown). See Chapter 11.

Lymphatic System

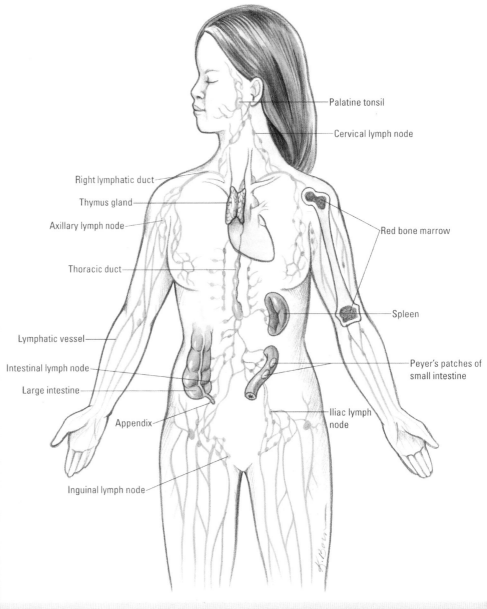

- Palatine tonsil
- Cervical lymph node
- Right lymphatic duct
- Thymus gland
- Axillary lymph node
- Red bone marrow
- Thoracic duct
- Spleen
- Lymphatic vessel
- Intestinal lymph node
- Peyer's patches of small intestine
- Large intestine
- Appendix
- Iliac lymph node
- Inguinal lymph node

The lymphatic system forms the infrastructure of the body's immune surveillance system. See Chapter 12.

Respiratory System

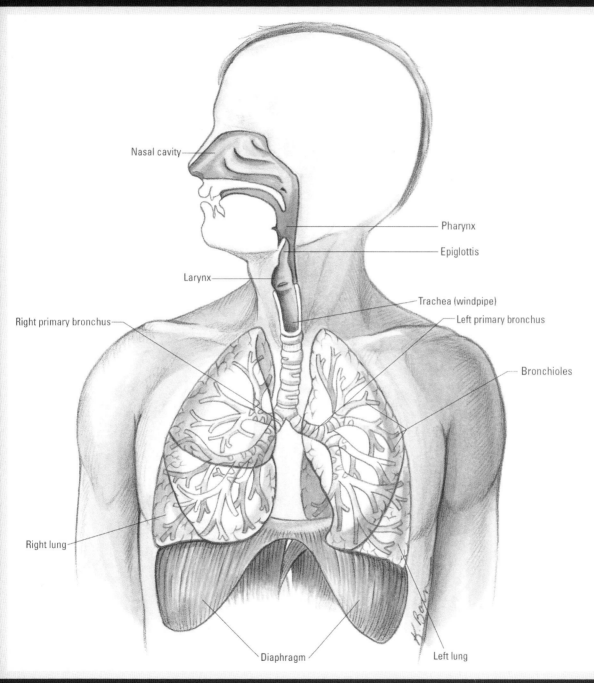

Nasal cavity

Pharynx

Epiglottis

Larynx

Trachea (windpipe)

Right primary bronchus

Left primary bronchus

Bronchioles

Right lung

Diaphragm

Left lung

Contraction and release of the diaphragm alternately decreases and increases air pressure in the lungs. Air is drawn in and expelled through the airway. See Chapter 13.

Structures of the Respiratory Membrane

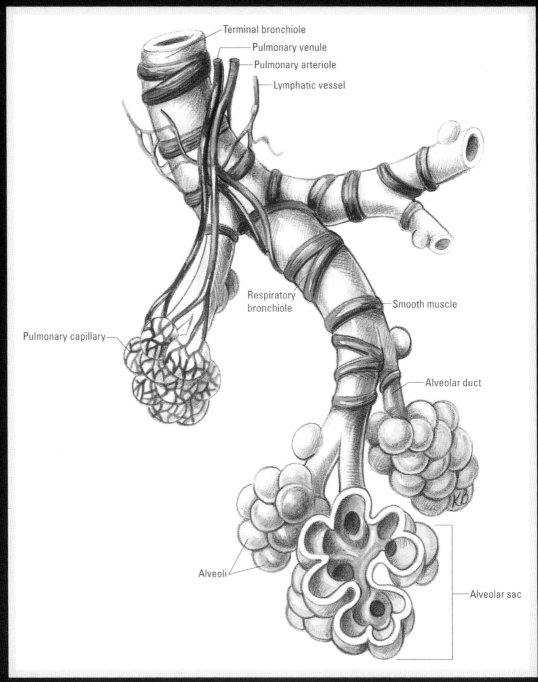

The pulmonary capillary is the specific site of gas exchange. See Chapter 13.

Digestive System

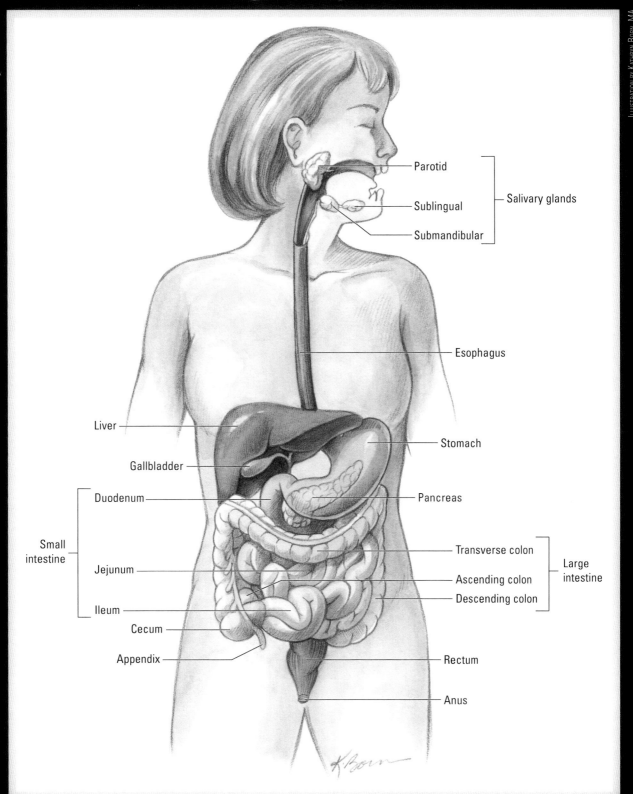

The transformation of food into physiologically available nutrients involves the participation of many organs. See Chapter 14.

Urinary System

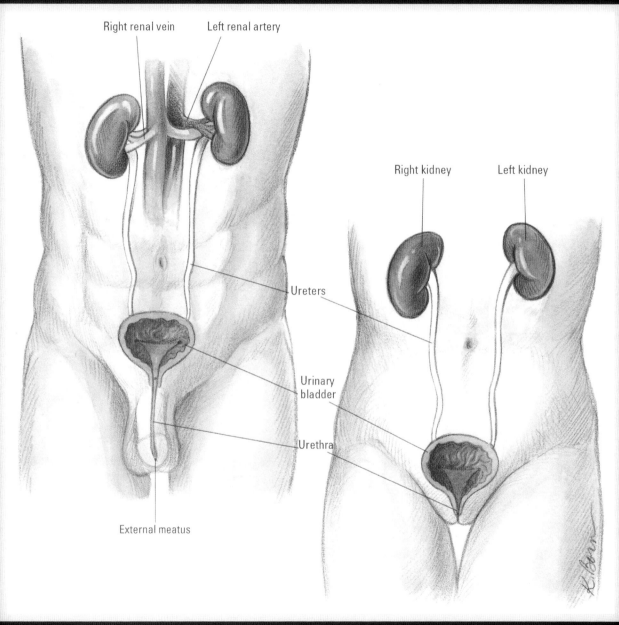

Right renal vein   Left renal artery

Right kidney   Left kidney

Ureters

Urinary bladder

Urethra

External meatus

The urinary system is specialized for the elimination of wastes and toxins. Its most complex organ, the kidney, also performs many other high-level homeostatic functions. See Chapter 15.

Kidney and Nephron

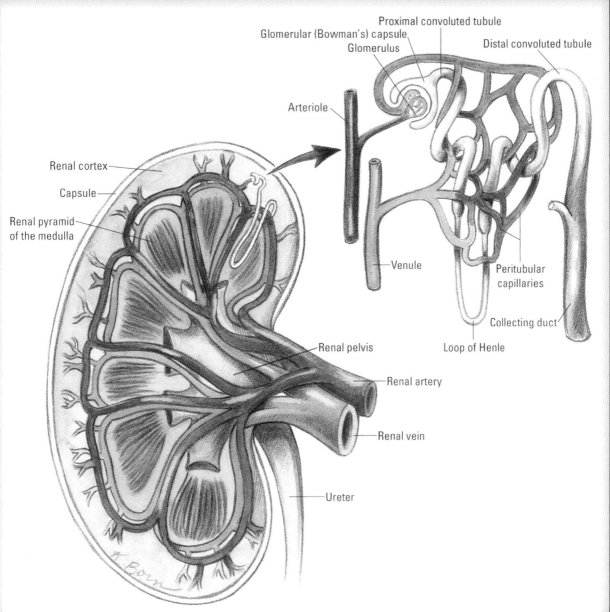

Illustration by Kathryn Born, MA

Proximal convoluted tubule

Glomerular (Bowman's) capsule

Glomerulus

Distal convoluted tubule

Arteriole

Renal cortex

Capsule

Renal pyramid of the medulla

Venule

Peritubular capillaries

Renal pelvis

Collecting duct

Loop of Henle

Renal artery

Renal vein

Ureter

The kidney is specialized for chemical processing, distribution, and disposal.
The nephron is the kidney's filtering unit. See Chapter 15.

Male Reproductive System

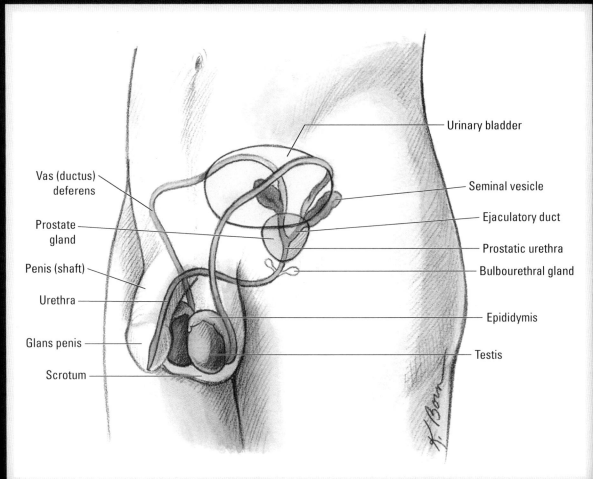

- Urinary bladder
- Vas (ductus) deferens
- Seminal vesicle
- Ejaculatory duct
- Prostate gland
- Prostatic urethra
- Penis (shaft)
- Bulbourethral gland
- Urethra
- Epididymis
- Glans penis
- Testis
- Scrotum

The organs of the male reproductive system are specialized to generate and deliver the sperm into the female reproductive tract. See Chapter 16.

Female Reproductive System

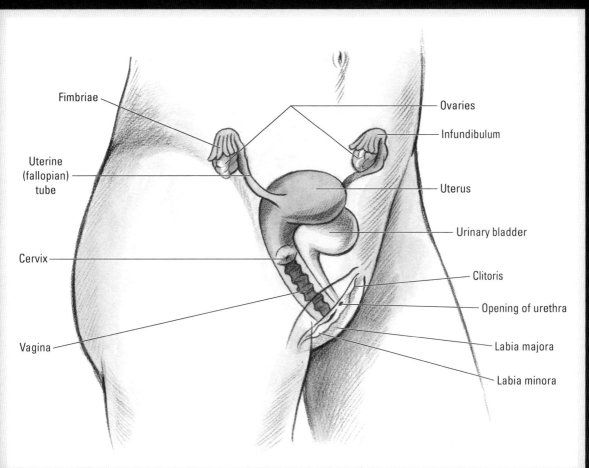

Fimbriae

Uterine (fallopian) tube

Cervix

Vagina

Ovaries

Infundibulum

Uterus

Urinary bladder

Clitoris

Opening of urethra

Labia majora

Labia minora

ILLUSTRATION BY KATHRYN BORN, MA

The female reproductive organs produce the ovum, supporting hormones, and serve to a house a developing fetus until childbirth. See Chapter 16.

Prenatal Development

Embryo at 5 Weeks
Length: Less than $\frac{1}{2}$ inch

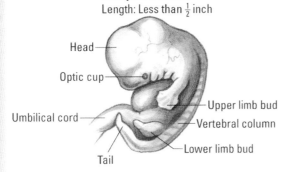

Head
Optic cup
Umbilical cord
Tail
Upper limb bud
Vertebral column
Lower limb bud

Embryo at 7 Weeks
Length: About 1 inch

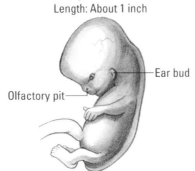

Ear bud
Olfactory pit

Fetus at 9 Weeks
Length: About 3 inches

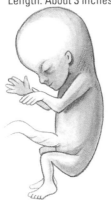

Fetus at 12 Weeks
Length: About $3\frac{1}{2}$ inches

Fetus at 21–25 Weeks
Length: 11–15 inches

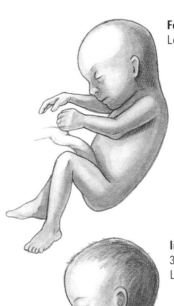

Infant at Birth
35–38 weeks
Length: 21 inches

Rapid growth and differentiation in the fetus are fueled through the placenta. See Chapter 17.

Thriving with the thyroid

The largest of the endocrine glands, the *thyroid* is shaped like a large butterfly, with two lobes connected by a fleshy *isthmus*. It is positioned in the front of the cervical region (neck), just below the larynx and straddling the trachea. (You can see the thyroid in Figure 9-1 earlier in this chapter.) In response to TSH from the anterior pituitary, the thyroid releases its two primary hormones: *thyroxine (T4)* and *triiodothyronine (T3)*. These hormones regulate the body's metabolic rate by acting on nearly every cell in the body.

TECHNICAL STUFF

The thyroid produces both T3 and T4, but only T3 is the active form of the hormone. Because these hormones are lipid-soluble, they enter the cell where T4 can be converted to T3 by enzymes. So why does the thyroid even bother making T4? T4 has a longer half-life than T3, meaning that it circulates in the bloodstream longer.

Thyroid hormones are unique in that they are nonsteroid hormones that are able to enter a cell. Steroid hormones by definition are made from lipid molecules (most often cholesterol). T3 and T4 are made from amino acids. The arrangement of their atoms, though, creates an overall polar molecule; thus, it can pass through the cell membrane. These two hormones are a big reason why we need iron in our diets and why we feel so lethargic when we don't have enough. The conversion of T4 to T3 removes one of the four iron atoms, which explains the numbers in the abbreviations. Here's a rundown of all the things T3 can start doing when it's in a cell:

>> Control the body's *basal metabolic rate* (the amount of energy needed to keep the body functioning at rest)

>> Increase the rate at which cells use glucose for energy

>> Help maintain body temperature by increasing or decreasing metabolic rate

>> Regulate growth and differentiation of tissues in children and teens

>> Increase the amount of certain enzymes in the mitochondria that are involved in cellular respiration

>> Influence the breakdown rate of proteins, fats, carbohydrates, vitamins, and minerals

>> Influence mood (high levels can cause anxiety; low levels, depression)

>> Increase the rate of protein synthesis

Parafollicular cells (also called *C cells*) in the thyroid secrete another hormone: *calcitonin*. This hormone promotes bone growth by stimulating osteoblasts (bone-building cells) to take up calcium from the bloodstream and deposit it into the matrix. High blood calcium levels stimulate the secretion of more calcitonin, triggering calcium ion storage.

Pairing up with the parathyroid

The parathyroid consists of four pea-size glands that are attached to the posterior side of the thyroid gland. These glands secrete *parathyroid hormone* (PTH), which does the opposite of calcitonin. When blood calcium levels dip, the parathyroid secretes PTH, which triggers osteoclasts to release the calcium ions stored in the bone by breaking down the matrix.

(See Chapter 6 to review the storage and release of calcium in the bones.) It also communicates with the intestines, stimulating more calcium absorption from the food we eat. The homeostasis of blood calcium is critical to the conduction of nerve impulses, muscle contraction, and blood clotting.

Pinging the pineal gland

The pineal gland is a small oval gland situated between the two hemispheres of the cerebrum, posterior to (behind) the diencephalon. This little guy's job is to regulate the body's circadian rhythm. As bedtime approaches, the pineal gland releases melatonin to signal your brain to begin the preparations for your sweet slumber.

REMEMBER

Because it both secretes a hormone and receives visual nerve stimuli, the pineal gland is considered to be part of both the nervous system and the endocrine system. The pineal gland is affected by changes in light; when its photoreceptors begin being stimulated less and less, it secretes melatonin, making you feel sleepy. This is also why you hear so much about the blue light from electronic devices and why you should avoid them at night. Your pineal gland doesn't get the message about decreasing light, so it doesn't begin its slow buildup of melatonin, and you have trouble falling and staying asleep.

Thumping the thymus

The thymus secretes a peptide hormone called *thymosin*, which promotes the production and maturation of T lymphocytes. T cells are a type of white blood cell that play a key role in immunity, mounting a targeted attack against invading germs. (For more on this topic, check out Chapter 12.) The gland is large in children and atrophies (degenerates) with age because the training of T cells is mostly complete by the onset of puberty.

Pressing the pancreas

The *pancreas* is both an exocrine and an endocrine gland, which means that it secretes some substances through ducts, whereas others go directly into the bloodstream. (I cover its exocrine functions in Chapter 14.) The pancreas is the long, flat, fibrous thing (I think it looks like a bag of tiny marbles) positioned posterior to the stomach, joining the beginning of the small intestine via a duct. If you weren't looking for it, you'd likely dismiss it as being fat and connective tissue.

The pancreatic endocrine glands are clusters of cells called the *islets of Langerhans*. Within the islets are a variety of cells, including

>> *A cells (alpha cells)* that secrete the hormone *glucagon* to increase blood sugar by triggering *glycogen* breakdown in the liver. (Glycogen is the storage form of glucose.)

>> *B cells (beta cells)* that secrete *insulin* to decrease blood sugar levels by triggering the liver to store glucose, increase lipid synthesis, and stimulate protein synthesis.

>> *D cells (delta cells)* that secrete *somatostatin* to inhibit the secretion of both insulin and glucagon.

WARNING

We tend to think of insulin as being a necessity to keep blood sugar levels in the normal range, which it certainly does. But its mechanism in doing so is more important. When insulin binds to receptors on a cell's surface, it triggers the opening of glucose channels, allowing glucose to diffuse into the cell. Without insulin, glucose doesn't get in; cells cannot obtain enough to perform cellular respiration and thus can't produce enough ATP to carry out their functions, so they not only become useless, but also might die. Cells in the brain are the only exceptions. Because they don't require insulin to bring in glucose, they get first dibs on the sugar.

Getting the gonads going

Your *gonads* (*ovaries* if you're female or *testes* if you're male) produce and secrete the steroid sex hormones: *estrogen* and *progesterone* in females and *testosterone* in males. Your body secretes sex hormones throughout your lifetime at different levels. Their production increases at puberty and normally decreases as you age. They are essential for preparing the body for reproductive function as well as the creation of *gametes*, or sex cells (egg and sperm). Here is a quick-run-down of these three reproductive hormones; more detailed information can be found in Chapter 16.

WARNING

Though high levels of the sex-specific hormones are necessary for the development and functioning of the reproductive organs, they're not exclusive. We now know that estrogen is found in males, for example; but surprisingly, it's also in growing plants! And estrogen isn't a hormone at all; it's actually a category, referring to the steroid hormones made in mass quantities by the ovaries (and also in the testes). Estrogens are associated with many nervous functions, such as learning and memory, mood regulation, and pain perception.

Estrogen

For ease in understanding, when the term *estrogen* is used in the context of reproductive functions, it refers to *estradiol*, so I'll follow that convention. In women, the increased production of estrogen at puberty is responsible for initiating development of the secondary sex characteristics, such as enlargement of the breasts and widening of the pelvis. Bone tissue grows rapidly, and height increases. Estrogen helps this process by stimulating the activity of osteoblasts and causing calcium to be transported in the bloodstream so it can be used for bone growth, particularly in the pelvis.

Progesterone

Progesterone works to prepare the uterus for implantation of a *blastocyst*. This tissue isn't called an *embryo* until it *implants* (takes up residence) in the uterus. It triggers the softening of the uterine lining by influencing secretions and plays a key role in maintaining pregnancy as well as contributing to breast development. Outside its reproductive functions, progesterone promotes the strengthening of skeletal muscle and has been found to have a protective effect on heart muscle cells.

Testosterone

Testosterone is essential for the production of sperm. It also causes the development of secondary sex characteristics in males. As a boy hits puberty, his sex organs enlarge, hair develops on his chest and face, and the hair on his arms and legs becomes darker and coarser. Regardless of sex, testosterone stimulates protein synthesis in muscle cells and triggers elongation of the growth plate in long bones. (This is why males undergo particularly dramatic growth spurts during puberty.) Testosterone also increases the production of red blood cells.

Does all this information have your hormones raging? Try answering some hormone-specific questions before tackling the chapter quiz.

PRACTICE **9** Why do we feel so tired when our iron levels are low?

_____

_____

_____

For questions 10–20, identify the structures of the endocrine system shown in Figure 9-2.

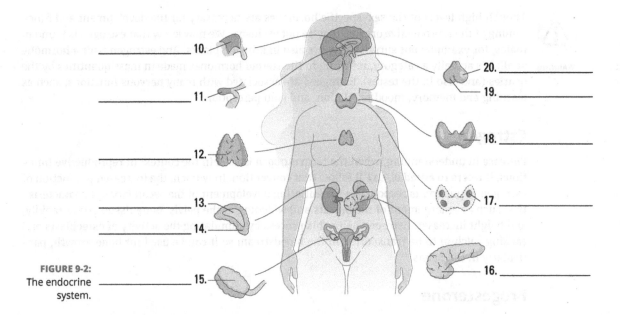

FIGURE 9-2:
The endocrine
system.

For questions 21–26, choose true or false. When the statement is false, identify the error.

21 Adrenaline is functional in the absorption of stored carbohydrates and fat.

_____

22 As part of the body's parasympathetic response, the pineal gland produces thyroxine.

_____

23 Aldosterone plays a role in regulating the amount of insulin in the body.

_____

24 The parafollicular cells of the thyroid produce hormones that affect the metabolic rate of the body.

_____

25 Cortisol is released in reaction to stress to provide an extra energy source.

_____

26 Calcitonin helps regulate the concentration of sodium and potassium.

_____

Practice Questions Answers and Explanations

1. **True.** Metabolism is one of the areas influenced by hormones.

2. **True.** In many ways, the system is self-regulating; just enough hormones are distributed to balance everything else.

3. **False.** The endocrine system is a key component of reproduction, as hormones from the anterior pituitary stimulate the ovaries and testes to carry out their reproductive functions in addition to releasing their reproductive hormones.

4. **True.** You could argue with me on this answer but hear me out. Steroid hormones are made from cholesterol, which is a lipid. All the nonsteroid hormones are themselves amino acids or are combinations of amino acids, because that's what proteins are made of.

5. **(d) Exocrine; endocrine**

6. **Steroid hormones are derived from cholesterol, making them lipid-soluble. Cell membranes are also made of lipids. As a result, steroid hormones can easily enter a cell, find their way to the nucleus, and directly stimulate (or inhibit) protein production. Nonsteroid hormones are water-soluble, so they don't have it so easy. A cell would have to invest a great deal of energy to bring a nonsteroid in, just so it could trigger a response. Instead, these hormones bind to receptors, triggering a chain reaction of events to elicit the target cell's response.**

7. **The anterior pituitary is filled with clusters of cells that make hormones, which are released into the bloodstream. The cells are stimulated to make and release their secretions in response to the releasing hormones from the hypothalamus. The posterior pituitary doesn't contain any endocrine cells; it's made of connective tissue and houses the axons of neurons that begin in the hypothalamus.** So these hormones (ADH and oxytocin) enter the bloodstream from the posterior pituitary, but they were manufactured in the neurons of the hypothalamus.

8. **(c) It tells the pituitary gland what to release and when.** If you remember that the pituitary gland is attached to the bottom of the hypothalamus, you can see the importance of the relationship.

9. **Iron is an essential component of the thyroid hormones T3 and T4, which help regulate your metabolism.** Your metabolism is all the chemical reactions that your body must do to keep you functioning, not the least of which is energy production. Not having a lot of T3/T4 circulating means that not as many cells are doing cellular respiration, leaving you fatigued.

10. **Hypothalamus**

11. **Pineal gland**

12. **Thymus**

13. **Adrenal medulla**

14. **Adrenal cortex**

15. **Ovary**

(16) **Pancreas**

(17) **Parathyroid gland**

(18) **Thyroid gland**

(19) **Posterior pituitary.** You have to look at its position in the brain to know whether it's the anterior or posterior pituitary.

(20) **Anterior pituitary**

(21) **False.** Adrenaline does lots of things, but not that. Adrenaline would allow access to the stored carbohydrates and fats to use for energy.

(22) **False.** First, the pineal gland releases melatonin, and the thyroid makes thyroxine. (The *thy-* prefix is on purpose!) Second, thyroxine boosts energy production, so it wouldn't be parasympathetic, which is our rest-and-digest system.

(23) **False.** This hormone regulates the uptake of mineral salts and, thus, the conservation of water by the kidneys.

(24) **False.** Parafollicular cells (or C cells) are indeed found in the thyroid, but these cells produce calcitonin.

(25) **True.** It's part of how the body gears up to manage an impending crisis. I suppose that if you were thinking that cortisol itself is the energy source, you answered false. But cortisol is released during our stress response, and it taps into several energy stores.

(26) **False.** Calcitonin lowers plasma calcium (and phosphate) levels.

If you're ready to test your skills a bit more, take the following chapter quiz, which incorporates all the chapter topics.

Whaddya Know? Chapter 9 Quiz

Do you have all this hormone knowledge regulated? It's time to test yourself. Answers and explanations are in the following section.

1. Which of the following is not a pituitary hormone?

   (a) Prolactin (PRL)

   (b) Thyroid-stimulating hormone (TSH)

   (c) Growth hormone (GH)

   (d) Epinephrine (EPI)

   (e) Luteinizing hormone (LH)

2. Fill in the blanks:

   Clusters of cells can be found within the islets of Langerhans in the pancreas. One such cluster, called _____, releases insulin after a meal to _____ blood glucose.

3. The body's initial reaction to a stressor is

   (a) activation of the parasympathetic response.

   (b) a surge of hormones from posterior pituitary.

   (c) prioritization of rapid wound healing.

   (d) promotion of normal metabolism.

   (e) centered on access to stored biomolecules.

For questions 4–13, match each hormone with its origin.

4. _____ Adrenocorticotropic hormone   (a) Adrenal cortex

5. _____ Somatostatin   (b) Adrenal medulla

6. _____ Thymosin   (c) Anterior pituitary

7. _____ Melatonin   (d) Hypothalamus

8. _____ Prolactin   (e) Pancreas

9. _____ Thyroid-stimulating hormone   (f) Parathyroid gland

10. _____ Aldosterone   (g) Pineal gland

11. _____ Glucagon   (h) Posterior pituitary

12. _____ Antidiuretic hormone   (i) Thymus

13. _____ Growth hormone   (j) Thyroid gland

14 The hormone that stimulates ovulation is

(a) follicle-stimulating hormone (FSH).

(b) antidiuretic hormone (ADH).

(c) oxytocin (OT).

(d) thyroid-stimulating hormone (TSH).

(e) luteinizing hormone (LH).

15 What would happen if the parathyroid glands stopped working?

_____

_____

_____

For questions 16–25, answer true or false.

16 _____ The pituitary gland is found in the sella turcica of the temporal bone.

17 _____ The anterior pituitary is called the master gland because of its influence on all the body's tissues.

18 _____ Nonsteroid hormones use a second messenger system to trigger a target cell's response.

19 _____ ADH causes constriction of smooth-muscle tissue in the blood vessels, which elevates the blood pressure.

20 _____ Thyroxine (T4) is normally secreted in lower quantity than triiodothyronine (T3).

21 _____ Homeostatic imbalance has primarily positive effects on the body.

22 _____ The parathyroid gland contains cells that secrete parathyroid hormone (PTH), which works to increase blood calcium levels.

23 _____ Exocrine glands secrete endogenous substances to carry out functions on our body's surfaces.

24 _____ Testosterone has such wide-ranging effects because its structure is derived from carbohydrates, so it's nonpolar and can move easily through the membrane's lipids.

25 _____ The adrenal glands are located in the cortex of the kidneys.

26 Which statement is *not* true of the pineal gland?

(a) It secretes melatonin.

(a) Nerve fibers stimulate the pineal cells.

(b) As light decreases, secretion increases.

(c) As part of our parasympathetic nervous system, it promotes immunity.

(d) It's considered to be part of both the endocrine and nervous systems.

For questions 27–32, fill in the blanks.

The challenge many of us face when that alarm rings each morning is going from groggy and wanting to hide under the covers to alert and ready to take on the day. Until we get that surge of metabolism-boosting hormones, though, we have to draw on sheer willpower and, for some of us, caffeine. This process begins with the (27) _____ communicating with the rest of the brain and unleashing (28) _____ to initiate the chain of commands that wakes us up the rest of the way. These hormones embark on their short journey to the (29) _____, where cells immediately release (30) _____, which is/are pulled right into the bloodstream. After circulating for a bit, these molecules find their way to the place that holds the key to our willingness to interact with the rest of the world: the (31) _____. As a result, (32) _____ is/are released, which kick-starts our cells' energy production, and we're ready to leave the warm, comfy bed behind until the day's work is done.

(33) The adrenal medulla serves one primary function. What is it?
(a) Producing erythrocyte cells
(b) Absorbing excess iodine from the bloodstream
(c) Producing epinephrine
(d) Secreting polypeptides
(e) Regulating water retention in the kidneys

(34) The thymus produces thymosin, which
(a) stimulates the pineal gland.
(b) inhibits the pituitary gland.
(c) initiates lymphocyte development.
(d) stimulates the thyroid gland.
(e) controls the hypothalamus.

(35) Which hormones are produced by the adrenal cortex? Choose all that apply.
(a) Cortisol
(b) Norepinephrine
(c) Testosterone
(d) Somatostatin
(e) Aldosterone
(f) Progesterone

For questions 36–40, match the releasing hormone with the eventual response of the body.

36 Production of sex hormones

37 Increase in glucose availability

38 Protein synthesis

39 Gamete production

40 Reduced inflammation

GnRH (gonadotropin-releasing hormone)

GHRH (growth hormone-releasing hormone)

CRH (corticotropin-releasing hormone)

Answers to Chapter 9 Quiz

1. **(d) Epinephrine (EPI).** This hormone comes from the adrenal medulla.

2. **beta cells; decrease**

3. **(e) centered on access to stored biomolecules.** Stressors of any kind cause a need for energy. Carbohydrates, lipids, and proteins can be broken down into molecules that can be used, and we store them in lots of cell types all over the body.

4. **(c) Anterior pituitary**

5. **(e) Pancreas**

6. **(i) Thymus**

7. **(g) Pineal gland**

8. **(c) Anterior pituitary**

9. **(j) Thyroid gland**

10. **(a) Adrenal cortex**

11. **(e) Pancreas**

12. **(h) Posterior pituitary**

13. **(c) Anterior pituitary**

14. **(e) luteinizing hormone (LH).** Don't be fooled into thinking that it's FSH; that hormone does its job earlier, when it encourages a follicle to mature.

15. **The parathyroid glands secrete parathyroid hormone, which is the only hormone that can access the calcium stored in our bones. Without it, when the level of calcium in blood flow drops, we won't have a way to bring it back up.** The implications of this are far more serious than just bone structure. Without the proper concentration of calcium ions, nervous communication would halt, as they are required for the neurons to release neurotransmitters.

16. **False.** The pituitary gland *is* found in the sella turcica, but that's part of the sphenoid bone, not the temporal bone.

17. **False.** It earned the title master gland because of its influence on the other endocrine glands.

18. **True.**

19. **True.** It's understandable that constriction increases pressure. This function, however, is secondary to its influence on the kidneys to retain water.

20. **False.** More T4 is secreted than T3 because it lasts longer.

21. **False.** Imbalance is a bad thing — so bad that it's capable of killing.

22. **True.**

(23) **False.** Exocrine glands release their secretions onto our body surfaces, but those secretions aren't endogenous substances because they don't have target cells.

(24) **False.** Molecules must be able to interact with the lipids to move through a cell membrane without a protein channel, which means that they have to be polar. Also, like all other steroid hormones, testosterone is made from cholesterol, not carbohydrates.

(25) **False.** They sit atop the kidneys.

(26) **(d) As a part of our parasympathetic nervous system, it promotes immunity.** It's more of a biological-clock kind of gland.

(27) **hypothalamus**

(28) **thyrotropin-releasing hormone (TRH)**

(29) **anterior pituitary**

(30) **thyroid-stimulating hormone (TSH)**

(31) **thyroid**

(32) **T3/T4**

(33) **(c) Producing epinephrine.** That's about all the adrenal medulla does. (Actually, it produces 80 percent epinephrine and 20 percent norepinephrine.)

(34) **(c) initiates lymphocyte development.** Specifically, it initiates the production and training of T cells.

(35) **(a) Cortisol, (c) Testosterone, (e) Aldosterone, and (f) Progesterone.** The sex hormones are made in small amounts in the adrenal gland.

(36) **GnRH (gonadotropin-releasing hormone)**

(37) **CRH (corticotropin-releasing hormone)**

(38) **GHRH (growth hormone-releasing hormone)**

(39) **GnRH (gonadotropin-releasing hormone)**

(40) **CRH (corticotropin-releasing hormone)**

4

Traveling the World

In This Unit . . .

IN THIS CHAPTER

» Looking at your blood and what's
  in it

» Inspecting interstitial fluid

» Learning about lymph

Chapter **10**

The River(s) of Life

All the body systems are dependent on one another. Each part fulfills a different role essential to the success of the whole . . . us. Communication between the parts is essential (as I discuss in Chapters 8 and 9), but so is logistics. We require a method to transport resources and wastes to and from even farthest reaches of our pinky toes. "I know!" you're thinking. "Blood!" And you're mostly right. But here's the problem: Blood never leaves its tubes, which isn't helpful for every other tissue in the body. Without our interstitial fluid and another fluid called lymph, each system would operate in isolation — an organizational plan that wouldn't be conducive to successful life.

So before I cover the next two body systems (cardiovascular and lymphatic), let's take a look at the rivers that help sustain our life, from blood to interstitial fluid to lymph.

Blood Is Thicker than Water

Blood — that deep-red, body-temperature, metallic-tasting liquid that courses through your body — is a vitally important, life-supporting, life-giving, life-saving substance that everybody needs. Every adult-size body contains about 1.3 gallons (5 liters) of the precious stuff.

Blood is a connective tissue (flip back to Chapter 4 if you need a reminder of why) that consists of many types of cells in a matrix called *plasma*. The different types of cells — *red blood cells, white blood cells,* and *platelets* — are referred to as *formed elements*. These elements are manufactured in the bone marrow by *hematopoietic stem cells*. The various types of blood cells combine to create 45 percent of the blood volume; the remaining 55 percent is plasma.

No river runs dry

Plasma is about 92 percent water. The remaining 8 percent or so is made up of plasma proteins, salts, ions, oxygen and carbon dioxide gases, nutrients (glucose, fats, and amino acids) from the foods you take in, and other substances carried in the bloodstream (such as wastes, hormones, and enzymes).

The plasma proteins, produced in the liver, are made for the plasma. That is, they aren't being transported anywhere. These proteins include

>> **Albumin:** The smallest plasma protein and the most abundant, albumin maintains the *osmotic pressure* (tendency to pull water in) of the bloodstream. Additionally, it serves as a carrier molecule, transporting substances such as hormones and calcium.

>> **Fibrinogen:** During clot formation, fibrinogen is converted into threads of *fibrin,* which form a meshlike structure that traps blood cells to form a clot.

>> **Immunoglobulin:** *Immunoglobulin* is another word for *antibody* — proteins that are created in response to an invading microbe. (See Chapter 12 for more on immunity.)

Moving trucks: Red blood cells

Red blood cells (RBCs), or *erythrocytes* (*erythro* is the Greek word for *red*), aren't just the most numerous blood cells; they're the most numerous cells in your body. About one quarter of the body's approximately 3 trillion cells are RBCs. They are among the cell types that must be regenerated and disposed of constantly. In fact, you produce and destroy around 2 million RBCs every second!

The cytoplasm of RBCs is full to the brim with an iron-containing biomolecule called *hemoglobin.* The iron-containing *heme group* in hemoglobin binds oxygen (O_2) molecules that have diffused into the cell; 98 percent of the oxygen in the bloodstream is transported this way. Hemoglobin that's carrying its maximum load of four molecules of O_2 is bright red, giving the oxygenated blood of the arteries its characteristic color. In veins, the RBCs have less heme-bound oxygen and thus are darker red.

TECHNICAL STUFF

Contrary to the belief of many people, the blood with low oxygen levels — the blood in your veins — isn't blue. It's a deeper shade of the fire-engine-red blood in the arteries; it's certainly not blue. The veins that are visible through the surface of some people's skin appear to be blue because of the properties of light. Remember prisms? Rainbows? ROY G BIV? Visible light is made of a range of wavelengths, each of which corresponds to a different color. When we see a color, we do so because that wavelength of light is bounced back to our eyes by the object. The red light doesn't make it back to our eyes through the skin, but the blue light does. If the skin weren't there, the veins would look red; if the blood wasn't there, both the veins and arteries would look white.

A red blood cell has about a three-month life span, at the end of which it's destroyed by a *phagocyte* (a large white blood cell with cleanup responsibilities) in the liver or spleen. The iron is removed from the heme group and transferred to either the liver (for storage) or the bone marrow (for use in the production of new hemoglobin). The rest of the heme group is converted to *bilirubin* and released into the plasma, giving plasma its characteristic straw-yellow color. The liver uses the bilirubin to form bile, which helps with the digestion of fats.

REMEMBER

RBCs have such short life spans for two reasons: They're constantly in motion, crashing into walls and other cells, which understandably causes some wear and tear, and they're full of hemoglobin and very little else. During differentiation (when a newly created cell matures and becomes specialized), an RBC loses its organelles, even its nucleus. Only the cell membrane retains its function.

The carbon dioxide (CO_2) waste that is generated when our cells use oxygen to generate ATP (cellular respiration) is also transported by the blood. Now with extra room (because it released some of the oxygen it was transporting), hemoglobin can take up CO_2 to form *carboxyhemoglobin*. At the lungs, carboxyhemoglobin releases the CO_2 and takes up oxygen again. Only 10 percent of the CO_2 in the blood is carried this way, though; most of it is converted to bicarbonate compounds that serve as buffers, maintaining the blood's pH.

Waiting on a war: White blood cells

White blood cells (WBCs) are derived from the same type of hematopoietic stem cells as RBCs. However, since their role is immune defense, they take different paths early in the process of differentiation. The WBCs, also called *leukocytes*, contain no hemoglobin and no iron. Unlike RBCs, all leukocytes retain their nuclei, organelles, and cytoplasm through their life cycle. Fewer leukocytes are produced than RBCs, by a factor of around 700.

Although WBCs are prevalent surfing the bloodstream, they don't actually function there. As our body's defenders against germs, they fight their battles in the interstitial fluid and lymph. It's fortunate, then, that they're the only cells that can exit blood flow. Leukocytes are active and most everywhere in our tissues all the time. (I discuss WBCs in greater detail in Chapter 12.) You only notice their presence, though, in the acute phase of certain diseases; the immune response, not the invader directly, produces the well-known symptoms of flu. But these cells are never far from a site of injury or infection, because they're everywhere. When a splinter pierces your finger, a contingent of local WBCs arrives at the site instantaneously.

Plugging up with platelets

Platelets are tiny pieces of cells. Large cells in the red bone marrow called *megakaryocytes* break into fragments, which are the platelets. Their job is to begin the clotting process and plug up injured blood vessels. Platelets, also called *thrombocytes* (*thrombus* means *clot*), have a short life span; they live only about ten days.

Not the least amazing thing about blood is its ability to stop flowing. The term for this ability is *hemostasis* (literally, *blood stopping*, and not to be confused with *homeostasis*). Hemostasis is the reason why you didn't bleed to death the first time you cut yourself. When vessels are cut (in the normal, run-of-the-mill sort of way), blood flows, but only for a moment. As you watch, the blood stops flowing, and a plug, called a *clot*, forms. Within a day or so, the clot has dried and hardened into a scaly scab. Cells along the injury path are stimulated to divide and do protein synthesis to repair the damage underneath the clot. Eventually, the scab falls off, revealing fresh new skin.

A blood clot consists of a plug of platelets enmeshed in a network of insoluble *fibrin* molecules. The *clotting cascade* is a physiological pathway that involves numerous components of the blood itself, called *clotting factors,* which interact to create a barrier to the flow. Each step in the process triggers the next, so the whole mechanism is an example of positive feedback. (See Chapter 2 for more on that topic.)

When a blood vessel is injured, collagen fibers within its walls are exposed. As the blood spills past, thrombocytes get stuck because of their rough edges, and they send out chemical signals that trigger the clotting cascade. Clotting factors in the blood are activated, and the platelets become sticky. The plasma protein fibrinogen is converted into fibrin. These long, sticky proteins create a mesh over the hole to which the platelets and blood cells get stuck, sealing it off. Within minutes, your blood is safe in your body, where it belongs.

The In-Between

Since the vast majority of the cells that need all the resources that the blood transports don't, in fact, come into contact with the blood, they must get their supplies elsewhere. Fortunately, all our tissues are bathed in *interstitial fluid,* and in the fluid, the resources are just floating around, waiting to be used. As I discuss in Chapter 3, cells pull their needed supplies in through their cell membranes, using methods such as diffusion and active transport. The same methods are used for cells to release their metabolic wastes into the interstitial fluid, where they can be pulled into blood flow and carried to their appropriate processing center: the liver, kidneys, or lungs.

So how do the resources get out of the blood? And while we're at it, where does all this interstitial fluid come from? The answers lie in the process of *capillary exchange.*

The stealthy swap: Capillary exchange

Capillaries are the smallest of the blood vessels. Their diameter is not much larger than that of an RBC, and their walls are simple squamous epithelium. The flattened cells are linked together with *tight junctions,* keeping the blood in the tube even though only a single layer of them creates the wall. There are, however, gaps between the cells where water and small molecules can pass through. Blood flowing through the capillaries applies force to the inside of the walls, which pushes out plasma and anything else that will fit through those gaps. This process is illustrated in Figure 10-1. Some capillaries also have purposeful pores called *fenestrations* to let more plasma out. The harder the blood pushes, the more plasma and resources come out.

The outflow of water in the first half of the capillary drops the *hydrostatic pressure* inside. As a result, interstitial fluid is pulled back into the capillaries the same way it left: between the cells. Any unused resources and the wastes added by the cells come along for the ride. This process ensures that the interstitial fluid stays fresh, with wastes removed and new resources added.

REMEMBER

Because the walls are cells, molecules can enter and leave the capillaries *through* them as well as *between* them. As a result, the standard cell-transport methods (such as osmosis, diffusion, active transport, and transcytosis) also contribute to the process of capillary exchange.

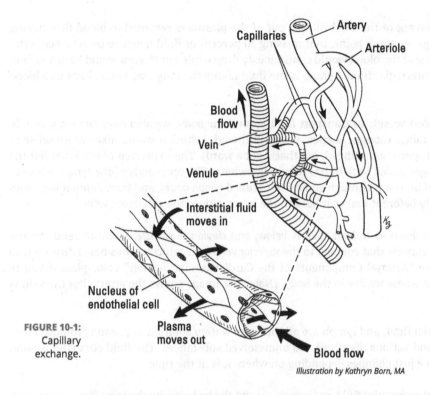

Labels on figure:
- Artery
- Capillaries
- Arteriole
- Blood flow
- Vein
- Venule
- Interstitial fluid moves in
- Nucleus of endothelial cell
- Plasma moves out
- Blood flow

FIGURE 10-1:
Capillary
exchange.

Illustration by Kathryn Born, MA

A new organ? The interstitium

The connective tissues within and between all our organs contain varying amounts of space. No space within our tissues remains empty; the constant outflow of plasma ensures that interstitial fluid fills it all. Consequently, we end up with an interconnected network of fluid called the *interstitium*. Many scientists argue that the interstitium should itself be classified as an organ, because it supplies needed resources to all tissues from the blood, aids in thermoregulation (water is quite effective in transferring heat), carries out the inflammatory response, and cushions organs from the movements of their neighbors. Should the medical and research communities support the classification of the interstitium as an organ, it would unseat the skin as our largest organ. (Perhaps we should put the question to a vote. Are you Team Interstitium or Team Skin?)

Of Sirens and Lymphs

Since the beginning of recorded history, water has been considered to be one of the four basic elements of life (along with air, earth, and fire). By then, it was already understood that to survive, we need to consume it in some form. With the great depths of bodies of water still unexplored, mystical properties were imparted on this life-giving resource. In Greek mythology, nymphs and sirens were water spirits — sort of the good and evil in the water. When sailors ran into trouble, the unseen underwater nymphs provided aid (whereas the sirens distracted them, leading them to their doom). This is the role of our *lymph* as well, playing a vital, unseen role in our survival (as opposed to leading us to our doom).

Only about 90 percent of the fluid filtered out of the plasma is returned to blood flow during capillary exchange. At some point, that missing 10 percent of fluid needs to be returned; otherwise, the volume of the blood would continuously drop while our tissues would keep swelling from the added interstitial fluid. Lymph is the fluid taking the long way home, back into blood flow.

In addition to blood vessels to transport fluid around the body, we also have *lymphatic vessels.* This network of tubes, though, is open-ended (although I think it would make more sense to describe them as open-beginninged, but that's not a word). The 10 percent of fluid that left the blood but didn't get pulled back in has to go somewhere. The open ends of the lymphatic vessels pull the fluid in, run it through some filters called *lymph nodes,* and then dump it back into blood flow shortly before it makes it back to the heart — into the *subclavian veins.*

TIP

Subclavian breaks down as follows: *Sub* is *below,* and *clavian* is *clavicle.* So these veins are the ones below each clavicle that connect to the *superior vena cava,* which carries blood directly into the heart. See the "Arterial Components of the Cardiovascular System" color plate if you're disoriented about where we are in the body. (*Note:* The image shows the artery, but the vein is right there too.)

REMEMBER

Plasma, interstitial fluid, and lymph are all the same watery solution of plasma proteins, electrolytes (ions), and various dissolved and undissolved substances. The fluid circulates around the body; its name just changes depending on where it is at the time.

The movement of interstitial fluid and lymph around the body is why the term *circulatory system* has fallen out of fashion. Instead, the cardiovascular system covers the blood, its vessels, and the heart, and the lymphatic system covers the path that the rest of our circulating body fluids take. Chapter 12 is devoted to the lymphatic system.

Now it's time to see how all this new information is flowing through you. Take the following chapter quiz, which incorporates all the chapter topics.

Whaddya Know? Chapter 10 Quiz

It's quiz time! Answer each question to test your knowledge of the various topics covered in this chapter. You can find the solutions and explanations in the next section.

1 Explain how plasma becomes interstitial fluid.

_____

_____

_____

2 Leukocytes are also known as

(a) red blood cells.

(b) lymphatic cells.

(c) platelets.

(d) interstitial cells.

(e) white blood cells.

3 Which of the following processes are involved in the movement of resources and wastes during capillary exchange? Choose all that apply.

(a) Osmosis

(b) Diffusion

(c) Filtration

(d) Active transport

(e) Endo/exocytosis

4 How is CO_2 most often carried through the blood?

(a) It's attached to hemoglobin.

(b) It's transported as bicarbonate molecules.

(c) It's transported as carbonic acid.

(d) It's dissolved in the plasma.

5 What are plasma proteins?

(a) All the proteins found in the plasma at a given time

(b) The proteins dissolved in plasma that are carried to interstitial fluid

(c) The proteins that carry out functions within the plasma

(d) All the proteins that contribute to the production of plasma

6 What do hematopoietic stem cells create? Choose all that apply.

(a) Red blood cells

(b) White blood cells

(c) Platelets

(d) Plasma

(e) Formed elements

(f) Plasma proteins

7 True or false: Although RBCs don't exit the capillaries during capillary exchange, WBCs and platelets do.

8 How is the clotting cascade an example of positive feedback?

_____

_____

_____

9 Which of the following statements best describes why any veins that are visible through the skin are blue?

(a) The capillaries in the skin appear to be blue, not the veins.

(b) Veins carry deoxygenated blood, which is more blue than arterial blood, which is red.

(c) The structure of the surrounding skin makes them appear that way.

(d) The walls of the veins are blue, but the blood isn't.

10 Blood vessels are often described as providing two-way flow, whereas lymphatic vessels flow parallel but only one-way. What does this mean?

_____

_____

_____

_____

Answers to Chapter 10 Quiz

1. **As the blood moves through its vessels, it applies forces on the walls. When it enters a capillary, it is pushing on a wall that is made of a single layer of cells. As a result, the plasma, which is mostly water, is forced out between the cells. Anything that's dissolved in the plasma or is just small enough to fit between cells is forced out as well. Thus, the plasma is filtered through the walls of capillaries.** When the fluid is in the tissues, it's called *interstitial fluid.* The term *plasma* is reserved for the watery component of blood.

WARNING

An important distinction here is that the fluid exits *between the cells,* not through holes and not through the cells. Some capillaries have holes (the fenestrations), but this situation is an exception, not a rule. And although resources can exit the blood by traveling through a cell, that's not how interstitial fluid is created, and it's not how the bulk of the molecules transported get there.

2. **(e) white blood cells**

3. All these processes are used during capillary exchange.

4. **(b) It's transported as bicarbonate molecules.** About 80 to 85 percent of CO_2 is transported this way.

5. **(c) The proteins that carry out functions within the plasma**

6. **(a) Red blood cells, (b) White blood cells, (c) Platelets,** and **(e) Formed elements.** The formed elements are RBCs, WBCs, and platelets, which is why (e) should be included in the answer. Plasma is the fluid portion and isn't really created anywhere. The plasma proteins are made by the liver.

7. **False.** Although WBCs can, platelets don't leave during capillary exchange. Though they're often small enough, the way they flow through the blood doesn't allow for them to exit; the reasons for this are beyond the scope of this book.

8. **Due to how they were created, platelets have rough edges. When they come into contact with proteins, they get stuck. The walls of blood vessels have numerous proteins, such as collagen, which are exposed when they break so the platelets stick. Once stuck, platelets signal other platelets to become stickier, so more platelets get stuck. This process also triggers the production of more platelets. The chemical signals from the stuck platelets also trigger the clotting cascade, which generates more proteins for them to stick to. More sticking leads to more sticking, and so on.** The response is in the same direction as the change: the definition of positive feedback.

9. **(c) The structure of the surrounding skin makes them appear that way.**

10. **Blood flows in two directions: away from the heart and then back. Some of the fluid leaves at the capillaries, and most of it returns. The fluid that doesn't return is pulled into lymphatic vessels and flows back toward the heart.** Plasma leaves, and plasma returns; lymph only returns.

Chapter **11**

Spreading the Love: The Cardiovascular System

More than any other system, the cardiovascular system has contributed strong imagery to people's daily language. *Heart* is a metaphor for love and courage. People say they nearly had a heart attack to describe an experience of surprise or shock. Abstract but distinctive characteristic qualities are said to be in one's blood. The blood itself runs cold, runs hot, and runs all over the place in informal speech and poetry. Scientifically speaking, emotions are more a matter of hormones than the myocardium, and nobody's blood is any redder or any hotter than anyone else's. The heart is neither soft nor hard, but a muscular, fibrous pump, and blood is a complex biological fluid that must be kept moving through its specialized network of vessels.

Meet & Greet Your Pumping Station

From a month after you're conceived to the moment of your death, your heart, a phenomenal powerhouse, pushes a liquid connective tissue — blood — and its precious cargo of oxygen and nutrients to every nook and cranny of the body. Then it keeps things moving to take carbon dioxide (CO_2) and waste products back out again. In the first seven decades of human life, the heart beats roughly 2.5 billion times. Do the math: How many pulses has your ticker clocked if the average heart keeps up a pace of 72 beats per minute, 100,000 per day, or roughly 36 million per year?

The cardiovascular system includes the heart, all the blood vessels, and the blood that flows through them. It functions as a *closed double loop* — *closed* because the blood never leaves the vessels, and *double* because there are two distinct circuits. (See Figure 11-1.)

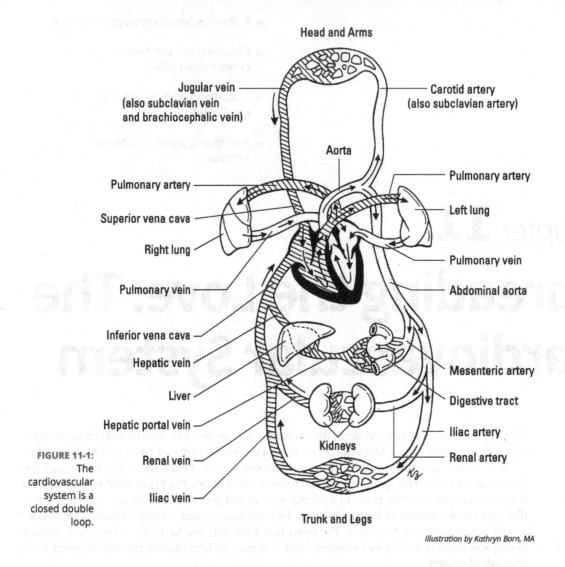

Head and Arms

Jugular vein
(also subclavian vein
and brachiocephalic vein)

Carotid artery
(also subclavian artery)

Aorta

Pulmonary artery

Pulmonary artery

Superior vena cava

Left lung

Right lung

Pulmonary vein

Pulmonary vein

Abdominal aorta

Inferior vena cava

Hepatic vein

Mesenteric artery

Liver

Digestive tract

Hepatic portal vein

Iliac artery

Kidneys

Renal vein

Renal artery

Iliac vein

Trunk and Legs

Illustration by Kathryn Born, MA

FIGURE 11-1:
The cardiovascular system is a closed double loop.

The heart is separated into right and left halves by the *septum*, allowing blood to flow through each half without mixing. With each *cardiac cycle* (a single set of muscle contractions), blood moves through the right and left sides simultaneously, thus creating the two circuits — the right side destined for the lungs and the left for everywhere else.

These two circuits have distinct functions:

>> **Pulmonary:** The *pulmonary circuit* carries blood to and from the lungs for gas exchange. Deoxygenated blood saturated with CO_2 enters the right side of the heart from the *inferior* and *superior vena cavae*. From there, it is pumped out through the *pulmonary artery* to the lungs, where *gas exchange* occurs through the capillaries and air sacs of the lungs. (**Note:** This process is not capillary exchange; see Chapter 13 to find out how it differs.) CO_2 exits

as oxygen (O_2) rushes in. The blood, now oxygenated, returns to the left side of the heart through the *pulmonary veins,* where it enters the second circuit.

REMEMBER

Though we tend to associate arteries with oxygenated blood and veins with deoxygenated blood, the terms refer to their structure. Arteries are adapted to carry blood away from the heart under high pressure; veins are the opposite. As such, the pulmonary arteries carry deoxygenated blood, and the pulmonary veins carry oxygenated blood.

>> **Systemic:** The *systemic circuit* uses the oxygenated blood to maintain a constant internal pressure and flow rate around the body's tissues, giving them access to the oxygen and other vital nutrients in the blood. The oxygenated blood enters the left side of the heart through the *pulmonary veins* and is pumped out through the *aorta* to be distributed throughout the body. Branching off the aorta is the *coronary artery,* creating a side circuit to provide resources to the *myocardium* (the muscular wall of the heart). From there, the *cardiac veins* carry the deoxygenated blood to the right side. Blood from the rest of the body is brought back to the heart in the vena cavae.

Although it's cutely depicted in popular culture as being uniformly curvaceous, the heart actually looks more like a blunt, muscular cone resting on the diaphragm. While it varies greatly by height and sex, on average the heart is 4.5 to 5.5 inches (12 to 14 centimeters) from top to bottom, 3 to 3.5 inches (8 to 9 centimeters) from left to right, and about 2 inches (6 centimeters) from front to back. The *sternum* (breastbone) and third to sixth costal cartilages of the ribs provide protection in front of the heart. Behind it lie the fifth to eighth thoracic vertebrae. Two-thirds of the heart lies to the left of the body's center, nestled into the left lung.

The cells of the wall of the heart can't pull resources from the blood inside of it; they need blood vessels to carry supplies to their tissue as well. The blood vessels that perform this task are visible on the surface of the heart. You can see the major ones in Figure 11-2.

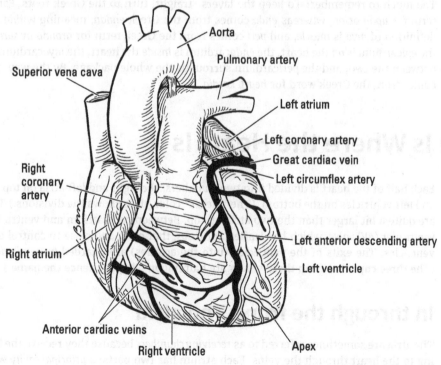

Aorta

Pulmonary artery

Superior vena cava

Left atrium

Left coronary artery

Great cardiac vein

Left circumflex artery

Right coronary artery

Left anterior descending artery

Right atrium

Left ventricle

Anterior cardiac veins

Apex

FIGURE 11-2:
A typical
healthy heart.

Right ventricle

Illustration by Kathryn Born, MA

In the body, the heart isn't positioned as you see it in illustrations; it's twisted slightly so the right side is anterior (more forward) to the left side. The *apex* of the heart is where it comes to a point, and the *base* is the opposite end where all the blood vessels attach.

TIP

So, the base is the top and the apex is the bottom? Isn't that backwards? Well, yes, sort of. But as you'll see by the time you finish this chapter, contraction starts at the apex to push the blood toward the base. If you think of it that way, you won't get the two terms confused.

The wall of the heart is composed of three layers:

>> **Epicardium:** On the outside lies the *epicardium,* which is composed of connective tissue containing adipose tissue (fat) to protect the underlying muscle. It forms a framework for nerves as well as the blood vessels that supply nutrients to the heart's tissues.

>> **Myocardium:** Beneath the epicardium is the *myocardium,* which is most of the wall's width. It is composed of organized layers and bundles of cardiac muscle tissue.

>> **Endocardium:** The *endocardium,* the heart's interior lining, is a layer of simple squamous endothelial cells. This layer is continuous with all the systems' blood vessels.

Because the heart beats with such gusto, it is isolated from all the other structures in the mediastinum in a protective sac called the *pericardium.* The pericardium is composed of multiple layers of connective tissue. The outermost layers comprise the *parietal pericardium,* and the innermost layer is the *visceral pericardium* (also known as the *epicardium*). Space is left between these layers, creating the *pericardial cavity.* This space, albeit tiny, is filled with a serous fluid to reduce friction, preventing damage to the heart's walls as they move with each set of contractions.

TIP

Too much to remember? To keep the layers straight, turn to the Greek roots. *Epi* is the Greek term for *upon* or *on,* whereas *endo* comes from the Greek *endon,* meaning *within.* The medical definition of *myo* is *muscle,* and *peri* comes from the Greek term for *around* or *surround.* Hence, the *epi*cardium is *on* the heart, the *endo*cardium is *inside* the heart, the *myo*cardium is the *muscle* between the two, and the *peri*cardium surrounds the whole package. By the way, the root *cardi* comes from the Greek word for *heart: kardia.*

Home Is Where the Heart Is

Each half of the heart is divided into two chambers: the right and left atria on top and the right and left ventricles on the bottom. (The "Heart" color plate shows the divisions.) The ventricles are quite a bit larger than the two atria up top. Between each atrium and ventricle is an *atrioventricular (AV) valve,* which has leaflets, or *cusps,* that open and close to control entry into the ventricles. The exits of the ventricles (into the arteries) are guarded by *semilunar (SL) valves.* (The three cups that form these valves are half-moon-shaped, hence the name.)

In through the front: The atria

The atria are sometimes referred to as *receiving chambers* because they receive the blood returning to the heart through the veins. Each atrium has two parts: a *principal cavity* with a smooth interior surface and an *auricle* (atrial appendage), a smaller, dog-ear-shaped pouch with muscular ridges inside called *pectinate muscles*. The atria have thin walls, as they only need to generate enough force to pump blood into the ventricles. The right atrium is slightly larger than the left, and the *interatrial septum* lies between them.

Since pressure is lower in the atrium than in the blood vessels, deoxygenated blood follows the path of least resistance and passively fills the right atrium through three openings:

>> The *superior vena cava*, which has no valve and returns blood from the head, thorax, and upper extremities

>> The *inferior vena cava*, which returns blood from the trunk and lower extremities

>> The *coronary sinus*, which returns blood from the heart

At the bottom of the right atrium is the *tricuspid valve*. Its three flaps open when the pressure in the right ventricle drops below that of the right atrium. This allows the blood to flow into the right ventricle. Attached to the tricuspid valve on the ventricular side (underneath) are strong strings called *chordae tendineae*, which fuse with *papillary muscles* in the walls of the ventricle. The chordae tendineae are pulled tight when the ventricle contracts to hold the tricuspid valve closed, preventing the backflow of blood.

The left atrium's principal cavity contains openings for the four *pulmonary veins*, two from each lung; none of these veins has a valve. Frequently, the two left veins have a common opening. The left auricle, a dog-ear-shaped blind pouch, is longer, narrower, and more curved than the right. The left atrium's AV opening is smaller than on the right and is guarded by the *mitral*, or *bicuspid*, *valve*. Though it has only two cusps, it contains chordae tendineae and functions the same way as the tricuspid valve.

Out through the back: The ventricles

The heart's ventricles are sometimes called the *pumping chambers* because it's their job to receive blood from the atria and pump it out through the arteries. More force is needed to move the blood great distances, so the myocardium of the ventricles is thicker than that of either atrium and the myocardium of the left ventricle is thicker than that of the right. (It takes greater pressure to maintain the systemic circuit.) The two ventricles are separated by a thick wall of muscle called the *interventricular septum*.

The right ventricle only has to move blood to the lungs, so its myocardium is one-third as thick as that of its neighbor to the left. Roughly triangular in shape, the right ventricle occupies much of the *sternocostal* (front) surface of the heart and joins the *pulmonary trunk* just left of the AV opening. The right ventricle extends downward toward where the heart rests against the diaphragm. Cardiac muscle in the ventricle's wall is in an irregular pattern of bundles and bands called the *trabeculae carneae*. The circular opening into the pulmonary trunk is covered by the *pulmonary valve*, which opens when the ventricle contracts (though not immediately). When the ventricle relaxes, the blood from the pulmonary artery tends to flow back toward the ventricle, filling the pockets of the cups and causing the valve to close. Openings into the ventricles are surrounded by strong, fibrous rings sometimes referred to as the *cardiac skeleton* (though there are definitely no bones here).

Longer and more conical in shape, the left ventricle's tip forms the apex of the heart. More ridges are packed more densely in the muscular trabeculae carneae. The opening to the aorta is protected by the *aortic valve*, which is larger, thicker, and stronger than the pulmonary valve's cups. The blood enters the ascending aorta into the aortic arch, where three vessels branch off:

>> The *brachiocephalic trunk,* which divides into the *right common carotid artery* and the *right subclavian artery*

>> The *left common carotid artery*

>> The *left subclavian artery*

It's time to pump up your knowledge with some practice.

PRACTICE

1 What's the difference between the pulmonary and systemic circuits?

(a) The pulmonary circuit supplies blood to the brain, and the systemic circuit supplies it everywhere else.

(b) The pulmonary circuit supplies blood to the internal organs, and the systemic circuit supplies blood to the extremities.

(c) The pulmonary circuit supplies blood to the skin, and the systemic circuit supplies blood to the underlying muscles.

(d) The pulmonary circuit supplies blood to the left side of the body, and the systemic circuit supplies blood to the right side.

(e) The pulmonary circuit travels to and from the lungs, and the systemic circuit keeps pressure and flow up for the body's tissues.

For questions 2–6, match the description with its anatomical term.

(a) Pericardium

(b) Pulmonary circuit

(c) Systemic circuit

(d) Epicardium

(e) Septum

2 _____ The system for gaseous exchange in the lungs

3 _____ The system for maintaining a constant internal environment in other tissues

4 _____ The membranous sac that surrounds the heart

5 _____ The wall that divides the heart into two cavities

6 _____ The layer providing nutrients to the heart's wall

7 Which blood vessel contains oxygenated blood?

(a) Pulmonary artery

(b) Pulmonary vein

(c) Pulmonary trunk

(d) Inferior vena cava

(e) Superior vena cava

8 Which of the following layers of the heart contains muscle? Choose all the apply.

(a) Pericardium

(b) Epicardium

(c) Myocardium

(d) Endocardium

(e) Endothelium

For questions 9–16, use the terms that follow to identify the heart's major vessels, shown in Figure 11-3. Not all the terms will be used.

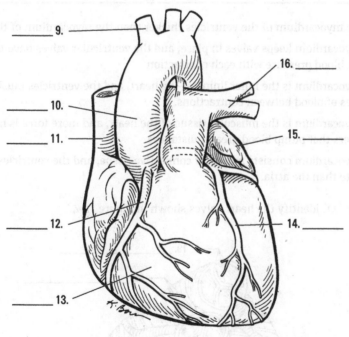

_____ 9. _____

_____ 10. _____

_____ 11. _____

_____ 12. _____

_____ 13. _____

16. _____

15. _____

14. _____

FIGURE 11-3: The heart and major vessels.

Illustration by Kathryn Born, MA

(a) Aorta

(b) Brachiocephalic trunk

(c) Inferior vena cava

(d) Left anterior descending
   artery

(e) Left atrium

(f) Left common carotid artery

(g) Left pulmonary arteries

(h) Left pulmonary veins

(i) Left subclavian artery

(j) Left ventricle

(k) Right anterior descending
   artery

(l) Right atrium

(m) Right coronary artery

(n) Right pulmonary arteries

(o) Right pulmonary veins

(p) Right ventricle

(q) Superior vena cava

17 Explain the role of the chordae tendineae and papillary muscles.

_____

_____

_____

_____

18 Why is the myocardium of the ventricles thicker than the myocardium of the atria?

(a) The myocardium keeps valves in place, and the ventricular valves have to hold back
   greater blood pressure with each contraction.

(b) The myocardium is the inner lining of the heart, and the ventricles must hold greater
   volumes of blood between contractions.

(c) The myocardium is the muscular tissue of the heart, and more force is needed in
   chambers that pump blood greater distances.

(d) The myocardium consists mostly of connective tissue, and the ventricles are more
   intricate than the atria.

For questions 19–22, identify the heart valves shown in Figure 11-4.

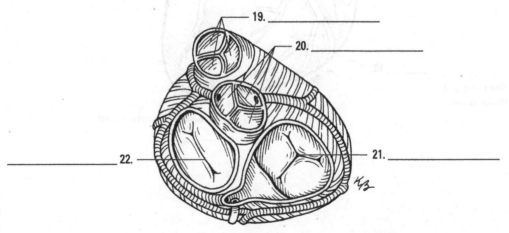

19. _____

20. _____

22. _____

21. _____

FIGURE 11-4:
The heart
valves.

Illustration by Kathryn Born, MA

Your Heart Song

To keep our blood flowing smoothly and consistently through our bodies, the contractions of the heart must be intricately timed. Both atria must contract at the same time, followed by both ventricles shortly thereafter, creating one complete heartbeat. The *cardiac cycle* is defined as the series of events required to generate a single heartbeat. This cycle is a series of closely coordinated electrical signals and pressure changes, like an intricate piece of music.

The orchestra

Certain structures of the heart, together called the *cardiac conduction system*, specialize in initiating and conducting the electrical impulses that induce your heartbeat, keeping it regular and strong in every part of the organ as it moves the blood through.

Unlike in skeletal muscle (see Chapter 7), cardiac muscle fibers (cells) connect end to end with *intercalated discs*. These gap junctions allow the electrical impulse to spread from cell to cell. Throughout the *myocardium* (muscle wall of the heart) are specialized myocardial fibers that, instead of being designed to contract, only carry impulse. These *conducting fibers* branch throughout their respective chambers so that the stimulus reaches the contracting fibers at the same time, creating a *syncytium* — a group of cells that function as one. The contracting fibers work in the same manner as the skeletal muscle fibers.

Five structures comprise the cardiac conduction system (see Figure 11-5):

>> **Sinoatrial (SA) node:** A small knot of conducting fibers embedded in connective tissue located on the back wall of the right atrium, near where the superior vena cava enters the heart. Once nervous input establishes the rhythm, the SA node becomes self-stimulating, which is why it's considered to be the heart's pacemaker. It does not require stimulus from a neuron to tell it to contract every time; it only needs to be told to speed up or slow down.

>> **Atrioventricular (AV) node:** A similar small mass of tissue located in the right atrium but near the septum. It divides the right and left atria from the ventricles and relays the impulses it receives from the SA node to the next part of the conduction system.

>> **Atrioventricular (AV) bundle (bundle of His):** A bundle of fibers that extends from the AV node into the *interventricular septum* (which divides the heart into right and left sides).

>> **Left and right bundle branches:** Where the septum widens, the AV bundle splits into the right and left bundle branches, each extending down toward the apex and then up along the outside of the ventricles.

>> **Purkinje fibers:** Fibers at the ends of the AV bundle branches that deliver the impulse up and around to the contracting fibers of the *ventricular syncytium*.

The conductor

Like a conductor guiding their orchestra through a song, the cardiac conduction system guides the myocardium through the cardiac cycle. The single heartbeat of a complete cycle is actually two contractions. First, the right and left atria contract together, pushing blood down into the ventricles. Moments later, the right and left ventricles contract together pushing the blood out of their respective arteries. The finale of the piece is blood of equal volume, called the *stroke volume*, bursting through both SL valves simultaneously.

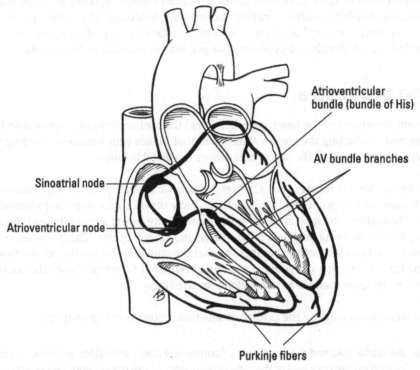

Atrioventricular bundle (bundle of His)

AV bundle branches

Sinoatrial node

Atrioventricular node

Purkinje fibers

FIGURE 11-5: The conduction system of the heart.

Illustration by Kathryn Born, MA

Here's how the cycle goes:

1. **The electrical impulse is initiated in the SA node.**

    The conducting fibers spread the impulse throughout the atrial syncytium. The right and left atria contract simultaneously, pumping blood into the right and left ventricles, respectively.

2. **The impulse passes to the AV node, which sends it to the AV bundle.**

    Some of the fibers connected to the SA node carry the impulse to the AV node. These fibers are narrower, providing the delay between the contraction of the atria and ventricles.

3. **The impulse passes into the right and left bundle branches and ultimately into the Purkinje fibers, causing the ventricles to contract.**

    By the time the ventricles begin to contract, the atria are relaxing. The impulse is carried to the apex first and then back up, for two reasons:

    -   To further ensure the delay before the ventricles contract
    -   To allow the ventricles to contract in a pattern

The Purkinje fibers weave into the myocardium of the ventricles in a spiraling pattern, causing the ventricular syncytium (walls of both ventricles) to contract in an upward, twisting motion.

4. The ventricles relax.

All chambers remain relaxed until the SA node generates the next impulse, starting the cycle over again.

The performance

For the blood to move on its well-timed path through the heart, the opening and closing of the valves must be coordinated as well. It's reasonable to conclude that the contractions are the cause of the valves opening and closing, but it's actually more complicated than that.

TECHNICAL STUFF

The movement of blood through the heart is driven by pressure changes, not by contractions. The relationship between pressure and volume is inverse: when volume increases, pressure decreases, and vice versa. When a chamber is relaxing, its volume is increasing. If fluids aren't blocked by, say, a valve, they flow to areas of lower pressure. Pressure changes are what drive the cardiac cycle; the contractions are used to generate those changes.

At the start of the cardiac cycle, all the chambers are relaxed and all the valves are closed. When a chamber is relaxed, it's said to be in *diastole*; the contraction phase is called *systole*. The sequence of events is as follows:

1. Blood trickles into the atria from the veins, increasing the internal pressure.

2. When the pressure in the atria rises above that of the ventricles (they would be relaxing at this time from the previous cardiac cycle, thus dropping the pressure), the AV valves are pushed open.

3. Blood passively fills the ventricles until they contain about 70 percent of the stroke volume. At this point, blood stops flowing because the pressures of the chambers (atria and ventricles) have equalized.

4. Atrial systole begins, pushing the remaining 30 percent of the blood into the ventricles.

5. The atria begin to relax as the ventricles begin systole. The relaxation drops the pressure of the atrium, creating a vacuum. The cusps of the AV valves are pulled upward as a result. However, the papillary muscles, which are continuous with the ventricular wall, are contracting, pulling on the chordae tendineae and preventing the cusps from being pulled into the atrium.

6. At this point, all valves are closed, the atria are in diastole, and the ventricles are in systole.

7. The ventricles continue contracting to increase their internal pressure, which isn't yet high enough. When ventricular pressure rises above the pressure inside the arteries, the semilunar valves are pushed open.

8. Blood enters the arteries to be carried elsewhere, and the ventricles begin diastole. The relaxation creates a vacuum pulling the blood backward, this time toward the ventricles.

9. Blood catches in the cups of the semilunar valves, sealing them shut (no chordae tendineae here). And now we're back to the beginning.

The recording

If you listen to your heart with a *stethoscope*, the device doctors and nurses use to do just that, you can hear the two sounds of a heartbeat. But as hard as it is to convince your brain otherwise, the sounds you hear aren't the contractions. A healthy heart makes a "lub-dub" sound as it beats because a single cardiac cycle contains two separate contractions. The first sound (the "lub") is heard clearest near the apex of the heart and comes at the beginning of ventricular systole; the sound you hear is the AV valves snapping shut. This sound is lower in pitch and longer in duration than the second sound (the "dub"), clearest over the second rib, which results from the SL valves closing during ventricular diastole.

If you're a fan of medical dramas, you're no doubt familiar with the heart-rate monitor — those squiggly lines moving across the screen with the corresponding beeps. The implication is that those waves correspond to the contraction of the chambers. But this isn't quite the case.

The heart's electrical signals can be measured and recorded digitally to produce a graph called an *electrocardiogram* (ECG or EKG). Three major waves appear on the EKG, each showing the electrical signals as they move through the heart. The first wave, called the *P wave*, records the spread of impulse through the right and left atria. The second and largest wave, the *QRS wave*, shows the spread of impulse through the right and left ventricles. The third wave, the *T wave*, records the recovery of the ventricles.

The conducting fibers spread impulse just as neurons do. (See Chapter 8 for more on that topic.) The waves on the EKG correspond to the movement of ions, measurable by electrodes on the skin because body fluids conduct this electricity. As the conducting fibers *depolarize* (spread the impulse), the EKG line moves up and back down. The same thing happens when they recover (or *repolarize*) because the flow of potassium ions out is measurable electricity just as the flow of sodium ions in is equally measurable. Because the repolarization of the atria occurs during the QRS, you don't see a wave for it.

The direction of the wave (up versus down) corresponds to the spread of impulse in relation to the lead electrode: up is toward and down is away. The length of the wave corresponds to the time it takes for the impulse to spread through the chambers. The QRS complex is a series of spikes rather than a wave because the impulse is covering a lot of ground very quickly.

REMEMBER

The waves on an EKG show the electrical conduction in the heart, not the contraction. The atria would be contracting in the period of time between the end of the P wave and the start of the QRS (roughly). The ventricles are contracting from the end of the QRS to the T wave.

PRACTICE

Give these practice questions a try to see if you've got the rhythm of this chapter.

23　Why is the sinoatrial (SA) node called the heart's pacemaker?

(a) It interrupts irregular heart rhythms to set a steadier pace.

(b) It initiates an electrical impulse that causes atrial walls to contract simultaneously.

(c) It speeds up and slows down the heart rate, depending on the body's activities.

(d) It tells the ventricles to contract simultaneously.

(e) It supplies a boost of energy to jump-start the heart when it slows.

24 Choose the correct conductive pattern.

(a) SA node → Bundle of His → Bundle branches → AV node → Purkinje fibers

(b) AV node → SA node → Purkinje fibers → Bundle of His → Bundle branches

(c) SA node → AV node → Bundle of His → Bundle branches → Purkinje fibers

(d) SA node → Purkinje fibers → Bundle of His → Bundle branches → AV node

(e) AV node → SA node → Bundle of His → Bundle branches → Purkinje fibers

25 Explain why ventricular filling is a two-part process.

_____

_____

_____

26 What is happening during ventricular diastole?

(a) The AV valves are being pulled open.

(b) The chordae tendineae are pulling the AV valves closed.

(c) The SL valves are being pushed open.

(d) The chordae tendineae are pulling the SL valves open.

(e) The SL valves are being pushed closed.

27 On Figure 11-6, draw brackets on the EKG to show atrial and ventricular systole, and label Q, R, and S.

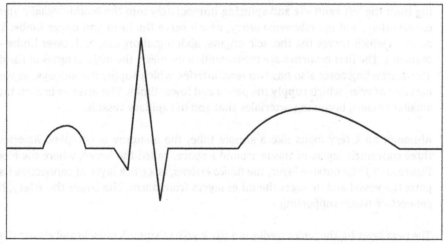

FIGURE 11-6:
A typical EKG.

Illustration by Kathryn Born, MA

Moving to the Beat of Your Own Drum

If you could do a ride-along with a single red blood cell, it wouldn't take you long to get a snap-shot of most of the body's tissues, though you'd spend such a brief moment in each area that you wouldn't be able to actually see anything. It takes less than a minute for a round trip, on average. You'd zip through the arteries, riding the waves of propulsion from ventricular systole and return on a leisurely ride through the veins' lazy rivers, with the destination of each trip determined by chance.

Your blood vessels comprise a network of channels through which your blood flows, but they are not passive tubes. Instead, they're very active organs that, when functioning properly, assist the heart in circulating the blood and influence the blood's constitution.

It's all in vein

The vessels that take blood away from the heart are *arteries* and *arterioles*. The vessels that bring blood toward the heart are *veins* and *venules*. Generally, arteries have veins of the same size running right alongside or near them, and they often have similar names. (See the "Arterial Components of the Cardiovascular System" color plate in the middle of the book for an illus-tration of the major arteries.) The arterial vessels (arteries and arterioles) decrease in diameter as they spread throughout the body. Eventually, they end in the *capillaries,* the tiny vessels that connect the arterial and venous systems. The venous vessels become larger as they converge on the heart. The smaller venules carry deoxygenated blood from a capillary to a vein, and the larger veins carry the deoxygenated blood from the venules back to the heart.

Starting with the arteries

Your arteries form a branching network of vessels, with the main trunk, called the *aorta,* aris-ing from the left ventricle and splitting immediately into the *brachiocephalic trunk, left common carotid artery,* and *left subclavian artery,* which serve the head and upper limbs. The *descending aorta* — which serves the thoracic organs, abdominal organs, and lower limbs — has several branches. The first branches are the *mesenteric arteries*— the main arteries of the digestive tract. The descending aorta also has two *renal arteries,* which supply the kidneys, as well as the *com-mon iliac arteries,* which supply the pelvis and lower limbs. The arteries branch into smaller and smaller vessels, becoming arterioles that end in capillary vessels.

Although an artery looks like a simple tube, the anatomy is complex. Arteries are made of three concentric layers of tissue around a space, called the *lumen,* where the blood flows. (See Figure 11-7.) The outside layer, the *tunica externa,* is a thick layer of connective tissue that sup-ports the vessel and protects the inner layers from harm. The larger the artery, the thicker the connective tissue supporting it.

The next layer in, the *tunica media,* is a thick wall of smooth muscle and elastic tissue. This layer expands with every *pulse* (wave of blood from the heartbeat). This layer controls *vasoconstriction* (decreasing diameter) and *vasodilation* (increasing diameter) of the vessels.

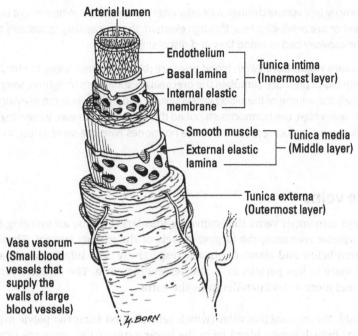

Arterial lumen

Endothelium
Basal lamina — Tunica intima (Innermost layer)
Internal elastic membrane

Smooth muscle — Tunica media (Middle layer)
External elastic lamina

Tunica externa (Outermost layer)

Vasa vasorum (Small blood vessels that supply the walls of large blood vessels)

FIGURE 11-7: The anatomy of an artery.

Illustration by Kathryn Born, MA

The inner layer, the *tunica interna*, is simple squamous epithelium that lines the lumen and is continuous through all the vessels and the heart (the endocardium). The *vascular endothelium*, as this tissue is also called, is active metabolically, releasing into the blood various substances that influence blood circulation and vascular health.

Cruising the capillaries

After passing through the arteries and arterioles, blood enters the capillaries, which lie between larger blood vessels in *capillary beds*. A capillary bed forms a bridge between the arterioles and the venules.

Your capillaries are your smallest vessels: only the single-cell-thick epithelial layer surrounds the lumen. The *precapillary sphincters* of the *metarterioles* (connectors of arteriole to capillary) can tighten or relax to control blood flow into the capillary bed, decreasing and increasing the pressure inside the capillary, respectively. Controlling the pressure within the capillary, or how hard the blood is pushing on the walls, provides control of *capillary exchange*. Higher pressure leads to more exchange of fluid as the blood passes through the capillary.

REMEMBER

Capillary beds are everywhere in your body, which is why you bleed anywhere that you even slightly cut your skin.

Besides aiding in the exchange of gases and nutrients throughout your body, capillaries serve two other important functions:

>> **Thermoregulation:** Precapillary sphincters tighten when you're in a cold environment to prevent heat loss from the blood at the skin surface. Then blood is diverted from an

arteriole directly to a venule through a nearby *arteriovenous shunt.* When you're in a warm environment or are producing heat through exertion, the precapillary sphincters relax, opening the capillary bed to blood flow and dispersing heat.

>> **Blood pressure regulation:** When blood pressure (blood volume) is low, the hormones that regulate blood pressure stimulate the precapillary sphincters to tighten, temporarily reducing the total volume of the blood vessel system and thus raising the pressure. When blood pressure is high, the hormones stimulate the sphincters to relax, increasing overall system volume and reducing pressure. These hormones have the same effect on the larger vessels.

Visiting the veins

Venules converge into small veins that converge into larger veins, all merging in the *inferior vena cava* and *superior vena cava,* the largest vessels of the venous system. These major veins return blood from below and above the heart, respectively. The inferior vena cava lies to the right of — and more or less parallel to — the descending aorta. The superior vena cava lies to the right of — and more or less parallel to — the aorta.

In the lower body, the *internal iliac veins* (which return blood from the pelvic organs) and the *external iliac veins* (which return blood from the lower extremities) converge into the *common iliac veins.* The *renal veins* return blood from the kidneys. Both these major veins flow into the inferior vena cava.

Blood from the digestive tract travels in the *hepatic portal vein* to the liver. Specialized cells in the liver move glucose molecules from the blood into storage. Phagocytic cells in the liver destroy bacterial cells that make it through the digestive process and remove toxins and other foreign material from the blood. Blood exits the liver through the *hepatic veins,* which flow into the inferior vena cava. The inferior vena cava empties into the right atrium.

Deoxygenated blood from the head and upper extremities drains into the *brachiocephalic veins.* The veins of the upper extremity — the *ulnar veins, radial veins,* and *subclavian veins* — also drain into the brachiocephalic veins. The *jugular veins* of the head and neck also drain into the brachiocephalic veins, which connect to the superior vena cava, which enters the right atrium.

After the blood from the right atrium has been pumped into the right ventricle, it's pumped to the lungs, where the blood is oxygenated; then it flows back to the heart in the *pulmonary veins,* the only veins that carry oxygenated blood.

Veins have a similar anatomy to arteries, although they tend to be wider, and their walls are thinner and less elastic. The tunica interna of a vein is also part of the continuous endothelial layer that lines the whole network. The tunica media has a much thinner layer of smooth muscle than in an artery. The veins have virtually no blood pressure, so they don't need a thick muscle layer to vary the vessel diameter or withstand fluid pressure. The outermost tunica externa is the thickest layer of a vein.

Because veins don't have a thick muscle layer to help push blood through them, and the momentum of the blood from the heart's contraction was used to power capillary exchange, they depend on contraction of skeletal muscles to move blood back to the heart. As you move your arms, legs, and torso, your muscles contract, and those movements massage the blood through your veins. The same occurs with each breath. The blood moves through the veins a little bit at a time and may occasionally try to sneak backward. Therefore, veins have semilunar valves embedded in the tunica interna that keep blood from flowing backward in the same way that the pulmonary and aortic valves do in the heart. The valves open in the same direction the blood is moving and shut after the blood passes through to keep the blood heading toward the heart.

REMEMBER

Since the push of the blood from the heart's contraction is gone by the time it enters the venous system, many veins have valves to prevent backflow. Although a pressure gradient still exists (lower in the heart), it's not enough to keep the blood flowing in that direction — especially against gravity. Muscle contractions from everyday movements squeeze the veins, pushing blood forward, and then they relax, pulling blood into the (temporarily) empty space. The process is much like how an eyedropper works.

Keep your finger on the pulse

You can feel the rhythmic pulsation of blood flow at certain spots around your body, most commonly on the radial artery on the inside of the wrist or on the carotid artery of the neck. What you feel as you touch these spots is your artery expanding as the blood rushes through it and then immediately returning to its normal size when the bulge of blood has passed. This pulsation corresponds to your heart rate and is usually recorded as beats per minute.

TECHNICAL
STUFF

The entire cardiac cycle takes about 0.86 seconds, based on the average of 70 heartbeats per minute. If your cardiac cycles take less time, your heart is beating too fast *(tachycardia)*; if there's too much time between your cardiac cycles, your heart is beating too slowly *(bradycardia)*.

Your cardiac output is determined by the heart rate and the *stroke volume* put out from a ventricle during one cycle (reported in liters per minute.). When either of these rises, the blood pressure rises. Heart rate is increased by physical exertion, the release of epinephrine (a hormone), and other factors. The blood volume is influenced by the action of antidiuretic hormone (ADH) and other mechanisms in the kidneys that control the amount of water removed from the urine and restored to the blood.

Under pressure

You've likely had your *blood pressure* taken before, the strap wrapped around your upper arm and inflated just to the point that you consider ripping it off from the pain. The purpose of this contraption is to block blood flow to your forearm. Then, as air is let out, blood begins to flow, creating the *sounds of Korotkoff* (which is why the stethoscope is positioned on your antecubital region, or inner elbow).

When the pressure around your arm matches the *systolic blood pressure* — the maximum force that the blood puts on the walls of the arteries due to ventricular contraction — some blood gets through, hitting the walls, which you can hear. Then, when the external pressure matches the *diastolic blood pressure* — the force applied to the walls between contractions — the sounds stop because the blood is now allowed to flow smoothly. Your blood pressure is recorded as the systolic pressure over the diastolic, or 120/80 mmHg on average.

REMEMBER

The importance of blood pressure isn't just about keeping the blood circulating; it's also the driving force behind capillary exchange, which I cover in Chapter 10.

Besides cardiac output, the *peripheral resistance* of blood vessels effects blood pressure. Peripheral resistance is an inverse measure of an artery wall's elasticity, or its give; high resistance means little elasticity. A blood vessel's diameter and length are also factors that contribute to peripheral resistance. Environmental agents such as smoking and plaque buildup (*atherosclerosis*) increase peripheral resistance, leading to an increase in blood pressure.

Blood pressure that's too high (*hypertension*) damages the artery walls, leading to a cascade of problems that can ultimately lead to heart failure. Unfortunately for us, the body doesn't seem to care much about this issue, as plenty of resources are being exchanged at the capillaries. There's no built-in mechanism for homeostasis of blood pressure if it's chronically high. On the other hand, low blood pressure (*hypotension*) means that the body's tissues aren't receiving enough oxygen and other nutrients — a problem that must be solved posthaste. A series of hormones is released via the *renin-angiotensin system*, (detailed in Chapter 15), which leads to *vasoconstriction* (decreasing the diameter of the arteries) and water retention in the kidneys, both of which lead to an increase in blood pressure.

How's your blood pressure after reading this section? Are you ready for some review questions?

PRACTICE

28) Number the structures in the correct sequence of blood flow from the heart to the radial artery for pulse. Start at the heart with the aortic semilunar valve.

_____ Axillary artery

_____ Subclavian artery

_____ Ascending aorta

_____ Brachial artery

_____ Aortic arch

29) Which layer of an artery is responsible for vasoconstriction?

(a) Basal lamina

(b) Lumen

(c) Tunica intima

(d) Tunica media

(e) Vasa vasorum

30) What are the sounds of Korotkoff?

(a) The pulsation of the blood on the walls of the veins during normal blood flow

(b) The turbulent sound of blood hitting the arteries when blood flow is restricted

(c) The sounds of contraction of the atria and ventricles

(d) The "lub-dub" sounds of the heart's valves closing

Practice Questions Answers and Explanations

(1) **(e) The pulmonary circuit travels to and from the lungs, and the systemic circuit keeps pressure and flow up for the body's tissues.** The Latin word *pulmon* means *lung.* None of the other answers mention the lungs.

(2) **(b) Pulmonary circuit**

(3) **(c) Systemic circuit**

(4) **(a) Pericardium**

(5) **(e) Septum**

(6) **(d) Epicardium**

(7) **(b) Pulmonary vein.** It's the only vein in the body that does so.

(8) **(c) Myocardium.** Only the middle layer contains muscle, and *myo* means *muscle.*

(9) **(b) Brachiocephalic trunk** (the first and largest branch of the aorta)

(10) **(a) Aorta**

(11) **(q) Superior vena cava.** The inferior would come up from the bottom.

(12) **(m) Right coronary artery** (the first branch of cardiac circulation)

(13) **(p) Right ventricle.** The label line isn't touching a vessel, so it's referring to the wall of the chamber.

(14) **(d) Left anterior descending artery.** This vessel on the surface of the heart comes forward off the left coronary artery that connects to the aorta.

(15) **(e) Left atrium.** Technically, the line points to the auricle of the atrium.

(16) **(g) Left pulmonary arteries**

(17) **The chordae tendineae are attached to the underside of the AV valves on one end and the papillary muscles on their other end. The papillary muscles are little knobs that are extensions of the myocardium, the muscle layer. When the wall of the ventricle contracts, so do the papillary muscles, which pull on the chordae tendineae. They pull down on the flaps of the valves, counteracting the force pulling them toward the atrium.** (The atrium is relaxing, thus decreasing pressure, and the ventricle is contracting, raising pressure.) This process serves to hold the AV valves in position, keeping them from flipping backward into the atria, which would allow blood to flow the wrong way.

(18) **(c) The myocardium is the muscular tissue of the heart, and more force is needed in chambers that pump blood greater distances.** It doesn't take a lot of power to push blood into the ventricles, so the atria don't need a lot of muscle tissue in the wall.

(19) **Pulmonary valve**

(20) **Aortic valve**

(21) **Tricuspid valve**

(22) **Mitral valve**

(23) **(b) It initiates an electrical impulse that causes atrial walls to contract simultaneously.** The other answers address rhythm, but the correct answer shows how the SA node coordinates the heart's contractions.

(24) **(c) SA node → AV node → Bundle of His → Bundle branches → Purkinje fibers**

(25) **With the atrium relaxed and slowly filling with blood, the ventricle ends its contraction. At some point as it relaxes, the pressure in the ventricle drops below what it is in the atrium. Because the valves open downward, the pressure shift causes the blood to push them open. There have been no contractions (in the new cardiac cycle) yet. The pressures of the atrium and ventricle equalize, so blood stops flowing. The amount of the stroke volume in the ventricle at this time is about 70 percent. To get the remaining blood out of the atrium, the chamber must contract to raise the atrial pressure over the ventricular.** The first part is called passive ventricular filling because no contraction means that no energy was used.

(26) **(e) The SL valves are being pushed closed.** The drop in pressure below that of artery causes the blood to briefly flow backwards, but it catches in the cups of the SL valves, pushing them shut.

(27) When you're looking at the EKG, you see a wave, three spikes, and then another wave. The point of the first spike is the Q, the R is the peak, and the S is the last downward spike. Contraction occurs in the period after the waves. Your bracket for atrial systole should span from where the first (P) wave ends to about the Q. For ventricular systole, it should start at around the R and begin as the final (T) wave is starting.

(28) **1. Ascending aorta; 2. Aortic arch; 3. Subclavian artery; 4. Axillary artery; 5. Brachial artery**

(29) **(d) Tunica media.** This layer contains the smooth muscle.

(30) **(b) The turbulent sound of blood hitting the arteries when blood flow is restricted**

If you're ready to test your skills a bit more, take the following chapter quiz that incorporates all the chapter topics.

Whaddya Know? Chapter 11 Quiz

Can you keep up with the pacemaker? It's time to find out! You can find the solutions and explanations in the next section.

1 Starting with a drop of blood returning to the heart from the legs, outline its path to its eventual return to the legs.

2 How and why do the conducting fibers spreading from the SA node to the atrial syncytium differ from those that go to the AV node?

_____

_____

3 Number the structures in the correct sequence of blood flow from the great saphenous vein back to the heart.

_____ External iliac vein

_____ Right atrium

_____ Common iliac vein

_____ Femoral vein

_____ Inferior vena cava

4 In an EKG, why don't you see a wave of repolarization for the atria?

_____

_____

5 What holds the cusps of the mitral valve in place?

(a) Supporting ligaments

(b) Chordae tendineae

(c) Trabeculae carneae

(d) Pectinate muscles

(e) Papillary muscles

For questions 6–9, Match the valve action to the event that caused it. Not all the letters will be used.

(a) Atria contract.

(b) Ventricle pressure is less than atria pressure.

(c) Ventricle pressure is greater than atria pressure.

(d) Ventricles contract.

(e) Artery pressure is less than ventricle pressure.

(f) Artery pressure is greater than ventricle pressure.

6   _____ SL valves open

7   _____ AV valves open

8   _____ AV valves close

9   _____ SL valves close

10 What covers the atrioventricular opening between the right atrium and the right ventricle?

(a) Bicuspid valve

(b) Tricuspid valve

(c) Pulmonary valve

(d) Mitral valve

(e) Aortic valve

For questions 11–19, fill in the blanks of the following paragraph.

The 11 _____ rhythm, or the heart's beats per minutes at rest, is controlled by the cardiac 12 _____ system. This system is comprised of interconnected conducting fibers that spread through the myocardium of the chambers. During each 13 _____, or single set of contractions, the impulse is initiated at the 14 _____. Then it spreads through the 15 _____, creating a coordinated contraction of both atria at the same time. Smaller fibers spread the impulse to the 16 _____, which creates the delay between contractions by slowing it down. Now speeding back up, the impulse moves through the 17 _____, heads down the interventricular septum to the apex, and finally branches into the 18 _____, which carry the impulse through the 19 _____, allowing both chambers to contract simultaneously in an upward, twisting motion.

For questions 20–25, choose true or false. When the statement is false, identify the error.

20 The heart is centrally located in the chest, directly behind the sternum.

_____

21 The chordae tendineae open the AV valves.

_____

22 Blood pressure is recorded as a fraction: systolic pressure over the diastolic pressure.

_____

23 The atrial syncytium is comprised of the conducting fibers of the atrium, which ensure that the atria contract before the ventricles.

_____

24 The renin-angiotensin system is activated to decrease blood pressure when it begins damaging arterial walls.

_____

25 The aortic valve does not get pulled backward into the chamber due to its cup-shaped structure.

_____

26 The superior vena cava enters the heart by way of the
(a) left ventricle.
(b) right ventricle.
(c) left atrium.
(d) right atrium.

27 Why does ventricular systole last longer than atrial systole?

_____

_____

28 In an EKG, which measurement would tell you how quickly impulse spread from the AV node to the apex?
(a) P–Q
(b) P–S
(c) P–P
(d) Q–S
(e) Q–T

29 How do you determine the measurements when taking a blood pressure?

_____

_____

_____

_____

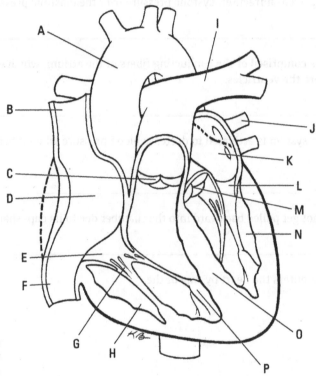

FIGURE 11-8:
The anatomy
of an artery.

Illustration by Kathryn Born, MA

For questions 30–36, use the descriptions to identify the structure identified in the heart in Figure 11-8. Not all the structures will be used.

30 _____ Returns blood to the heart from the head, thorax, and upper extremities

31 _____ Contracts to hold AV valves closed

32 _____ Valve located between the right ventricle and pulmonary artery

33 _____ Returns blood to the heart from the trunk and lower extremities

34 _____ Valve located between the left atrium and left ventricle

35 _____ Sends oxygenated blood out to the body

36 _____ Returns blood from the lungs

Answers to Chapter 11 Quiz

1 The connections of your outline should be as follows:

Inferior vena cava → Right atrium → Tricuspid valve → Right ventricle → Pulmonary valve → Pulmonary trunk → Pulmonary arteries → Lungs → Pulmonary veins → Left atrium → Mitral valve → Left ventricle → Aortic valve → Aorta → Body

2 **The conducting fibers that spread through the atria are thicker than those traveling to the AV node, allowing for a rapid spread of impulse.** To ensure that the atria finish contracting before the ventricles do, we need to slow down that spread between the nodes, so the conducting fibers connecting the SA and AV nodes are narrower. On leaving the AV node, the conducting fibers get progressively bigger, spreading the impulse faster. The Purkinje fibers spread impulse faster than any of the heart's other conducting fibers.

3 **1. Femoral vein; 2. External iliac vein; 3. Common iliac vein; 4. Inferior vena cava; 5. Right atrium**

4 **The atria are repolarizing at the same time that the ventricles are depolarizing.** The spread of impulse through the ventricles traverses a lot of space in a short period of time, so the QRS complex masks the atria's recovery wave.

5 **(b) Chordae tendineae.** Although the papillary muscles aid in the process, it's the chordae tendineae that keep them from flapping up into the atrium.

6 **(e) Artery pressure is less than ventricle pressure.**

7 **(b) Ventricle pressure is less than atria pressure.**

8 **(c) Ventricle pressure is greater than atria pressure.**

9 **(f) Artery pressure is greater than ventricle pressure.**

10 **(b) Tricuspid valve**

11 **sinus**

12 **conduction**

13 **cardiac cycle**

14 **SA node**

15 **atrial syncytium**

16 **AV node**

17 **bundle of His**

18 **Purkinje fibers**

19 **ventricular syncytium**

20 **False.** Two thirds of the heart lies to the left of the body's midline.

21 **False.** They don't close the valves, either. They prevent the valves from opening backward.

22 True

23 **False.** The AV node ensures the delay. Although the first part of the statement is true, the purpose of the syncytium is to make both atria contract at the same time.

24 **False.** This system is used to increase blood pressure.

25 **True.** This fact is a good thing, because there isn't anywhere for the chordae tendineae to attach without obstructing blood flow.

26 **(d) right atrium.**

27 When the ventricle begins to contract, it immediately increases the pressure to a level higher than the atrium. But even when the AV valve closes, the pressure isn't higher than the pressure inside the artery that's on the other side of the SL valve. Thus, it has to *keep contracting*, further increasing the pressure until it raises above arterial pressure, and the SL valve pops open.

28 **(d) Q–S.** The first spike, the Q, indicates impulse hitting the AV node. As it heads down the septum, towards the lead, the EKG line shoots up to the peak, the R. As it spreads back up and around the ventricle walls, the line dives down to the S.

29 The purpose of the blood pressure cuff (called a *sphygmomanometer*) is to cut off circulation to your brachial artery. So at first, you hear nothing when you listen with a stethoscope. As the pressure around the arm is gradually released, blood starts to come through in little spurts. These spurts hit the sides of the empty artery, making a sound you can hear: the sounds of Korotkoff. The level at which you first hear these tapping sounds is the systolic pressure. When the pressure around the artery matches how hard the blood pushes on the walls when the heart isn't contracting, the flow is smooth, and sounds are no longer heard. The pressure where you heard the last sound is the diastolic pressure.

30 **B** (superior vena cava)

31 **P** (papillary muscle). The chordae tendineae don't contract.

32 **C** (pulmonary valve)

33 **F** (inferior vena cava)

34 **L** (mitral valve)

35 **A** (aorta)

36 **J** (pulmonary veins)

Chapter **12**

Underground Defenses: The Lymphatic System

You see it every rainy day: water, water everywhere, rushing along gutters and down storm drains into a complex underground system that most would rather not give a second thought. You see the water in your shower and sink spiral down the drain — out of sight, out of mind. Well, it's time to give hidden drainage systems a second thought, because your body has one: the *lymphatic system*.

In addition to its vital role in circulation, your lymphatic system is all that stands between you and a planetful of invasive microorganisms that regard you, metaphorically speaking, as a large serving of biological molecules that could satisfy them indefinitely. Your body calls on the immune processes of the lymphatic system to protect itself from invading microbes, such as bacteria and viruses, other foreign cells, and your own cells that have gone bad (such as those that have become cancerous).

 We still have much to learn about the lymphatic system. It was long thought that the central nervous system didn't have lymphatic flow; it wasn't until 2015 that we learned it had lymphatic vessels. In 2017, researchers were finally able to use advances in imaging technology to confirm **WARNING** that there are, in fact, lymphatic vessels in and around the brain.

Meet & Greet Your Body's Customs and Border Control

Like the interstitial fluid getting pulled back into plasma during capillary exchange (see Chapter 11), storm water finds its way back into local waterways. That isn't so for household water, which gets processed in a wastewater treatment plant first. Likewise, your lymphatic system drains the excess interstitial fluid and filters it before putting it back into the supply.

Because lymph is pulled out of the circulating fluids that provide resources to cells, it is the ideal battleground for our immune defense:

» **Confronting marauders:** Whether you're well or ill, your immune system is always alert and active. That's why none of the bacteria, fungi, parasites, and viruses that exist in uncountable millions in the air you breathe, on the surfaces you touch, and on your food, are eating you. Sure, you have lots of ways of keeping the invaders from entering your body in the first place. But germs have evolved their own devious means to get past your defenses. Plenty get through. When your immune system is functioning properly, most germs don't stay for long. The immune system hunts them, destroys them, and wraps their cellular remains with a neat little bow for elimination from the body.

» **Stopping renegades:** The second major function of the immune system is to recognize and destroy cells of your own body that have gone rogue and have become potential seeds for cancerous growth. Cells go rogue every day. Most cancers can take hold only when the immune system malfunctions and fails to eliminate them. Unfortunately, the effectiveness of our immune cells declines with age.

The structures of the lymphatic system resemble those of other organs and systems that function to move fluids around. The lymphatic system has its own tubes, pipes, connectors, reservoirs, and filters, as shown in Figure 12-1. It lacks its own pumping organ, but like venous circulation, it uses valves, skeletal muscle action, and an overall pressure gradient for this purpose.

» **Lymphatic vessels:** The lymphatic vessels are the tubes that carry lymph. They are similar in structure to veins but with a larger lumen, a layer of elastic tissue, and less smooth muscle. They form a network similar to that of the venous system, starting small (the *lymphatic capillaries*) and then getting larger *(lymphatic vessels)* and even larger *(lymphatic ducts)*. Lymphatic vessels are distributed throughout the body, more or less alongside the blood vessels.

» **Lymphatic ducts:** The largest of the lymphatic vessels, the lymphatic ducts, drain into two large veins. The *right lymphatic duct,* located on the right side of your cervical region (neck) near your right clavicle, drains lymph from the right arm and the right half of the body above the diaphragm into the *right subclavian vein.* The *thoracic duct,* which runs through the middle of your thorax, drains lymph from everywhere else into the *left subclavian vein.*

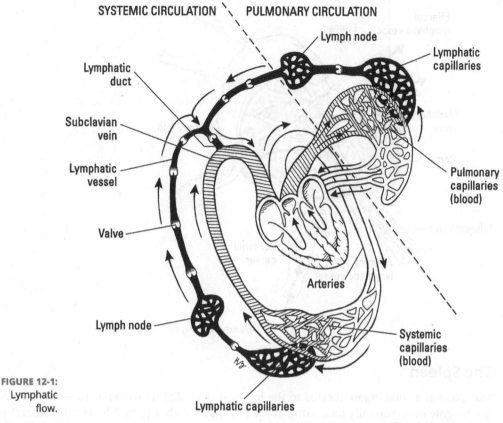

SYSTEMIC CIRCULATION   PULMONARY CIRCULATION

Lymph node

Lymphatic
capillaries

Lymphatic
duct

Subclavian
vein

Lymphatic
vessel

Pulmonary
capillaries
(blood)

Valve

Arteries

Lymph node

Systemic
capillaries
(blood)

FIGURE 12-1:
Lymphatic
flow.

Lymphatic capillaries

Illustration by Kathryn Born, MA

>> **Lymph nodes:** The little, bean-shaped structures located along the pathway of lymph are
the *lymph nodes*. Dense clusters of lymph nodes are located in the mouth, pharynx, armpit,
groin, all through the digestive system, and elsewhere, as you can see in the "Lymphatic
System" color plate in the center of the book. The *capsule* (fibrous outer membrane) folds
in, creating sections called *nodules*. *Afferent lymphatic vessels* carry the lymph in, pooling it
into the nodules. White blood cells (WBCs) patrol for germs as the lymph is drained through
the *efferent lymphatic vessels* on the opposite side; the process is much like passing through
border patrol upon entering a country. The structure of a lymph node is illustrated in
Figure 12-2.

TIP

If you have trouble remembering your afferent from your efferent, think of the *a* as standing
for *approach* or *access* and the *e* as standing for *exit*.

Having a Spleendid time

This detour to the fluid circulation of the blood is supported by two main organs: the spleen
and thymus. And with the help of a few smaller organs, they keep constant watch on what's
traveling in our bodies, primed for a battle when necessary.

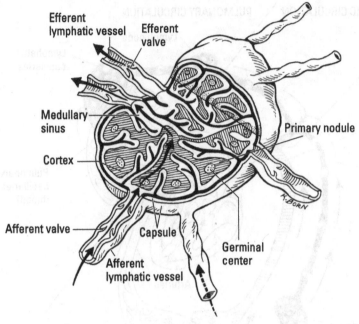

Efferent lymphatic vessel
Efferent valve
Medullary sinus
Cortex
Afferent valve
Afferent lymphatic vessel
Capsule
Germinal center
Primary nodule

K. BORN

FIGURE 12-2:
A lymph node.

Illustration by Kathryn Born, MA

The Spleen

The *spleen* is a solid organ, located to the left of — and slightly posterior to — the stomach. It's roughly oval, normally measuring about 1 by 3 by 5 inches (3 by 8 by 13 centimeters) and weighing about 8 ounces (23 grams). Essentially, its structure is that of a really large lymph node, and it filters blood in much the same way that the lymph nodes filter lymph, with WBCs patrolling the pool of fluid.

The spleen is enveloped by a fibrous capsule that folds inward, creating chambers called *lobules.* It has a *hilum* (concave area or bend) containing the *splenic artery, splenic vein,* and an efferent lymphatic vessel, in a similar configuration to a lymph node. Arterioles leading into each lobule are surrounded by masses of developing lymphocytes that give those areas of so-called *white pulp* their appearance. On the outer edges of each compartment, tissue called *red pulp* consists of blood-filled cavities. Unlike lymph nodes, the spleen doesn't have any afferent (access) lymph vessels, which means that it doesn't filter lymph — only blood.

Blood flows slowly through the spleen to allow it to remove microorganisms, exhausted erythrocytes (red blood cells [RBCs]), and any foreign material that may be in the stream. Among its various functions, the spleen can be a blood reservoir. When blood circulation drops while the body is at rest, the spleen's vessels can dilate to store any excess volume. Later, during exercise or if oxygen concentrations in the blood begin to drop, the spleen's blood vessels constrict and push any stored blood back into circulation.

REMEMBER

The spleen's primary role is as a biological recycling unit, capturing and breaking down defective and aged RBCs to reuse their components. Iron stored by the spleen's macrophages goes to the bone marrow, where it's turned into hemoglobin for new RBCs. By the same token, bilirubin for the liver is generated during breakdown of hemoglobin. The spleen produces RBCs during embryonic development but shuts down that process after birth; in cases of severe anemia, however, the spleen can start production of RBCs again. Nonetheless, the spleen isn't considered to be a vital organ; if it's damaged or has to be removed, the liver and bone marrow can pick up where the spleen leaves off.

The thymus gland

Tucked just behind the sternum and between the lungs in the upper thorax (the *superior mediastinum*, if you want to be technical), the *thymus gland* was a medical mystery until recent decades. Its two oblong lobes are largest at puberty, when they weigh around 40 grams (somewhat less than an adult mouse). Through a process called *involution*, however, the gland atrophies and shrinks to roughly 6 grams by the time an adult is 65. (You can remember that term as the *inverse* of *evolution*.)

The thymus gland serves its most critical role as a nursery for immature T lymphocytes, or T cells, during fetal development and the first few years of a human's life. Before birth, fetal bone marrow produces *lymphoblasts* (early-stage lymphocytes) that migrate to the thymus. Shortly after birth and until adolescence, the thymus secretes several hormones, collectively called *thymosin*, that prompt the early cells to mature into full-grown T cells that are *immunocompetent*, ready to go forth and conquer invading microorganisms. (These hormones are the reason why the thymus is considered to be part of the endocrine system too; check out Chapter 9 for details on this system.)

The tonsils, Peyer's patches, and appendix

The *tonsils* are the usual culprits of childhood sore throats. Like the thymus, they are largest around puberty and tend to atrophy with age. *Invaginations* (ridges) in the tonsils form pockets called *crypts*, which trap bacteria and other foreign matter.

Unlike the thymus, the tonsils don't secrete hormones, but they do produce lymphocytes and antibodies to protect against microorganisms that are inhaled or eaten. Although only two are visible on either side of the pharynx, there are five tonsils:

>> The two you can see, which are called *palatine tonsils*

>> The *adenoid* or *pharyngeal tonsil* in the posterior wall of the nasopharynx

>> The *lingual tonsils,* which are round masses of lymphatic tissue arranged in two approximately symmetrical collections that cover the posterior third of the tongue

Peyer's patches are masses of lymph nodules just below the surface of the ileum, the last section of the small intestine. When harmful microorganisms try to migrate into the small intestine, Peyer's patches can mobilize an army of B cells and macrophages to fight off infection.

Just after the ileocecal sphincter (where the small and large intestines join) lies a dead-end tube: the poor misunderstood *appendix*. Regarded by some people as being a useless (*vestigial*) organ, the appendix is a small, muscular tube of lymphatic tissue. It serves as a center of immune response, producing lymphocytes when we're battling a germ in the intestines. The appendix also seems to play a role in housing the good bacteria that are normally present to help with digestion, repopulating the large intestine when they've all been flushed away by a course of antibiotics or a bout of lower-intestinal distress.

Picking Your Nodes

As the afferent lymphatic vessels pour lymph into the lymph nodes, the lymph spreads out and pools in the nodules. The space is filled with *reticular connective tissue* — the fibers creating a net to support the hundreds of leukocytes (WBCs) that monitor the fluid. Also spread throughout the node are germinal centers for the production of B cells.

The lymph is slowly pulled through the nodules and out of the efferent lymphatic vessel at the hilum. In the process, pathogens, cancerous cells, and broken cells are destroyed by macrophages or targeted lymphocytes. (Stay tuned; I cover how these WBCs attack shortly.)

In this way, the fluid is filtered before being returned to the blood. It's important to note, though, that we're filtering germs and damaged cells, not toxins or wastes; that job is handled by the liver and kidneys, respectively.

WARNING

Lymph nodes are often mistakenly called *lymph glands*. They don't secrete anything, however, so that term is a misnomer. When we're fighting off a germ, the lymph nodes often swell due to an influx of fluid — the result of inflammation. Swollen lymph nodes (swollen glands) aren't themselves a disease or pathological condition; they're a sign that the immune system is doing its job.

Think you have a *node*tion (sorry, not sorry) about what's happening here? Test your knowledge.

PRACTICE   For questions 1–5, match the lymphatic structure with its function. Choices can be used more than once or not at all.

    (a) Bone marrow

    (b) Lymph nodes

    (c) Peyer's patches

    (d) Spleen

    (e) Thymus

    (f) Tonsils

1 _____ House lymphocytes to remove foreign particles from lymph

2 _____ Is the largest lymphatic organ

3 _____ Develops and trains T cells

4 _____ Produces and stores lymphocytes and monocytes, removes damaged erythrocytes

5 _____ Control the bacterial composition of the intestines

6 What's the significance of the fact that the spleen has no afferent vessels?

(a) Foreign materials can't get inside and cause problems.

(b) The spleen doesn't filter lymph.

(c) It's why humans can survive removal of the spleen.

(d) The spleen evolved that way to prevent phagocytosis.

(e) It allows for a direction connection with lymph nodes.

7 The two lymphatic ducts empty into the

(a) vena cavae.

(b) spleen.

(c) thymus.

(d) right atrium.

(e) subclavian veins.

For questions 8 and 9, fill in the blanks to complete the following sentence.

Lymph moves from a/an **8** _____ vessel through a lymph node and back out into the blood through a/an **9** _____ vessel.

10 What is the largest lymphatic vessel in the body?

(a) Right lymphatic duct

(b) Spleen

(c) Thoracic duct

(d) Lymph node

(e) Bone marrow

Immunity: A Three-Act Play

The lymphatic system is the battleground for our immune processes, but nearly every tissue in our bodies has some role to play (which is why you no longer see the immune system as one of the body's listed systems). Defense against threats, both foreign and domestic, is a collaborative process, not the function of a single body system. It functions like a long-running, fine-tuned Broadway show that's been rehearsed until the director no longer has to worry. That's Immunity.

Our numerous immune cells are constantly patrolling our bodies, floating in the bloodstream, hanging out in the interstitial fluid, and moving through the lymph. Many of these cells post up in the lymph nodes acting like guards, checking everything that comes through a security gate. Most often, our bodies are able to destroy pathogens before they can unpack their bags and start making us sick. This scenario is the strategy of the *innate immune system*, Act 1 of our immunity play. If the pathogen is especially prolific or finds a good hiding spot, we start to feel the characteristic symptoms of an infection. The battle isn't lost, though. We have a secret weapon: Act 2, the *adaptive immune system*. We enlist specialized WBCs to mount a targeted attack on the pathogen, minimizing the collateral damage to our own cells that the innate strategies inevitably cause. Finally, when the battle is (presumably) won, we have Act 3, true *immunity*. That particular pathogen won't be able to take up residence in our bodies for the foreseeable future.

The set

The success of our immune processes relies on one critical factor that I haven't mentioned yet: recognition. If our cells can't recognize *self* from *nonself*, they don't know what to leave alone and what to attack. Molecules called *antigens* serve this purpose. Antigens are molecules, usually proteins, that stick up off the surfaces of cells. All cells have them, including our own. Some nonliving things (such as pollen, dust, and toxins) have them too. When immune cells are patrolling the body, looking for invaders, they're looking for unfamiliar antigens. When the cells locate these antigens, our immune response begins, resulting in the destruction of the invaders.

To prevent cases of mistaken identity, our cells have human leukocyte antigens (HLA, for short) that identify them as self. These cells get left alone; the antigens don't trigger an immune response.

TIP

The proper definition of an antigen is a molecule that elicits an immune response in the body. Many texts talk about how immune cells and chemicals recognize these antigens as foreign and destroy them. We recognize any germs that enter our bodies by their unfamiliar antigens and mount our attack. But if we only destroy the antigens, the cell is left intact. It's a case of one word taking on too many meanings. When you come across a statement about "attacking antigens," try to think "attaching the antigens" instead. When they're attached, whatever protein they're sticking out of can be destroyed too.

The stars of the show: Leukocytes

The players in our immunity show are the *leukocytes*, the WBCs. These cells are unique in many ways besides their ability to use the bloodstream as their own personal transportation service. In shape and size, they're far from the compact epithelial or muscle cell types. They have about a dozen distinctive shapes and many sizes, and some have the ability to transform themselves into other, even weirder forms and to multiply extremely rapidly. See Table 12-1 for an overview of immune-system cells.

WARNING

Over the past few decades, immunologists and cell biologists have discussed immune-system cells in terms of several classification systems that haven't held up well on further investigation. Forming and discarding theories and classification systems based on new knowledge is expected and welcome in immunology, as in any science. The structure and physiology of immune-system cells will be under study for the foreseeable future. The following discussion of specific types of immune-system cells gives you some idea of some established concepts. Keep this limited goal in mind, especially when checking out Table 12-1.

Table 12-1   Cells of the Immune System

Cell Type	Function	Prevalence in Circulation
Neutrophil	First responders that phagocytize bacteria; live for only a few hours.	50–70% of total WBCs
Basophil	Trigger inflammation; prevent blood clots at infection site.	<1% of total WBCs
Eosinophil	Destroy cells marked as foreign; specialized for parasites.	1–4% of total WBCs
Monocyte	Mature into macrophages, which phagocytize bacteria and viruses.	4–8% of total WBCs; largest of WBCs
Lymphocyte	Carry out targeted attacks. There are three types:	20–45% of total WBCs; relatively few circulate
	B cells: Produce antibodies; especially useful for bacteria	Mature in bone marrow
	T cells: Attack infected cells; especially useful for viruses	Mature in thymus
	NK cells: Attack any unrecognized cells	Mature in bone marrow

Their talents: Mechanisms of action

The variety of leukocytes gives us a variety of methods to counter any potential troublemakers. Some simply eat the unwelcome cell via phagocytosis. Others use chemicals to inflict damage through a process called *degranulation*. Throughout, communication is critical. Most cells release signaling chemicals when performing their actions to call for reinforcements, for example.

Phagocytosis

Phagocytosis is the simplest mechanism and the most frequently used to deal with invaders. An immune cell engulfs its target by wrapping its membrane around it; then it destroys the target.

Neutrophils are the most numerous of the WBCs and are the leading candidate for the Cell with the Shortest Life Span Award. They circulate for about a day and then undergo *apoptosis* (programmed cell death). If they're signaled to an infection site, they function for another day or two, but no more. They're sometimes referred to as *microphages* because although they perform phagocytosis, they're much smaller than the familiar macrophages that come from monocytes.

Another type of WBC, *monocytes,* are themselves inactive. When signaled, they migrate into the infected tissue and divide into a macrophage and a *dendritic cell.* The macrophages are large, cell-eating machines that target not only germs, but also all the broken cells and debris that inevitably result from the battle. Dendritic cells perform phagocytosis of a single cell and then work to signal our adaptive immune response (more on that momentarily).

WARNING

Pus, like mucus, is yucky. But pus is proof that your body is fighting an invader and that your immune system is doing its job. You could say that pus is a product of the immune system. This thick, white-to-yellow goo, sometimes shot with blood, appears at a site of injury or infection. (What would adolescence be without at least a few pus-filled pimples on the skin?) Its color and texture come from the dead phagocytes that make up most of it — WBCs that have done their job and then died with the cellular remains of thousands and thousands of invaders wrapped up inside.

Degranulation

Neutrophils, basophils, and *eosinophils* have *granules* in their cytoplasm. Granules themselves aren't chemicals; they're little packets of chemicals, such as cellular toxins, enzymes, and other proteins. Cell biologists have identified several types of granules with specific chemical contents. *Degranulation* is the process that moves the granules out of the cell, releasing the contents into the interstitial space. This process is the opposite of phagocytosis (*exocytosis;* see Chapter 3). The chemicals released into the interstitial fluid carry out a range of specialized immune functions. Some destroy invaders directly; others regulate the immune system's processes.

Granules in *eosinophils* play a crucial part in the immune response to *enteric* (intestinal) parasites through their release of toxic proteins (our own natural pesticides!). Eosinophil numbers increase during allergic reactions and parasitic infections.

The degranulation of *basophils* releases histamine and *heparin.* Histamine is the main chemical that triggers inflammation so that more resources can be delivered to the site of infection. Heparin prevents clotting to keep the blood flowing smoothly in the area; because where there's infection, there's damage.

Inflammation

Histamines are also released by *mast cells.* Although these cells aren't WBCs, they are present in most tissues (characteristically surrounding blood vessels and nerves) and are especially numerous near the boundaries between you and the outside world, such as in the skin and in the mucosa of the respiratory and digestive systems. Mast cells are granular cells that play a key role in the inflammatory process. When activated, a mast cell rapidly releases the contents of its granules and various hormonal mediators into the extracellular space.

The last time you got a splinter in your hand or foot, you may have noticed that the site of entry became red, swollen, and tender. These reactions are signs and symptoms of an *inflammation response,* a basic way that the body reacts to infection, irritation, or injury. Think of it as a mechanism for removing the injurious object and initiating the healing process. When the splinter punctures your skin, the injured cells release numerous signaling chemicals, particularly histamine.

TECHNICAL STUFF

Histamine triggers vasodilation in the area where it's released, increasing the diameter of the arterioles but not the capillaries (they don't have any smooth muscle to respond). The change in volume as the blood leaves the arterioles and enters the capillaries is more dramatic, thus increasing the pressure. Blood pushes harder on the walls of the capillaries, releasing more plasma (with its accompanying resources) into the interstitial fluid.

As annoying as it is, triggering pain and itch receptors as it does, inflammation is a critical part of both the healing and innate immunity processes.

Act 1: Innate Immunity

Innate, or *nonspecific*, *defenses* are the tools our bodies use to attack foreign invaders regardless of their ilk. Germs can be bacteria, viruses, fungi, or other microorganisms; other foreign particles (such as pollen and toxins) can be problematic as well. Our innate defenses target all these substances.

First and foremost, we have our skin — the body's largest organ and our first line of defense. Along with our other *mechanical barriers,* such as mucus and tears, most of the potential invaders are never allowed entry. Should one make it into the body, we have other innate strategies for our second line of defense:

>> **Phagocytosis:** Consumes foreign invaders by specialized WBCs

>> **Chemical barriers:**

  - Enzymes (in saliva and gastric juices) break down cell walls.

  - *Interferons* block replication (especially of virus and tumor cells).

  - *Defensins* rip large holes in cell membranes.

  - *Collectins* group pathogens for easier phagocytosis.

  - *Perforins* poke tiny holes in cell membranes.

>> **Inflammation:** Dilates blood vessels, sending more resources to the area where the pathogen was identified

>> **Fever:** Weakens microorganisms, stimulates phagocytosis, and triggers mobilization of WBCs

>> **Natural killer cells (NKs):** Secrete *perforins* to perforate cell membranes

TECHNICAL STUFF

NKs are the only lymphocytes involved in innate immunity. The other lymphocytes (B and T cells) are pathogen-specific; they respond only to one particular antigen. NKs are far less specific, responding not to the presence of a nonself antigen but to the lack of a self antigen.

Unfortunately, the occasional pathogen gets past these defenses, so our bodies must mount a targeted attack: our adaptive defenses. Furthermore, if we relied solely on our innate defenses, there would be massive amounts of collateral damage to our own cells (which is responsible for many of our symptoms of illness in the first place).

PRACTICE

Your mechanical barriers aren't keeping all this knowledge out, are they? It's time to find out.

11 Explain why inflammation is an essential event in immunity.

_____

_____

_____

For questions 12–16, identify the innate strategy described.

12. _____ Disrupt replication of pathogen

13. _____ Gather pathogens for destruction

14. _____ Tear large holes in cell membrane

15. _____ Released by NKs to poke tiny holes in cell membranes

16. _____ WBC engulfs the entire pathogen

17. A high level of these cells indicates a parasitic infection.

    (a) Basophil

    (b) Eosinophil

    (c) Neutrophil

    (d) Monocyte

    (e) Lymphocyte

18. How do lymphocytes differ from the other leukocytes?

    (a) They can perform phagocytosis.

    (b) They release chemicals via degranulation.

    (c) They're triggered only by specific antigens.

    (d) They produce histamines for inflammation.

    (e) They can exit blood flow.

Act 2: Adaptive Immunity

When the mechanisms of our innate immunity don't eliminate the pathogen quickly on their own, our adaptive mechanisms join in the battle. The trigger requires an *antigen-presenting cell* (APC), usually a dendritic cell. When a dendritic cell comes across an unrecognized pathogen (meaning that it hasn't been affected by any sort of immune response), it engulfs and digests it. But it preserves the antigens and displays them on its own cell membrane, thus becoming an APC. Then it flows around our fluids, hunting for a T cell that has a receptor that can bind to the antigen. When this link occurs (probably in a lymph node, because that's where lymphocytes tend to hang out and pass the time), the adaptive immune response is activated.

Our body has at least one lymphocyte that can attach to any pathogen that enters it; researchers estimate that the odds of that not being the case are practically nonexistent. The issue is how long the APC will take to find that pathogen. The more unique an antigen is, the longer it will take our body to mount this targeted attack. In the meantime, the innate system is doing its best, but the collateral damage to our own tissues starts to add up. Add in the ability of many pathogens to produce molecules that damage our tissue themselves, and you have the recipe for a pathogen-caused disease with a high mortality rate.

Cell-mediated immunity

When the APC binds to a *helper T cell*, it triggers the helper T to *proliferate* — make a bunch of copies of itself, in other words. The new helper T cells release *interleukins*, which activate the matching *cytotoxic T cells*. These cytotoxic Ts (sometimes called *killer Ts*) bind with antigens on the invader and release perforins, killing the pathogen. This process is illustrated in Figure 12-3.

When the battle has waned, *suppressor T cells* signal the adaptive immune process to stop. Some T cells remain as *memory T cells* when the pathogen has been defeated for a quick response if that germ ever returns.

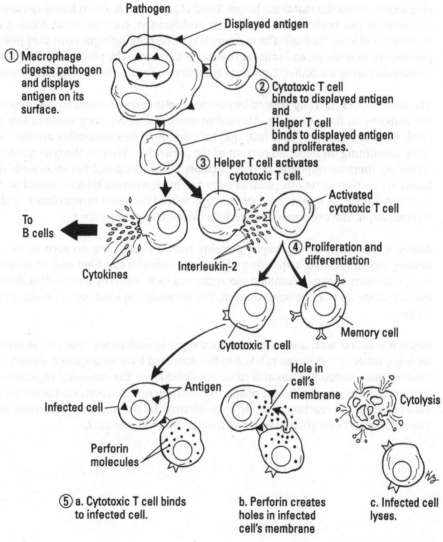

Pathogen

Displayed antigen

① Macrophage digests pathogen and displays antigen on its surface.

② Cytotoxic T cell binds to displayed antigen and
Helper T cell binds to displayed antigen and proliferates.

③ Helper T cell activates cytotoxic T cell.

To B cells

Cytokines

Interleukin-2

Activated cytotoxic T cell

④ Proliferation and differentiation

Cytotoxic T cell

Memory cell

Antigen

Infected cell

Hole in cell's membrane

Cytolysis

Perforin molecules

FIGURE 12-3: Cell-mediated immunity.

⑤ a. Cytotoxic T cell binds to infected cell.

b. Perforin creates holes in infected cell's membrane

c. Infected cell lyses.

Illustration by Kathryn Born, MA

The T-cell process, or *cell-mediated immunity*, creates an army of cells that attack the pathogen that initiated the process. Their mode of attack is much like sending soldiers to the front lines to battle the enemy with weapons and is especially effective against viruses and cancerous cells.

Humoral immunity

Humoral immunity, also called *antibody-mediated immunity*, is our adaptive defense that uses the B cells. It works in conjunction with (rather than after or instead of) the cell-mediated process. The strategy here, to continue the army metaphor, is to turn the B cells into bomb-making factories.

Naïve B cells bind to the matching antigen of a pathogen, but until they're activated, nothing else occurs. When the matching helper T cell is activated, it also releases *cytokines*. The chemicals activate this bound B cell, leading to proliferation. Each round of division creates both a *memory B* and a *plasma B cell*. The memory B cells do nothing right now; they just circulate, find somewhere to settle in, and wait for that germ to return. The plasma B cells go about their task of manufacturing *antibodies*. Figure 12-4 illustrates the process of humoral immunity.

The *antibodies*, exclusively created by plasma B cells in the thousands, are proteins that bind to the antigens on the pathogens. Also called *immunoglobulins*, they come in five classes named (with great imagination) IgA, IgD, IgE, IgG, and IgM. IgG antibodies are the most important class, accounting for about 80 percent of the antibodies. They're the type most involved in the secondary immune response, as they circulate in the blood and the other body fluids. IgAs are found in exocrine secretions (such as tears and bile), whereas IgDs are found on the surfaces of B cells (forming the receptors). IgMs are specialized for blood compatibility, and IgEs promote inflammation. The overproduction of IgEs causes allergic reactions.

Antibody molecules are Y-shaped proteins with a *binding site* on each of the short arms. A binding site in turn has a specific and intricate shape; it can bind only to an antigen with the complementary shape. A common metaphor is a lock-and-key mechanism. Just as a key has a specific shape and can fit only one lock, the antibody can bind only to antigens that match its shape.

When it's bound with antibodies, the pathogen is *neutralized*. The cell is unable to interact with any other cell; it is also marked to be consumed by a macrophage. Because they have two binding sites, antibodies can also cause *agglutination* — the clumping together of the neutralized invaders for more efficient phagocytosis. They can also activate the *complement cascade*, a series of chemical reactions that directly destroy the pathogen by effectively ripping it open. The antibody actions are illustrated at the bottom of Figure 12-4.

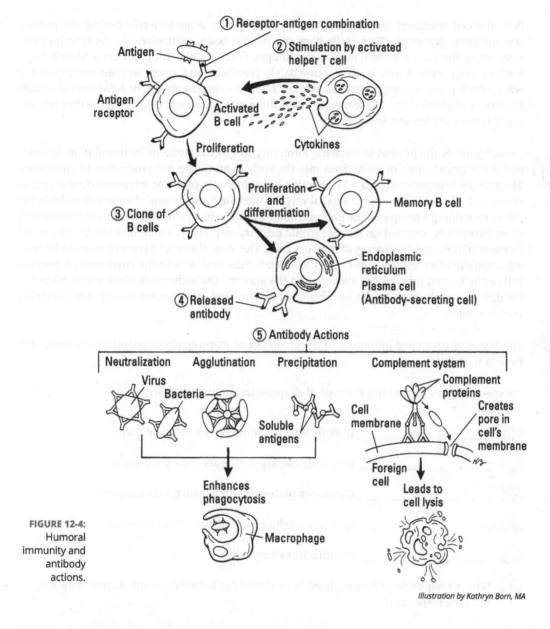

① Receptor-antigen combination

② Stimulation by activated helper T cell

Antigen

Antigen receptor

Activated B cell

Cytokines

Proliferation

Proliferation and differentiation

③ Clone of B cells

Memory B cell

Endoplasmic reticulum

Plasma cell (Antibody-secreting cell)

④ Released antibody

⑤ Antibody Actions

Neutralization   Agglutination   Precipitation   Complement system

Virus

Bacteria

Soluble antigens

Cell membrane

Complement proteins

Creates pore in cell's membrane

Foreign cell

Enhances phagocytosis

Leads to cell lysis

Macrophage

FIGURE 12-4:
Humoral immunity and antibody actions.

Illustration by Kathryn Born, MA

Act 3: Becoming Immune

Immunity is the upside of infection and disease (or at least some infections and diseases). After your immune system has dealt with certain pathogens, it produces the condition of immunity, allowing you to resist infection by a particular pathogen because your body has defeated it before.

Both the cell-mediated and humoral mechanisms produce *memory cells* during the *primary response* (first exposure). These cells hang out in your body, particularly in the lymph nodes, monitoring the fluid for their matching pathogen (with its matching antigen). Should these memory cells come across it, they immediately reactivate and attack in their corresponding ways, which is the *secondary response*. We don't have to wait for an APC or for chemical signals from the site of infection. These memory cells jump into action and take down the invader, and you'll never even know it was there.

REMEMBER

Immunization is the process of inducing immunity to specific antigens by inoculation. *Inoculation* is the introduction of an antigen into the body to stimulate the production of antibodies. The antigen is introduced, often by injection but sometimes orally or intranasally, in a preparation called a *vaccine.* There are several ways to introduce an antigen, the most familiar being a dose containing a sample of the pathogen that's dead or in an *attenuated* (alive but weakened) form. Essentially, vaccination induces a mild primary response (so mild that ideally, you won't be aware of it) to create an army of memory cells. This way, the secondary response can be activated rapidly when you encounter the pathogen for the first time in the environment, because it'll be the second time you've encountered the antigen. Different methods of vaccination and for different pathogens are being developed all the time in response to our ever-changing environment.

PRACTICE

Think you've developed immunity to the challenge of learning about adaptive immunity? It's time to find out!

For questions 19–23, identify the type of adaptive immunity cell described.

19 _____ Directly attacks pathogens by using perforins

20 _____ Responds quickly during secondary exposure

21 _____ Consumes pathogens then displays its antigens

22 _____ Releases cytokines to activate humoral response

23 _____ Manufactures antibodies

24 Why are dendritic cells considered to be the bridge between our innate and adaptive immune responses?

_____

_____

_____

25 One mechanism used by antibodies is to clump together pathogens for more efficient phagocytosis. This process is called

(a) inflammation.

(b) neutralization.

(c) precipitation.

(d) agglutination.

(e) degranulation.

26 This type of antibody forms receptors on B cells.

(a) IgA

(b) IgD

(c) IgE

(d) IgG

(e) IgM

27 Which statement best describes the role of cytotoxic T cells?

(a) They're a critical component of the humoral immune response.

(b) They attach to antigens and release perforins to destroy cells.

(c) They signal to other immune cells that the germ has been defeated.

(d) They bind to a pathogen and kill it by using cytokines.

(e) They activate the cell-mediated response by releasing interleukins.

28 Explain what the term *immune* means.

_____

_____

_____

_____

Practice Questions Answers and Explanations

(1) **(b) Lymph nodes**

(2) **(d) Spleen**

(3) **(e) Thymus**

(4) **(d) Spleen**

(5) **(c) Peyer's patches**

(6) **(b) The spleen doesn't filter lymph.** No afferent vessels means no access for lymph.

(7) **(e) subclavian veins**

(8) **afferent**

(9) **efferent**

(10) **(c) Thoracic duct.** Yes, this duct is the largest lymphatic vessel. Answer (b) isn't correct, because the spleen is the largest lymphatic *organ*.

(11) **Inflammation is the result of histamine release, causing vasodilation in the area of infection. Widening the supply lines (the arterioles) brings more blood to the area. That increased blood volume enters the capillaries, which have unchangeable diameters. This process causes increased pressure; the blood pushes harder on the walls, squeezing more plasma out into the interstitial fluid.** This process is essential because it's from the interstitial fluid that all cells get their supplies, including the WBCs that fight off germs.

(12) **Interferons.** They interfere with replication.

(13) **Collectins.** *Collect* is another word for *gather*.

(14) **Defensins.**

(15) **Perforins.** *Perforations* are tiny holes, like those in perforated paper.

(16) **Phagocytosis.**

(17) **(b) Eosinophil**

(18) **(c) They're triggered only by specific antigens.** NKs are lymphocytes that don't respond only to a specific antigen, but none of those in the other categories do.

(19) **Cytotoxic T.** Also known as killer Ts, so it makes sense that they're the ones that attack directly.

(20) **Memory.** Both B and T cells have memory versions.

(21) **Dendritic.** It's not an adaptive immune cell itself, but it's the trigger for the process.

(22) **Helper T.** It *helps* both cell-mediated and humoral processes get started.

(23) **Plasma B.** B cells make antibodies.

(24) **Dendritic cells are like macrophages in that they attack foreign cells by swallowing them whole (*phagocytosis*). The difference is that before destroying the germ, dendritic cells remove the antigens. Then they display the antigens on their cell membrane and set off to find a helper T cell with a receptor that can bind the antigen. When they do, chemicals are released to activate our adaptive immunity pathways.**

(25) **(d) agglutination.** Agglutination occurs when antibodies *glue* the cells together.

(26) **(a) IgA**

(27) **(b) They attach to antigens and release perforins to destroy cells.** T cells run the cell-mediated response, but they're the ones that are activated by interleukins. Cytokines are the signaling chemical that start the humoral, or B-cell, process. The regulatory Ts signal victory.

(28) **When you're immune to a pathogen, you have enough memory cells and antibodies that are circulating and stored in your lymph nodes that if you're exposed to that germ again, your lymphocytes would destroy it before you started having symptoms.**

Hopefully you're not lymphing along and are ready for the chapter quiz!

Whaddya Know? Chapter 12 Quiz

I hope your confidence inflamed because it's quiz time! Complete each question to test your knowledge of the various topics covered in this chapter. You can find the solutions and explanations in the next section.

1. The germinal center of lymph nodules produces what type of cells?

    (a) Monocytes

    (b) Lymphocytes

    (c) Phagocytes

    (d) Neutrophils

    (e) Basophils

2. What role does the spleen play besides that of a biological recycling unit?

    (a) It undergoes involution as a person ages to make way for enlarged lymph nodes.

    (b) It routes the lymph from lymphatic vessels to the subclavian veins.

    (c) It produces neutrophils to maintain immunocompetent T cells.

    (d) It has vessels that dilate to store excess blood volume that can be released when needed.

    (e) It houses and activates T lymphocytes during the cell-mediated response.

For questions 3–7, fill in the blanks to complete the following paragraph.

Pathogens, like our own cells, are covered in 3 _____ which enable our WBCs to distinguish between self and nonself. When activated, 4 _____ that have the matching receptors release perforins to destroy the specific pathogen. Meanwhile, B cells synthesize 5 _____ that target the pathogen as well. Both types of lymphocyte keep 6 _____ around for a faster response on the next exposure, granting you 7 _____ from that particular pathogen.

8. Besides immune defense, what is the major role of the lymphatic system?

    _____

    _____

    _____

9 Lymphatic tissue in the ileum creates the

(a) Peyer's patches.

(b) lymph nodes.

(c) appendix.

(d) adenoids.

(e) lymphatic duct.

10 Which are the least common of the circulating WBCs?

(a) Basophils

(b) Eosinophils

(c) Lymphocytes

(d) Monocytes

(e) Neutrophils

11 Which of the following are examples of mechanical barriers? Choose all that apply.

(a) Skin

(b) Stomach acid

(c) Mucus

(d) Hair

(e) Inflammation

12 How does lymph move around inside its vessels?

(a) Rapidly, through the action of tiny pumps within lymph nodes

(b) Slowly, and only as additional lymph enters the system

(c) Slowly, through the squeezing of surrounding skeletal muscles

(d) Steadily and rhythmically, mirroring the heart rate

13 Differentiate among monocytes, macrophages, and dendritic cells.

_____

_____

_____

For questions 14–19, choose true or false. When the statement is false, identify the error.

14 Lymph nodes use Peyer's patches to destroy any bacteria that comes through them.

_____

15. Lymph nodes sometimes swell when an infection is present to make room for additional lymph fluid.

_____

16. IgEs are the antibodies overproduced in allergies.

_____

17. When helper T cells proliferate, one cell becomes a cytotoxic T, and the other becomes a memory T.

_____

18. Fevers work by killing bacteria and weakening viruses.

_____

19. Neutrophils are also called microphages because they're the smaller version of cells that can perform phagocytosis.

_____

20. Lymph nodes have two primary functions. What are they?

    (a) To conserve WBCs and to produce lymphocytes

    (b) To produce bilirubin and to cleanse lymph fluid

    (c) To produce lymphocytes and antibodies and to filter lymph fluid

    (d) To conserve iron and to remove erythrocytes

    (e) To produce erythrocytes and to conserve iron

21. Which method(s) of antibody action directly destroy(s) a pathogen? Choose all that apply.

    (a) Agglutination

    (b) Precipitation

    (c) Complement fixation

    (d) Neutralization

    (e) Phagocytosis

22. The thymus is also considered to be an endocrine gland because it secretes hormones that

    (a) make T lymphocytes immunocompetent.

    (b) stimulate the tonsils to make more T lymphocytes.

    (c) signals the Peyer's patches as they produce B lymphocytes.

    (d) activate the lymph nodes.

    (e) triggers emptying in the lymphatic ducts.

23 This chemical released by WBCs disrupts cell division.

(a) Collectins

(b) Defensins

(c) Interferons

(d) Interleukins

(e) Perforins

24 Why is the spleen's white pulp that color?

(a) It is adipose tissue, storing lipids to provide energy to immune cells.

(b) Blood has drained from the red pulp back into the bloodstream.

(c) It is a concentration of lymph nodes housing memory cells.

(d) Lymphocytes clump there and give it its color.

(e) Reticular tissue in these areas hides the color of the blood.

25 Explain how each division of our adaptive defenses is activated.

_____

_____

_____

_____

Answers to Chapter 12 Quiz

(1) **(b) Lymphocytes.** Your answer hint is in the question: *Lymph* nodules produce *lymph*ocytes.

(2) **(d) It has vessels that dilate to store excess blood volume that can be released when needed.** Enlarged lymph nodes don't travel, so you know that the first answer is wrong. Also, you know the spleen is in the abdomen, not the throat, so toss out the second answer. Did you waver between the third and fourth answers? Just remember all those TV shows in which internal bleeding is chalked up to a damaged spleen, and you've got it!

(3) **antigens**

(4) **cytotoxic T cells**

(5) **antibodies**

(6) **memory cells**

(7) **immunity**

(8) Besides germ patrol, the lymphatic system rounds out our circulation. Blood isn't the only fluid that circulates around our bodies; interstitial fluid flows slowly through the tissues as well. This fluid is continuously refreshed (supplied with new resources) by pushing some plasma out of blood at the capillaries (capillary exchange). Much of the lost fluid is replaced, drawing interstitial fluid back into the vessels as they near the veins, but not all of it. **The excess fluid in the tissue is drained through the lymphatic vessels and put back into blood flow just before it returns to the heart.**

(9) **(a) Peyer's patches.** The ileum is the final segment of the small intestine, which is why appendix can't be the answer.

(10) **(a) Basophils.** These WBCs account for less than 1 percent of all the WBCs floating in the bloodstream.

(11) **(a) Skin, (c) Mucus, and (d) Hair.** Although mucus is a chemical, its mode of action isn't a chemical reaction, such as stomach acid. Its goal is to cause a foreign particle to get stuck, which is what the hair does too.

(12) **(c) Slowly, through the squeezing of surrounding skeletal muscles.** You know that the first answer is wrong because lymph nodes aren't pumps. Because you know that the lymphatic system doesn't have pumps, you know the last answer is wrong, because there isn't anything steady or rhythmic about it. You're left with one of the "slowly" answers. The second answer can't be right, because the system wouldn't have room to take in more lymph if what was already there wasn't moving through some kind of action. Voilà! You have the right answer.

(13) **Monocytes are circulating WBCs that haven't fully differentiated, which means that they don't do anything themselves. When they exit the bloodstream, they divide into macro-phages and dendritic cells, both of which are innate WBCs that perform phagocytosis of germs. But the dendritic cells display the antigens, becoming an APC.**

(14) **False.** Peyer's patches are bundles of lymphatic tissue in the small intestine, far too big to fit into a lymph node.

15. **False.** This statement seems to be logical, but inflammation can bring more fluid through the lymphatic vessels, making them swell. Regardless, the influx of massive numbers of lymphocytes is the biggest contributor to swollen lymph nodes.

16. **True.** IgEs cause allerg*eeees*.

17. **False.** B cells proliferate this way.

18. **False.** Contrary to popular belief, a fever doesn't generate enough heat to kill a germ directly. Weaken, yes; kill, no. The biggest benefit of a fever is that it increases our cellular activities, effectively boosting our immune cells.

19. **True.** Macrophages are often twice as big.

20. **(c) To produce lymphocytes and antibodies and to filter lymph fluid.** Be careful that both parts of the answer are correct.

21. **(c) Complement fixation.** Although phagocytosis destroys pathogens directly, antibodies can't; they're proteins. Agglutination and precipitation make phagocytosis more efficient (a macrophage can consume many germs at the same time), and neutralization keeps a pathogen from causing problems until something destroys it.

22. **(a) make T lymphocytes immunocompetent.** That's just a fancy way of saying that the hormones are *activated*, or able to do their jobs. Also, the *T* of *T cells* comes from the thymus.

23. **(c) Interferons.** *Disrupt* is another way to say *interfere*.

24. **(c) Lymphocytes clump there and give it its color.** They're *white* blood cells, after all.

25. **As a whole, the adaptive process begins when an APC finds a helper T cell that has a receptor that can grab onto the foreign cell's antigens. When this occurs, two signaling chemicals are released. Interleukins trigger T cell proliferation and activate the cytotoxic Ts for their direct attack. Then cytokines stimulate B cells to divide, but only if they are bound to antigens on that same foreign cell.**

5

Supply and Demand

In This Unit . . .

Chapter **13**

Oxygenating the Machine: The Respiratory System

We need lots of things to survive, but the most urgent need from moment to moment is oxygen. Without a continual supply of this vital element, we don't last long. Though we have reserves of the other things we need — carbohydrates, lipids, and proteins — we don't have a storehouse of oxygen. Why not, you ask? Simple. We're walking around in it all the time!

Meet & Greet the Old Windbags

The air we inhale more than 20,000 times each day is made mostly of nitrogen. Lucky for us, though, it also contains the vital, irreplaceable gas that is oxygen (O_2). All the carbohydrates, lipids, and proteins we need to survive would be useless without oxygen; our bodies can't metabolize the energy they need from these substances without a constant stream of oxygen to keep things percolating along. And a key role of the respiratory system is to access that constant stream.

Ironically, we do need nitrogen, particularly to build proteins, but we aren't equipped to pull it out of the air we breathe. That job falls to the digestive system, which I discuss in Chapter 14.

All that metabolizing creates another, equally important need. We must have a means to get rid of our bodies' key gaseous waste: carbon dioxide (CO_2). If it builds up in our systems, we die, so it must be removed from our bodies almost as fast as it's formed. Conveniently, breathing in fulfills our need for oxygen, and breathing out fulfills our need to expel carbon dioxide.

As a system, the respiratory organs carry out the following functions:

>> **Ventilation** — the act of breathing — isn't the same as *respiration,* which refers to the exchange of gases. You don't have to think about performing either task, but *ventilation* is the process of bringing air into and out of the lungs. See "Breathe In the Good, Breathe Out the Bad" later in this chapter for more on that topic.

>> **Respiration** is the process of moving gases into and out of the blood. *External respiration* involves pulling O_2 out of the lungs and into the bloodstream, and vice versa with CO_2. *Internal respiration* happens when O_2 leaves the blood and enters the tissues while CO_2 leaves tissues and enters blood flow. This process is detailed in "All in One Breath" later in this chapter.

>> **Regulating blood pH** within the accepted range requires coordination between the respiratory and urinary systems, with help from the endocrine system and, of course, the cardiovascular system.

>> **Producing speech:** The ability of humans to control breathing consciously permits speech (among other sounds).

The major organs of the respiratory system include the nose, pharynx, larynx, trachea, lungs, and diaphragm. Flip to the color plate titled "Respiratory System," and take a moment to familiarize yourself with the labeled structures. The *respiratory tract* — the pathway air takes into and out of the lungs — is divided into two sections: the *upper respiratory tract* (from the beginning of the airway at the nostrils to the larynx, or throat) and the *lower respiratory tract* (from the larynx to the microscopic air sacs within the lungs).

Also crucial to the respiratory system, but not part of the actual pathway of air, is the *diaphragm.* This large, curved muscle provides the power for breathing. It contracts when we breathe in and relaxes when we breathe out.

Now use that diaphragm to take a deep breath and find out how you did it!

A Breath of Fresh Air

Before entering the trachea, air must pass through the nose, pharynx, and larynx — structures that help prepare and filter the air. Though we are able to breathe through our mouths, that's more of a back-up plan, so here, we focus on the nose.

Nose around

You may care a great deal about how your nose is shaped, but the external structure actually makes little difference to your breathing. The nose is simply the most visible part of your respiratory tract. Now, this is the one time you get to turn your nose up at anatomy. Really. Find a mirror and have a look!

Between those oh-so-familiar nostrils, which are formally called *external nares*, is a band of cartilage called the *septum*. Beyond where you can see, the *vomer* (a skull bone) extends the septum, creating the two *nasal cavities*. Near the entry, you see some hairs (that is, if you haven't buzzed or plucked them off lately). As much as these hairs can annoy us, they do serve a purpose: They trap dirt, dust particles, germs, and any other particulates floating invisibly in the air. Numerous things hitch a ride on the air that we definitely don't want to take into the depths of our lungs; these hairs are our first defense in stopping them.

Assisting in this filtration is the lining of the nasal cavities themselves. This lining, the *nasal mucosa*, is a *ciliated pseudostratified epithelium* (flip back to Chapter 4 if you need a reminder of what that is) that is packed with *goblet cells* and tiny glands, all of which produce mucus. Poor mucus gets a bad rep. We only realize its existence when it's annoying us, usually because we're fighting off germs or have breathed in some particularly dirty air. (Trapping particles tends to make the mucus thicker.) Mucus is in fact quite essential. The cilia on the cells sweeps the mucus up and back to be swallowed (which usually occurs unnoticed) along with anything trapped in it.

As an added bonus, the mucus also serves to moisten the air — an essential step if you want gas exchange to occur deep in the lungs. Curled bones called the *nasal conchae* increase the surface area of the nasal cavities, providing more opportunity for filtration, moistening, and warming of the air as it makes its way to the lungs.

The respiratory region of the nasal cavity performs several important functions:

>> Filtering the air with hairs and mucus.

>> Moistening the air with mucus. As the air flows by, it "pulls" water from the mucus if necessary. See Chapter 2 for a review of how this works.

>> Warming the air. Numerous blood vessels run very close to the epithelium, allowing the heat from the blood to be transferred to the air.

Draining into the nasal cavities are the paranasal sinuses. Ah, the sinuses. They can be such headaches. These cavernous spaces come in four pairs, named for the skull bone they're found in: frontal, maxillary, ethmoidal, and sphenoidal. (See Chapter 6 to review these particular bones.) As hollow, air-filled spaces, the sinuses perform the following functions:

>> **Providing filtration:** The lining of the sinuses is also ciliated pseudostratified; air loops through the sinuses when we breathe. Mucus and the particles trapped within it are swept into the nasal cavity.

>> **Reducing the weight of the skull:** Without these spaces, there would be so much bone that your skull would be far too heavy for you to hold your head up.

>> **Enhancing sound resonance:** The sinuses contribute to the sound qualities of your voice.

>> **Performing drainage:** Ever notice how your nose always runs when you cry? That's because ducts in the corners drain those excess *lacrimal gland* secretions (also known as *tears)* through the sinuses and into the nasal cavity.

Jump down your throat

In laymen's terms, it's the throat. But you know better, right? The *pharynx*, as it's properly called, is a hollow tube running from your nasal cavity down to your voice box. (Yeah, it's not called that either, but we'll get to that shortly.)

As air moves along the pharynx, it passes through three regions:

>> **Nasopharynx:** The nasal cavities merge into the *nasopharynx*. If you press your tongue to the roof of your mouth, you can feel your *hard palate*. This bony plate separates your mouth *(oral cavity)* from your nose *(nasal cavities)*. If you move your tongue backward along the roof of your mouth, you reach a soft spot. This spot is the *soft palate*. Superior to (above) that is the nasopharynx. Across from the soft palate, along the back wall of the pharynx here, are the *adenoids,* a patch of lymphatic tissues that plays an important role in immunity. (For a review of lymphatic tissue, check out Chapter 12). Opening into the nasopharynx on each side are the *eustachian tubes,* which provide drainage to the middle ear.

>> **Oropharynx:** The soft palate narrows into the *uvula,* that odd ball of tissue hanging down in the back of your mouth. Its job is to close off the nasopharynx when you swallow. Posterior to (behind) that is the *oropharynx,* or back of the throat, which extends down to the level of the *hyoid bone.* (I mention the special characteristics of the hyoid bone in Chapter 6.) Contained in the oropharynx, where the oral cavity joins, are your tonsils, one on each side. Like the adenoids, the tonsils are lymphatic tissue.

>> **Laryngopharynx:** The final portion, the *laryngopharynx,* leads to the pharynx (voice box). Here, you find the *epiglottis,* a flap of cartilage that folds over the top of the larynx when you swallow, preventing anything except air from following that path.

TECHNICAL STUFF Normally, the soft palate raises when you swallow, blocking off the nasopharynx and keeping food from going up into your nose. But when you're laughing and eating or drinking at the same time, your soft palate gets confused. When you start to swallow, it starts to move back, but when you laugh suddenly, it thrusts forward, allowing whatever's in your mouth to flow up into your nasal cavities and immediately fly out of your nostrils, to the delight of everyone around you.

Connecting the pharynx to the *trachea* (windpipe) is the *larynx*. A hollow tube as well, it's surrounded by nine pieces of cartilage, the largest and most prominent of which is the thyroid cartilage. Place your fingertips gently along the front of your neck (in a straight line, down the center). Tilt your chin up slightly, and swallow. Feel that? You can feel the *thyroid cartilage* rise and fall with the larynx. Just superior to the thyroid cartilage is the *hyoid bone;* just inferior is the *cricoid cartilage.* Numerous laryngeal muscles attach here and to the cartilages; as just demonstrated, one of the functions is to aid in swallowing.

We call the larynx by its more familiar name, the voice box, because that's the function we generally think of: creating our voice. Within all those cartilages, inside the hollow tube, are two sets of folds. These folds, made of smooth muscle and connective tissue, are relaxed and the larynx is fully open when we breathe. But if there's muscle, muscle contraction can't be too far behind, implying that these folds can close off the tube, which is exactly what happens. Seems counterproductive, right? The first set of folds are the *false vocal cords*, which you can guess by the name aren't the ones that produce sound. These cords fold over when we swallow, collapsing the opening through the larynx, which right here is dubbed the *glottis*. This is our last chance to keep anything besides air from making it into the lungs. Unfortunately, it also blocks the airway, which causes choking. The glottis is the narrowest part of the tube and is surrounded by the *true vocal cords*. When exhaling, we can contract the many laryngeal muscles to control the true vocal cords, creating different pitches and intensities of our voice. (The words themselves come from our lips, cheeks, and tongue.)

Like the nasal cavity, the pharynx and larynx are lined with ciliated cells and produce mucus to continue the filtration of particulates from the air. From here, the air enters the lower respiratory tract — which means that it's a great time to stop and check your understanding.

PRACTICE

For questions 1–5, provide the name of the structure described, followed by the corresponding letter from Figure 13-1. Not all the letters will be used.

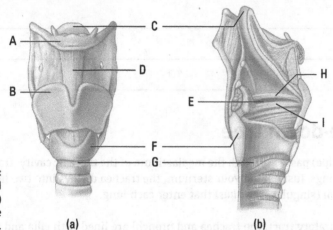

FIGURE 13-1:
Front (a) and
lateral (b)
views of the
larynx.

(a)   **(b)**

Illustration by Imagineering Media Services Inc.

1 Transports air to the lungs: _____ letter: _____

2 Keeps solids from entering the airway: _____ letter: _____

3 Provides a key attachment point for laryngeal muscles: _____
letter: _____

4 Encases the larynx, providing protection around the portion containing the vocal cords:
_____ letter: _____

5 Contracts over the narrowest part of the larynx when you swallow:
_____ letter: _____

6 Found in the back of the nasopharynx, these structures include lymphatic tissue that helps reduce the chance of infection from inhaled air.

(a) Uvula

(b) Tonsils

(c) Conchae

(d) Adenoids

(e) Glottis

7 Which of the following statements about the mucous membranes of the nasal cavity is *not* true?

(a) They contain an abundant blood supply.

(b) They moisten the air that flows over them.

(c) They're made of stratified squamous epithelium.

(d) They're able to trap particulates in the air.

(e) They contain numerous goblet cells and cilia.

8 What is the difference between ventilation and external respiration?

_____

_____

_____

The upside-down tree

Your *trachea* (windpipe) passes through the mediastinum of the thoracic cavity, from the larynx to just above your lungs. Just behind your sternum, the trachea divides into two large branches called *primary bronchi* (singular, *bronchus*) that enter each lung.

Like the upper respiratory tract, the trachea and bronchi are lined with cilia and mucus. Their thick walls contain smooth muscle used to dilate and constrict the tubes to let in more or less air, respectively. Surrounding the trachea and bronchi are rings of cartilage that prevent the airways from collapsing.

The lungs themselves are large, paired organs within your thoracic cavity on either side of your heart. They're spongy and, like the heart, are protected by the rib cage. The left lung is smaller than the right, having two *lobes* (sections) rather than three. This creates a depression between the lungs, called the *cardiac notch*, where the heart comfortably sits.

The lobes are further divided into segments and then into *lobules*, the smallest subdivision visible to the eye. The *bronchial tree*, the branching pathway of air, spreads throughout these divisions. The right or left primary bronchus enters the lung and quickly branches off into the *secondary bronchi* which then split further into the *tertiary bronchi*. If you turn the lungs upside down (the figure, of course), the pattern looks like an actual tree. The bronchi form the lower, thick branches, which then split off in all directions, filling the space of . . . well, the tree. These smaller tubes that branch off the tertiary bronchi are called *bronchioles*, which route the air to the *lobules*, the smallest divisions of the lungs. The bronchioles decrease in diameter as they approach the tiny air sacs (which would be the leaves of the tree). The cells lining these tubes become less ciliated as well, with increasingly less mucus production.

WARNING

Be careful with your spelling here! The smallest tubes are the bronchiOles, with an O. The entire tree is bronchiAl, with an A, the adjective form of bronchi, which are not the same thing. Don't label those tiny branches "the bronchials"; you're likely to get that marked wrong!

REMEMBER

When you look at 2D figures as part of your study ritual, it's a struggle to keep in mind that all these organs are in fact 3D. With the lungs especially, it's easy to fall into the trap of thinking that the air is carried to the bottom of the lungs, thus inflating and deflating them like balloons with each breath. The bronchial tree carries the air throughout both lungs, in all directions. It dead-ends in the air sacs, which makes the lungs feel solid but spongy — not like balloons at all.

The light at the end of the tunnel

Flip to the color plates, and take a moment to look at the "Structure of the Respiratory Membrane" figure, which shows the bronchioles leading to the air sacs via the *alveolar duct*. We've finally reached our destination, where the air pauses momentarily before turning back around to make the trip in reverse.

The *alveolar sacs*, or air sacs, look remarkably like clusters of grapes. The individual "grapes" are *alveoli* (singular *alveolus*). There are no cilia and mucus-making cells here, just the simple squamous walls. You can see the pulmonary capillaries covering the surface of each alveolus creating the surface for the swap of O_2 and CO_2. The areas of the alveoli that interact with the simple squamous walls of the capillaries form the *respiratory membrane*. Because they are covered in capillaries, we get a large surface area for maximum efficiency in exchange of the gases.

I hope you weren't thinking "Finally! Enough of the pathway details already; tell me how it works!" because there's one more thing. Remember the other job of mucus — not the filtration part that I keep bringing up, but the part about moistening the air? As I mention earlier, gases need a moist environment if they want to make travel a tad easier. And because we want them to travel through two layers of cells, we need a thin film on the inner surface of the alveoli, which we have. But here's the problem: That film is water. Water has surface tension. (It's strongly attracted to other molecules of water, creating barriers that are sometimes hard to break through.) To top it all off, alveoli are really tiny — microscopic, if you want to know the truth. All the water molecules want to hold hands (form hydrogen bonds), collapsing the sphere. Collapsed alveoli don't pass gas, which in this case is a bad thing. Luckily for us, cells in the walls of the alveoli secrete a compound called *surfactant* that breaks the surface tension of water — much like a chaperone, keeping the two young-and-in-love water molecules from holding hands but still letting them hang out together.

Now we're ready for gas exchange, but not before some practice questions!

PRACTICE

For the structures given in questions 9–15, write the corresponding letter from Figure 13-2. Not all the letters will be used.

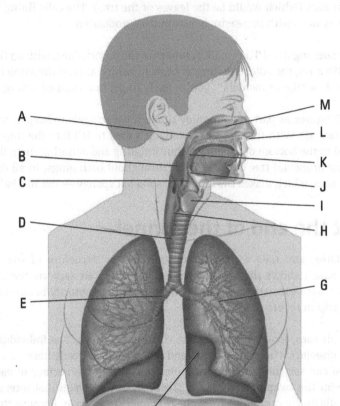

FIGURE 13-2: The respiratory tract.

A

B

C

D

E

M

L

K

J

I

H

G

F

9. Cartilage rings _____

10. Cricoid cartilage _____

11. Laryngopharynx _____

12. Nares _____

13. Nasal conchae _____

14. Tertiary bronchus _____

15. Uvula _____

16. In Figure 13-2, what does the letter F refer to? _____

17 The production of surfactant is one of the last events in fetal development. In fact, the lack of surfactant is what makes a premature birth dangerous. Why?

_____

_____

_____

18 How do the lungs relate to the mediastinum?

_____

_____

Breathe In the Good, Breathe Out the Bad

Breathing is essential to life, and fortunately, your body does it automatically. Instead of having to constantly suck air in and then blow it out again, air enters and leaves the lungs because of changes in pressure. A quick check of *Boyle's Law* tells us that an increase in volume leads to a decrease in pressure. And all molecules (including all those that make up air), flow to where pressure is lower. Your body takes advantage of this to breathe in and breathe out — that is, perform the process of *ventilation*.

Keep it under control

When you're sleeping, sitting still, and doing normal activity, your breathing rate is 12 to 20 breaths per minute (one breath being a complete cycle: inhale and then exhale). The pons and the medulla oblongata (of the brainstem) work together to control this rhythm. They initiate the regular, alternating contraction and relaxation of the diaphragm by signaling the diaphragm via the *phrenic nerves*.

TECHNICAL STUFF

Although your respiration rate is involuntarily controlled by the respiratory center — the pons and medulla oblongata — you can also voluntarily control your breathing. The diaphragm is still triggered to contract, but the order originates from the cortex of the brain instead.

Normal breathing

Inspiration (breathing in) is the result of the diaphragm's contraction. Seems backward, doesn't it? When the diaphragm contracts, it pushes downward into the abdomen, creating more space in the thoracic cavity. The *intercostal muscles* (the muscles between the ribs) may also be stimulated to contract, helping expand the lungs even more by pulling the ribs up and out. Air moves in at the top of the airway (through the nose or the mouth, if necessary) and all the way down to take up the expanded space in the alveoli.

During normal, restful breathing, *expiration* (breathing out) is a passive process. That is, it requires no energy or instruction. The brain simply stops sending impulses to the diaphragm (and any other muscles that are engaged), causing it to relax. As the diaphragm settles back into position, the volume of the thoracic cavity decreases, aided by *elastic recoil.* Much of the tissue in and around the lungs has stretchy elastic fibers. When the diaphragm stops contracting, they lengthen a little extra, or recoil. (Think about stretching and then letting go of a rubber band that flies across the room instead of just falling straight down.) This increases the pressure in the lungs, forcing the air out through the respiratory tract. (See Figure 13-3.) You can control your exhales as well, but this makes it an active process, as it requires the contraction of different muscles (such as the abdominals) to decrease the volume more.

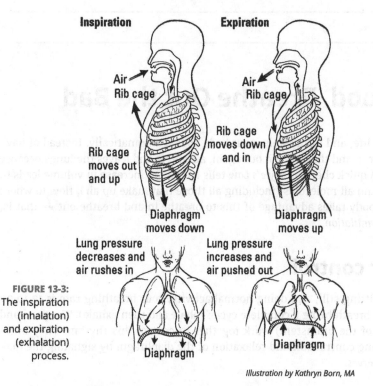

FIGURE 13-3:
The inspiration (inhalation) and expiration (exhalation) process.

Illustration by Kathryn Born, MA

Giving it that little extra

Because we can consciously amplify our ventilation, there are differing, measurable lung volumes, summarized in this list:

>> **Tidal volume (TV):** The amount of air moved in or out during normal, restful ventilation.

>> **Inspiratory reserve (IR):** The additional amount of air (above the tidal volume) that can be inhaled with a forceful inspiration.

>> **Expiratory reserve (ERV):** The additional amount of air that can be exhaled with a forceful expiration after a normal, restful breath (in and out).

>> **Inspiratory capacity (IC):** The maximum amount of air that can be drawn into the lungs during normal ventilation. This equals TV plus the IR.

>> **Vital capacity (VC):** The maximum amount of air that can be moved into (or out of) the lungs. VC differs from IC because it includes the extra exhale first (ERV). You have to get rid of more air to bring in more air.

>> **Residual volume (RV):** The amount of air that must remain in the lungs at all times. The lungs never deflate with exhalation, even if it's forceful.

>> **Total lung capacity (TLC):** The total amount of air the lungs can hold. TLC is equal to VC plus RV.

Mechanics

I point out in Chapter 1 that the lungs reside in their respective pleural cavities. This space is created by two membranes: the *parietal pleura*, which lines the chest wall, and the *visceral pleura*, which sits on the surface of the lungs. Both membranes secrete watery fluids that prevent friction so that the lungs aren't damaged by rubbing against the chest wall when you inhale. Seems like a silly thing to say, right? Watery? This is quite important because of that pesky surface tension. Inside our alveoli, the attraction between water molecules poses a problem. Here, we can't increase our lung volume without it.

When the diaphragm contracts, it's the volume of the thoracic cavity that increases, not the lungs. In fact, the diaphragm is pulled away from the lungs. But the parietal pleura, with its molecules of water, moves with the thoracic wall. The water molecules on the visceral layer really don't want to let go of their parietal partners, so they don't. Since the visceral pleura is the outer layer of the lungs, the volume of the lungs expands as the cavity does. More volume, less pressure and voilà! Air flows in.

There's not a lot of visible change outside the body during restful breathing. Contracting only the diaphragm generates enough of a pressure change for air to flow into the lungs. You don't even notice it. (Except now you do because you're thinking about it!) A small drop in pressure (1–2 mmHg) within the lungs leads to a gentle flow of fresh air. The same holds true for an exhale; the elastic recoil briefly increases the pressure slightly above the air pressure of your environment (the *atmospheric pressure*, which is on average around 760 mmHg). To move more air, we need to generate a bigger pressure difference (the atmosphere compared to inside the lungs). A full breath engages the external intercostal muscles, and a deep breath enlists help from muscles such as the *pectoralis, serrators, sternocleidomastoid,* and *trapezius* to pull the rib cage up and out even more. This process also generates increased pull on the surrounding connective tissues; so when you exhale (stop contractions), the elastic recoil force is even stronger.

To push out extra air — make a forceful expiration, in other words — you can engage muscles such as the *rectus abdominus* and the *external obliques* to push the diaphragm farther up into the thoracic cavity. Doing this, followed by a deep inspiration, allows for a greater exchange of air. That is, more air leaving the respiratory tract, with the excess CO_2 makes more room for fresh air loaded with O_2 to come in.

Now use that inspiratory capacity and see if you can answer some questions.

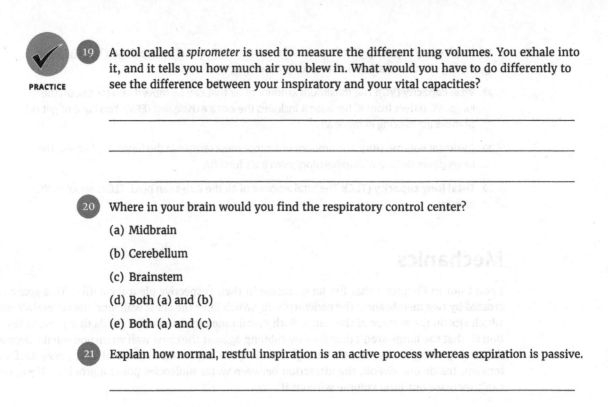

19 A tool called a *spirometer* is used to measure the different lung volumes. You exhale into it, and it tells you how much air you blew in. What would you have to do differently to see the difference between your inspiratory and your vital capacities?

PRACTICE

_____

_____

_____

20 Where in your brain would you find the respiratory control center?

(a) Midbrain

(b) Cerebellum

(c) Brainstem

(d) Both (a) and (b)

(e) Both (a) and (c)

21 Explain how normal, restful inspiration is an active process whereas expiration is passive.

_____

_____

_____

All in One Breath

So we've got the air in the lungs, which is great, but it's not air we want moving into the bloodstream; it's *oxygen*. How do we get the O_2 out of the air into the blood? This process is *gas exchange,* and it occurs only in the alveoli.

Passing gas

Each of the approximately 300 million alveoli is wrapped with *pulmonary capillaries,* whose walls, like those of the alveoli, contain simple squamous epithelium. Because each wall is only one cell layer thick, this tissue is well suited for the exchange of materials. (See the "Structures of the Respiratory Membrane" color plate to refresh your memory of what this crucial area looks like.)

REMEMBER

In Chapter 11, I talk about how blood circulates through two loops: the pulmonary (which goes to the lungs and back) and the systemic (which goes to the body's tissues and back). The *pulmonary circuit* allows us to exchange gases, whereas the job of the *systemic circuit* is tied to oxygen delivery and the pickup of carbon dioxide. Much like how the heart must supply oxygen to itself, so must the lungs. Although you see the alveoli covered in capillaries, these are not the supply route for lung's tissues. A network of systemic vessels spreads throughout the lungs as well. (These vessels are often excluded from figures, as they distract from the unique function of the lungs.)

Until now, we've cared only about air getting into and out of the alveoli. Now we have to focus on external respiration, or gas exchange. As it does during ventilation, movement happens via diffusion: from higher pressure to lower pressure. The basis here is the same but there isn't an air pressure gradient between the alveoli and the blood. There are, however, gradients for O_2 and CO_2.

TIP

We tend to discuss diffusion along a concentration gradient; any molecule that can, will move to where there is less of it. Because of their nature, though, gases don't have concentrations; they have partial pressures. Conceptually, it's the same thing. As *Dalton's Law* states, gases will travel to areas of lower partial pressure (of the same gas). If the term *partial pressure* nudges you out of your comfort zone, mentally change the term to *concentration*. Just realize that information won't be given to you that way, and you should probably change it back when answering questions.

Study Figure 13-4 carefully to gain an understanding of the events that occur during respiratory gas exchange.

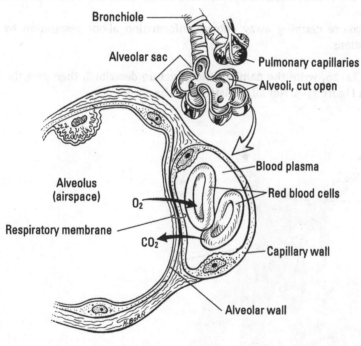

FIGURE 13-4: Respiratory gases are exchanged by diffusion across alveolar and capillary walls.

Illustration by Kathryn Born, MA

REMEMBER

All the tubes of the airway have connective tissue and smooth muscle surrounding them; only the alveolar walls are a thin layer of simple squamous epithelium. This means that gas exchange can occur only at the alveoli. The bronchial tree, though, still has air in it, which gives us our residual volume. All the structures of the upper respiratory tract contain air as well. This volume, along with the residual volume, is known as *dead space* because it doesn't participate in gas exchange.

To get into the blood, oxygen has to diffuse through only two cells. Because the blood is returning after having dropped off oxygen to the tissues, the partial pressure of oxygen (pO_2) will be much lower than it is in the air (which is now inside the alveoli). So out it goes into the plasma. Conversely, the pCO_2 (partial pressure of carbon dioxide) is higher in the blood, as the CO_2 was picked up when the O_2 was dropped off. Fortunately, there's not too much CO_2 in the air, so the pCO_2 inside the alveoli is lower, and the carbon dioxide diffuses out to be exhaled.

Once they're in the plasma, O_2 molecules are picked up by red blood cells (RBCs), because that's their job, after all. RBCs are packed full of the protein *hemoglobin*, so much that they contain little else — not even a nucleus. Each hemoglobin molecule can carry four oxygen molecules. The vast majority of O_2 is transported this way, which is good because unescorted oxygen wreaks havoc, much like unaccompanied middle-schoolers (that is, they tend to damage nearby structures). So once it diffuses into the plasma, the O_2 quickly diffuses into an RBC, where it binds to hemoglobin. When the blood approaches an area with a lower pO_2, like, say, our tissues, the hemoglobin loosens its hold. Some of the O_2 pops off (but not all) and diffuses out of the cell and back into the plasma, just in time to be pushed out during capillary exchange.

RBCs with fully saturated hemoglobin (all of them carrying four oxygens) appear bright red. This is why in figures, arteries are traditionally red; red is the color of oxygenated blood. Veins are usually blue, but not because deoxygenated blood is blue. You're not a lobster. Although it's a much deeper red (which makes a full vein look blue), the blood is most definitely not blue.

See whether you're carrying away enough information about respiration by tackling the following questions.

PRACTICE

For questions 22-26, write the name of the structure described, then give the letter of that structure from Figure 13-5. Not all the letters will be used.

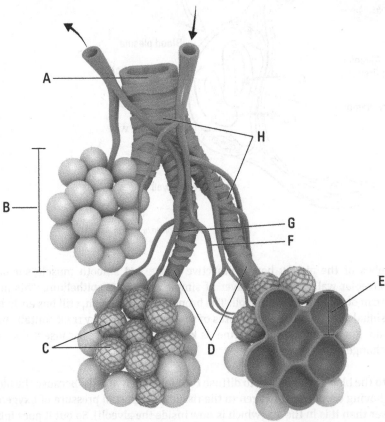

FIGURE 13-5:
The alveoli.

22. Carries deoxygenated blood through the lungs: _____ letter: _____

23. The final destination for inhaled air: _____ letter: _____

24. Controls the diameter of the airway: _____ letter: _____

25. Carries CO_2 to be exchanged: _____ letter: _____

26. Routes air into air sacs: _____ letter: _____

27. What's the difference between residual volume and dead space?

    _____

    _____

    _____

28. What is hemoglobin's primary role?

    (a) Filtration

    (b) Gaseous transport

    (c) Ventilation

    (d) Cellular respiration

    (e) Gas exchange

29. Blood in the pulmonary artery has a pO_2 of 40 mmHg and a pCO_2 of 46 mmHg. The air in the alveoli has a pO_2 of 105 mmHg and a pCO_2 of 40 mmHg. Which statement best describes what will occur once the capillaries are reached?

    (a) Both O_2 and CO_2 will exit the alveoli.

    (b) Both O_2 and CO_2 will enter the alveoli.

    (c) O_2 will enter the alveoli, and CO_2 will enter the blood.

    (d) O_2 will exit the alveoli, and CO_2 will exit the blood.

The carpool line

The oxygenated blood makes its way back to the heart to begin its journey to the rest of the body. With oxygen escorted by the hemoglobin, the blood moves through the ever-branching arteries until it reaches the capillaries. Now we're at the site of *internal respiration*. Unlike in external respiration (gas exchange at the alveoli), diffusion is not the driving force of this process. Here, we're dealing with capillary exchange. (See Chapter 10 if you need a refresher on that topic.) Plasma, which now contains oxygen (among other useful molecules), is forced out between the cells of the capillary wall. Notice my use of the phrase *forced out*? That's clear evidence that diffusion, which is passive, is not driving the exchange of gases. We're talking *filtration*, powered by the force of blood pressure. Now the oxygen is in the interstitial fluid of the tissue, where the cells can actually access it. The cells use diffusion to pull in the oxygen to use for cellular respiration (see Chapter 2) so that they can generate energy, creating carbon dioxide as a byproduct. Then the CO_2 diffuses out of the cell into the interstitial fluid and is pulled into blood flow as it moves through the end of the capillary.

On its way back to the lungs, the CO_2 can be carried by hemoglobin or dissolved in the plasma, but most of it is converted into *bicarbonate molecules*. The CO_2 left floating around in plasma leads to the formation of *carbonic acid* (that's how the CO_2 dissolves). It's interesting to note that the bicarbonate molecules used to transport CO_2 are buffers, which resist the drop in pH caused by the carbonic acid, keeping the blood neutral. When we're placing a higher demand on our cells (such as during exercise), they generate more carbon dioxide, and the buffers can't keep up, so the blood pH drops ever so slightly, signaling the respiratory center to increase respiration. It seems illogical, but it is, in fact, an increase in the amount of CO_2 that leads to faster breathing, not a drop in O_2 levels.

REMEMBER

The process of gas exchange in the lungs — CO_2 leaving the blood and O_2 entering it — happens independently of ventilation. It doesn't matter whether you're inhaling, exhaling, or neither. There's always air in the alveoli, and there's always blood flowing through the pulmonary capillaries. That's why we're able to hold our breath!

Practice Questions Answers and Explanations

(1) **Trachea: G.** D is a muscle called the sternohyoid. (The lines running along its length indicate that it's a muscle.)

(2) **Vestibular fold: H** or false vocal cords

(3) **Hyoid: A.** The thyroid cartilage also serves as an attachment point, but more important, it protects the glottis (E) and its surrounding folds. (The vocal folds are I.)

(4) **Thyroid cartilage: B.** F is the cricoid cartilage.

(5) **Epiglottis: C.** Although H (vestibular fold) would apply here, it's the answer to question 2.

(6) **(d) Adenoids.** The tonsils perform the same function, but they're found in the oropharynx.

(7) **(c) They're made of stratified squamous epithelium.** All the other statements are true, but the mucous membranes, or lining, of the nasal cavity is pseudostratified.

(8) **Ventilation is the process of moving air into and out of the lungs, while external respiration is the exchanging of gases (O_2 and CO_2) between the air found in the lungs and that found in the blood.** It's understandably confusing because external means outside. But that is where the air came from; it's not part of the body. Also recall that internal respiration occurs *inside* the body in both directions (between blood and interstitial fluid).

(9) **D**

(10) **H**

(11) **C**

(12) **L**

(13) **M**

(14) **G**

(15) **K**

(16) **Cardiac notch.** This space is created by the left lung for the heart to sit in (because it has one less lobe than the right one).

(17) **Without surfactant, the alveoli would collapse. Surfactant disrupts the surface tension of water. Without it, the molecules are attracted to one another, and because the alveoli are spherical, they would collapse in all directions.** No space means no air, which means no gas exchange.

(18) **The lungs surround the mediastinum. The mediastinum is the area within the thoracic cavity that isn't encased in the right and left pleural cavities.** The pericardial cavity lies within this space; the trachea, esophagus, and major blood vessels also pass through the mediastinum.

(19) **To find vital capacity, you'd have to forcefully exhale first. VC is the maximum amount of air you can physically bring into your lungs, which means you have to send out as much air as possible first. IC would be how much of a deep breath you can take when breathing**

normally, so that means a normal (not forced) exhale first. Don't let the mechanics of the tool distract you from the answer to this question, because it measures the volume exhaled, and the question is referencing inhaled volumes.

20. **(c) Brainstem.** Were you looking for the pons and medulla oblongata? Both are part of the brainstem.

21. **Normal, restful inspiration requires muscle contraction, which requires energy; thus, it's an active process. A normal, restful expiration relies on elastic recoil. No muscles are contracting; they're just relaxing. This process doesn't require energy, so it's passive.**

22. **Pulmonary arteriole: F**

23. **Alveoli: E**

24. **Smooth muscle: H**

25. **Pulmonary capillary: C**

26. **Alveolar duct: D**

27. **Dead space is all the air in the respiratory tract that isn't involved in gas exchange. The residual volume is what is left inside the lungs at all times.** Think of how residue is a film left behind; in this case, it's air.

28. **(b) Gaseous transport.** It carries oxygen and sometimes CO_2. That's it.

29. **(d) O_2 will exit the alveoli, and CO_2 will exit the blood.** Gases that are able to pass through whatever is in their way (in this case, two layers of flattened cells) will always move to the area where their partial pressure is the lowest. Note that the question gives you information about the blood in the artery; the levels wouldn't have changed as the vessels branched into the capillaries.

If you're ready to test your skills a bit more, take the following chapter quiz, which incorporates all the chapter topics.

Whaddya Know? Chapter 13 Quiz

Think this chapter is now a breeze? Complete each problem to test your knowledge on the various topics covered in this chapter. You can find the solutions and explanations in the following section.

1 What does the mediastinum have to do with the lungs?

(a) It's the bone that protects them from collapsing during blunt trauma.

(b) It's the region between the right and left lungs.

(c) It's the control mechanism that reduces minimal air when the body is stressed.

(d) It's the membrane surrounding each lung.

(e) It's the muscle structure that moves the diaphragm to inflate and deflate them.

For questions 2–6, identify the statement as true or false. When the statement is false, identify the error.

2 The lungs inflate with inspiration and deflate with expiration.

_____

3 Although we need water in our bodies, its surface tension disrupts respiratory function throughout the system and must be overcome.

_____

4 The left lung has two lobes rather than three to allow room for the heart.

_____

5 The hemoglobin found in red blood cells can also carry CO_2.

_____

6 Even though internal respiration keeps happening, we are limited in how long we can hold our breath, because the drop in O_2 levels signals the brain to increase respiration.

_____

For questions 7–12, fill in the blanks to complete the following sentences.

Upon inhalation, molecules of (7) _____ enter the plasma via (8) _____. From there, these molecules enter (9) _____, which contain a pigment called (10) _____. Simultaneously, a (11) _____ molecule, which was formed during (12) _____, moves into the lungs to be expelled during exhalation.

(13) In which part of the respiratory system is cartilage never found?

(a) Bronchioles

(b) Primary bronchi

(c) Secondary bronchi

(d) Tertiary bronchi

(e) Trachea

(14) Which of the following statements best describes the application of Dalton's Law in the respiratory system?

(a) A change in volume directly causes a change in pressure, allowing for the passive exchange of gases into and out of the blood.

(b) The air pressure in the lungs increases with the volume when the diaphragm contracts, allowing the air under atmospheric pressure to flow in.

(c) As a result of the cells generating ATP, the concentration of carbon dioxide in the tissues increases, causing diffusion into the blood.

(d) Because the partial pressure of oxygen in the tissues is lower than in the blood, hemoglobin lets go of the O_2 molecules so they can be pushed out during capillary exchange.

(e) Carbon dioxide has a higher partial pressure in the air due to its size, so when the blood reaches the surface of the alveoli, CO_2 easily diffuses out to be exhaled.

(15) Which of the following respiratory structures does *not* play a role in immunity?

(a) Tonsils

(b) Nasal conchae

(c) Goblet cells

(d) Adenoids

(e) Cilia

16 What is the purpose of mucus in the respiratory tract? Choose all that apply.

(a) To reduce friction of the air passing through

(b) To provide moisture to the incoming air

(c) To coat the alveoli to facilitate gas exchange

(d) To trap gas molecules for transport into the blood

(e) To trap dust and pollutants in the air for removal

For questions 17–22, fill in the blanks to complete the following sentences.

The trachea divides into two 17 _____, which then divide into 18
_____ with a branch going to each lobe of the lung. Upon entering the
lobe, each tube divides into 19 _____, then subdividing into many
smaller tubes called 20 _____. This last set of tubes routes the air to
the smallest subdivisions of the lungs, the 21 _____. Finally, ducts
supply the 22 _____, where gas exchange can occur.

23 If a pin completely pierced the body from the outside, into the thoracic region, which
three structures — in order — would it pass through?

(a) Lung, pleural cavity, and terminal bronchioles

(b) Visceral pleura, lung, and pleural cavity

(c) Parietal pleura, pleural cavity, and visceral pleura

(d) Pleural cavity, trachea, and lungs

(e) Primary bronchi, secondary bronchi, and terminal bronchioles

24 Which muscles would be contracted during a normal, restful inspiration? Choose all
that apply.

(a) Diaphragm

(b) Intercostal muscles

(c) Pectorals

(d) Abdominal muscles

(e) Trapezius

25 Veins appear to be blue through the skin because

(a) the combination of bright red oxygenated blood plus the vein walls creates a blue color.

(b) the fully saturated hemoglobin in the blood makes the walls appear to have a blue hue.

(c) the blood inside is blue because it lacks oxygen, which is visible through the walls of
the veins.

(d) the unsaturated hemoglobin creates dark red blood, which appears to be blue when
viewed through the vein walls.

(e) the vein walls contain a blue pigment; the blood color is actually irrelevant.

26 Which of the following statements best describes the role of surfactant in respiration?

(a) Alveoli remain inflated by disrupting the hydrogen bonding between water molecules.

(b) It provides a layer of moisture for the gases to pass through.

(c) The alveoli are prevented from collapse by enhancing the surface tension of water.

(d) It creates a connection with the thoracic cavity and the surfaces of the lungs so they can expand with inspiration.

(e) It disrupts the surface tension of water between the pleural membranes so the cavity can remain inflated.

27 This structure is responsible for the vital task of keeping solids out of our lungs. Choose all that apply.

(a) Uvula

(b) Epiglottis

(c) Bronchiole

(d) Oropharynx

(e) Glottis

For questions 28–31, fill in the blanks to complete the following sentences.

The maximum amount of air you can bring into your lungs is called 28

_____. This differs from 29 _____, which is the actual

volume of the lungs. The values differ because some air must be left in the lungs at all times.

This is known as the 30 _____. Normally, though, we don't come close to

moving the maximum amount of air in and out of the lungs. During normal, restful breath-

ing, the amount we bring either in or out of the lungs is called the 31 _____.

32 Which of the following processes of breathing is correctly described?

(a) Ventilation: Forcefully sucking air in through the nose and blowing it out through the nose or mouth

(b) External respiration: The exchange of air with the environment; inspiration and expiration

(c) Cellular respiration: Oxygen and carbon dioxide diffusing in and out (respectively) of a cell

(d) Internal respiration: The movement of O_2 and CO_2 out of and into the blood and body tissues

Answers to Chapter 13 Quiz

(1) **(b) It's the region between the right and left lungs.** It encases everything contained in the thoracic cavity except the lungs.

(2) **False.** The lungs never fully deflate. (Normally, that would be a collapsed lung and would require medical intervention.)

(3) **False.** Although this statement is true inside the alveoli, the surface tension of water works to our benefit inside the thoracic cavity.

(4) **True.**

(5) **True.**

(6) **False.** The first half of the statement is true — gas exchange does keep occurring when we hold our breath. But the signal to increase respiration, or the desire to inhale when holding one's breath, stems from the increase in CO_2, which causes a pH drop.

(7) **oxygen (O_2)**

(8) **diffusion**

(9) **red blood cells**

(10) **hemoglobin.** A pigment is just a substance that has a characteristic color, in this case a protein. Don't let that throw you off.

(11) **carbon dioxide (CO_2)**

(12) **cellular respiration.** Don't confuse cellular respiration with internal respiration, which happens via capillary exchange.

(13) **(a) Bronchioles.** While bronchioles do have smooth muscle bands, like the rest of the respiratory tract, they do not have cartilage.

(14) **(d) Because the partial pressure of oxygen in the tissues is lower than in the blood, hemoglobin lets go of the O_2 molecules so they can be pushed out during capillary exchange.** Boyle's Law deals with the relationship between volume and pressure. Dalton's Law says gases will move to where they have a lower *partial pressure*. Remember that gases don't have "concentrations."

(15) **(b) Nasal conchae.** Tonsils and adenoids are areas of lymphatic tissue that house numerous immune cells. Goblet cells produce mucus, which captures particulates in the air, which definitely includes germs, and the cilia sweeps the mucus away. The nasal conchae are the bones within the nasal cavity that increase surface area.

(16) **(b) To provide moisture to the incoming air** and **(e) To trap dust and pollutants in the air for removal**

(17) **primary bronchi**

(18) **secondary bronchi**

(19) **tertiary bronchi**

(20) **bronchioles**

(21) **lobules**

(22) **alveolar sac.** Technically, alveoli wouldn't be a correct answer. An alveolar duct supplies the alveolar sac, which is a bundle of individual alveoli.

(23) **(c) Parietal pleura, pleural cavity, and visceral pleura.** Note that the question asks you to choose among the lists provided, not from the entire structure of the body.

(24) **(a) Diaphragm.** No other answer is correct. The abdominal muscles would be engaged during forceful expiration; the others would work to increase inspiration above the normal, restful breath in.

(25) **(d) the unsaturated hemoglobin creates dark red blood, which appears to be blue when viewed through the vein walls.**

(26) **(a) Alveoli remain inflated by disrupting the hydrogen bonding between water molecules.** Surfactant is found only in the alveoli, and enhancing surface tension would pull the waters closer together, making collapse more likely. And although the surfactant is necessary to have the moisture for gases to pass though, it's the water doing that.

(27) **(b) Epiglottis** and **(e) Glottis.** The epiglottis closes off the larynx when we swallow, and the glottis is our last chance to keep solids out, as it's halfway down the larynx and is the narrowest portion.

(28) **vital capacity**

(29) **total lung capacity**

(30) **residual volume**

(31) **tidal volume**

(32) **(d) Internal respiration: the movement of O_2 and CO_2 out of and into the blood and body tissues.** Although (a) is close, ventilation doesn't have to be forceful. Answer (b) is the definition of ventilation, and (c) is just normal cell events.

Chapter 14

Fueling the Fire: The Digestive System

L iving systems constantly exchange energy. Physiological processes — anabolic, catabolic, and homeostatic — require energy. Ultimately, that energy comes from the light energy that plants use to transform carbon in the atmosphere (as CO_2) into biological matter (as carbohydrates) in the process of photosynthesis. Unfortunately, we're not able to do that. We have to eat. Although now that I think about it, maybe it is a fortunate thing.

It's time to feed your hunger for knowledge about how nutrients make their way into the human body. In this chapter, I help you swallow the basics about getting food into the system and digest the details about how nutrients move into blood flow. You also get plenty of practice following the nutritional trail from first bite to final elimination.

Meet & Greet the Alimentary School

We tend to think of digestion as beginning in the stomach, with its acid. But that's wrong — twice, actually. The process begins with the first bite of food, which doesn't have to be solid to be considered food from a digestive perspective. The food travels through the center of our bodies in a series of tubes known as the *alimentary canal* (digestive tract). These are the familiar organs of the digestive system. Flip to the "Digestive System" color plate to refresh your memory of which one is where.

As the food moves through the alimentary canal, the digestive system carries out its functions with the help of the *accessory organs*, including the liver, gallbladder, and pancreas. We can whittle down the functions of the digestive system into a four-step process:

>> **Ingestion:** *Ingestion* is eating, usually followed by *mastication,* or chewing, which is powered by the *masseter* muscles in your cheeks. The food is mixed with secretions from the salivary glands; mashed up; and prepared for *deglutition,* the fancy term for swallowing. Then food is transported to the stomach via the esophagus.

>> **Digestion:** *Digestion* is the breakdown of food. This needs to happen on two scales, though. You need to break the food into manageable pieces, in a process called *mechanical digestion.* You don't shove an entire slice of cake in your mouth; you take a bite, making the size manageable for your mouth. Also, you don't immediately swallow the bite whole; you chew it to make the size manageable for your esophagus.

Mechanical digestion continues in the stomach but is harder to visualize at this point. I suppose it's like the mashed-up bits of mushy food are then spread out into molecules. Even so, many of the molecules are still too large to move into your actual body (the tissue rather than the space), but there's no more mechanical digestion to be done. This stage is where the enzymes come in, performing *chemical digestion,* breaking bonds to create smaller molecules that can move through your cells.

Though we're quite familiar with the mechanical breakdown done in our mouths, chemical digestion begins in the mouth as well, with the help of saliva.

>> **Absorption:** *Absorption* is the process of getting the molecules — the nutrients, in other words — out of what you consumed and into your blood flow. They pass through the lining of the small intestine and into the interstitial fluid of the connective tissue on the other side. Because this tissue is chock-full of blood vessels, everything you absorbed now has access to blood flow. Rather than distribute it throughout the body, though, you send it all to the liver via a specialized system of veins called the *hepatic portal system* for processing.

>> **Defecation:** *Defecation* is the method you use to get rid of what's left of the food you ate. Note that this process is not the way you get rid of your metabolic wastes — the ones made by your cells. That's the function of the urinary system (which is covered in Chapter 15).

Unlike the respiratory tract, which is a two-way street — oxygen flows in and carbon dioxide flows out, as I describe in Chapter 13 — the digestive tract is designed to have a one-way flow (though we're all familiar with the body's override that can make the food come back up when it deems necessary). Between the organs along the digestive tract, you'll find a *sphincter,* a ring of smooth muscle whose resting state is tightly contracted to prevent backflow. Under normal conditions, food moves through your body in the following order:

Oral cavity ⇨ Pharynx ⇨ Esophagus ⇨ Stomach ⇨ Small intestine ⇨ Large intestine ⇨ Rectum ⇨ Anus

About 8 meters (about 26 feet) long, the alimentary canal is a muscular tube with a four-layered wall. The outermost layer, the *serosa,* is a thin membrane with cells that secrete a clear, watery fluid to reduce friction as the organs rub against one another. The serosa is also referred to as the *visceral peritoneum.* The next layer is the *muscularis,* which contains smooth muscle that contracts to push food through via *peristalsis*. (See "Swallow your pride" later in this chapter for the skinny on peristalsis.) The outer part of this layer consists of *longitudinal muscle* that runs parallel to the length of the tube. Beneath that muscle is a layer of *circular muscle* that wraps around the tube. The stomach has a third layer of *oblique muscle* to aid in its mixing movements. The next layer is the *submucosa*. The submucosa is mostly areolar (loose) connective tissue that serves to provide space for all the nerves, glands, and vessels that reside there. The innermost layer is the *mucosa*, creating the continuous lining of the canal. In the esophagus, this layer is smooth, but in the other organs, it folds on itself to varying degrees. The creation of little pits and valleys increases surface area for interaction with the food occupying the *lumen*, the term for the space within the tube. The epithelial tissue of the mucosa does all the secreting and absorbing of the alimentary canal.

Nothing to Spit At

Mucus and spit get such a bad rep; they're among the many gross things in gross anatomy. Their roles in digestion are so essential, though, that every organ of the alimentary canal produces them. Without them, food would repeatedly get stuck as it was pushed through the canal.

It all begins in the mouth

The mouth's anatomy begins with the lips, which are covered by a thin modified mucous membrane. That membrane is so thin that you can see the blood in the underlying capillaries (the unromantic reason for the lips' natural rosy glow). The inner surface of the lips is covered by a mucous membrane as well. Sickle-shaped pieces of tissue called *labial frenula* attach the lips to the gums. The cheeks are made up of *buccinator muscles* and a *buccal pad* (a subcutaneous layer of fat) and the *masseters*. The buccinator muscles keep the food between the teeth during the act of chewing; the masseters are the main power behind that act.

The roof of the oral cavity is defined by both the *hard palate* (formed from the *maxilla* and *palatine bones)* and the *soft palate* (a movable partition of fibromuscular tissue that prevents food and liquid from getting into the nasal cavity). As the soft palate nears the *nasopharynx*, it curves downward; where the soft palate ends, the oral cavity arches back into the *oropharynx*. Here, you'll find the *uvula* hanging down; its job is to help block off the nasal cavity when you swallow. On each side of the oropharynx are the *palatine tonsils,* a mass of lymphatic tissue that serve an immune function. That is, if a surgeon hasn't removed them because of frequent childhood infections.

We have 32 teeth — 16 on the top row and 16 on the bottom row — that we use to tear and grind food into pieces small enough to swallow. They come in four basic types: incisors for biting, canines for tearing (especially meat), and premolars and molars for grinding.

Teeth rise from openings in the jawbone called *sockets*, or *alveoli*, and are held in place by the *gingiva* (gums).

Regardless of type, each tooth has three primary parts (see Figure 14-1):

>> **Crown:** The part that projects above the gum

>> **Neck:** The region where the gum attaches to the tooth

>> **Root:** The internal structure that firmly fixes the tooth in the alveolus (socket)

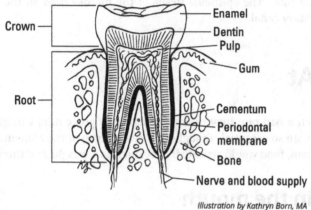

Crown —
Enamel
Dentin
Pulp
Gum
Root —
Cementum
Periodontal membrane
Bone
Nerve and blood supply

FIGURE 14-1:
The composition of a tooth.

Illustration by Kathryn Born, MA

Teeth consist primarily of yellowish *dentin* with a layer of *enamel* over the crown and a layer of *cementum* over the root and neck. The root and neck are connected to the bone by the *periodontal ligament*, or *periodontal membrane*. Cementum and dentin are nearly identical to bone in mineral composition; enamel consists of 94 percent calcium phosphate and calcium carbonate, and is thickest over the chewing surface of the tooth. That's all they have in common, though; teeth are not bones.

In addition, each tooth has a *pulp cavity* in the center that's filled with connective tissue, nerves, and blood vessels that enter the tooth through the *root canal* via an opening at the apex of the root of the tooth called the *apical foramen*. Now you know why it hurts so much when dentists have to drill down and take out that part of an infected tooth!

Swallow your pride

As you chew, three pairs of salivary glands release secretions that combine to create *saliva*, which contains water, electrolytes, mucus, and enzymes. Saliva begins the chemical digestion of the ingested food as well as softens it for ease in swallowing. The pairs of salivary glands include:

>> The *parotid* salivary glands are the largest salivary glands, found in the posterior region of the mandible, just in front of and below each ear.

>> The smallest pair of the trio, the *sublingual* salivary glands, lie near the tongue below the oral cavity's floor.

>> The *submandibular* (or *submaxillary*) salivary glands are at the back of your mouth, under the mandible. They're about the size of walnuts and release fluid onto the floor of the mouth, under the tongue. They produce approximately 70 percent of the saliva that enters the oral cavity.

Those secretions are nothing to spit at. Saliva does the following things:

>> Dissolves and lubricates food to make it easier to swallow

>> Contains *amylase,* an enzyme that breaks down certain carbohydrates and *lipase,* which begins lipid (fat) breakdown

>> Contains *lysozymes* that break down the cell wall of some bacteria, killing them

>> Moistens and lubricates the mouth and lips, keeping them pliable and resilient for speech and chewing

>> Frees the mouth and teeth of food, foreign particles, and loose epithelial cells

The upper surface of the tongue is covered by little bumps called *papillae,* which contain your taste buds. These bundles of *gustatory receptors* allow us to enjoy our food by contributing to the sensation of taste. (See Chapter 8 for more on taste buds.) The papillae also contain *mechanoreceptors* to provide touch information so the tongue can sense where the food is in your mouth. Because it's made of skeletal muscle, the tongue moves the food around in your mouth to assist in chewing. The tongue muscles attach to the oral cavity's floor, anchored by the stringy piece you see when you touch your tongue's tip to the roof of your mouth: the *lingual frenulum.* The tongue's muscles also attach to the *hyoid bone,* which allows for swallowing.

Deglutition, better known as swallowing, occurs in three phases:

1. Food is pushed into the pharynx.

    Following mastication, the tongue rolls the food mass into a ball called the *bolus.* Then it pushes against the hard palate, propelling the food into the oropharynx. This is the only step in deglutition that's voluntary.

2. The swallowing reflex occurs.

    Tensor and *levator* muscles contract, causing the uvula and soft palate to raise, blocking off the *nasopharynx.* The hyoid bone and larynx elevate, forcing the *epiglottis* over the *trachea.* This step prevents food from entering the airway but momentarily stops breathing — hence, you risk choking if you gasp or someone makes you laugh. Next, the tongue pushes on the soft palate, sealing off the oral cavity. Now the only place left for the food to go is into the *esophagus.* The longitudinal muscles of the pharynx contract, pulling the top of the esophagus up to the bolus. (And you thought swallowing pushed the food down!) The *upper esophageal sphincter* relaxes, and the swallowing reflex is complete.

3. The *bolus* heads down the hatch.

The hatch is borrowed nautical slang for the esophagus. The esophagus performs a series of contractions called *peristalsis* to push the food down to the stomach. When the bolus enters the esophagus, it causes the tube to stretch. This stretch triggers the *circular muscles* in the previous ring (the section of esophagus just above the food) to contract and the next ring (the section just below the food) to relax. This *receptive relaxation*, along with the contraction of the *longitudinal muscles* around the food, pushes the food down into the next segment, and the process repeats until the bolus reaches the end of the esophagus. The *cardiac (lower esophageal) sphincter* is triggered to relax, and the bolus enters the stomach.

Chew on these sample questions about the alimentary canal.

PRACTICE

1. What is the difference between chemical and mechanical digestion?

_____

_____

2. Indicate the order of the layered walls of the alimentary canal, from the innermost (1) proceeding outward (6).

_____ Muscularis, longitudinal

_____ Muscularis, circular

_____ Serosa

_____ Mucosa

_____ Lumen

_____ Submucosa

3. Place the following events of the digestive process in order: Absorption, Defecation, Deglutition, Digestion (enzymatic), Ingestion, Mastication

For questions 4–14, use Figure 14-2 to identify the structures of the mouth and throat.

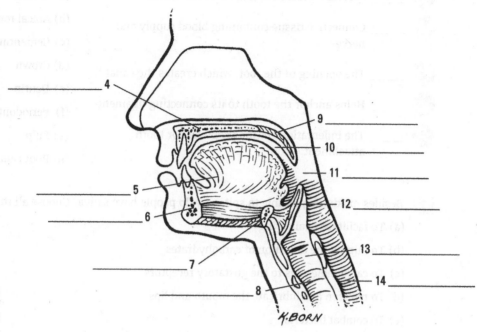

Illustration by Kathryn Born, MA

FIGURE 14-2:
Structures of
the mouth and
throat.

15 What's the function of the soft palate, which vibrates when you snore?

(a) It separates the uvula from the periodontal ligament.

(b) It prevents you from swallowing food before it's thoroughly masticated.

(c) It forces saliva back over the tongue.

(d) It prevents food and liquid from getting into the nasal cavity.

(e) It secretes a watery fluid, making the bolus easy to swallow.

16 What structure anchors the tongue?

(a) The hard palate

(b) The labial frenulum

(c) The lingual frenulum

(d) The gingiva

(e) The mandibular arch

For questions 17–21, match tooth structure to function. You won't use all the choices.

17 _____ Creates a protective layer over the crown

18 _____ Connective tissue containing blood supply and nerves

19 _____ The opening of the root, which creates the canal

20 _____ Helps anchor the tooth to its connecting ligament

21 _____ The indentations of the jawbones for tooth attachment

(a) Alveoli

(b) Apical foramen

(c) Cementum

(d) Crown

(e) Dentin

(f) Periodontal ligament

(g) Pulp

(h) Root canal

22 Besides making it possible to spit, why do people have saliva? Choose all that apply.

(a) To facilitate swallowing

(b) To initiate the digestion of carbohydrates

(c) To carry molecules to the gustatory receptors

(d) To moisten and lubricate the mouth and lips

(e) To combat bacteria

The Stomach: More Than a Vat of Acid

Once the bolus completes its eight-second journey (or one-second journey if it's liquid) through the esophagus, the *cardiac sphincter* relaxes, allowing entry into the stomach. The widest and most flexible part of the alimentary canal, the stomach lies inferior to (beneath) the diaphragm, just left of the body's midline. Food spends anywhere from two to six hours here as it's both mechanically and chemically digested into *chyme*.

When empty, the *mucosa* of the stomach lies in folds called *rugae*, which allow expansion of the stomach when you gorge and shrink when the stomach is empty to decrease the surface area exposed to any lingering acid. Food enters the upper end of the stomach, called the *cardiac region*, through the *cardiac sphincter*, which generally remains closed to prevent gastric acids from moving up into the esophagus. (When gastric acids do move into places they don't belong, the painful sensation is referred to as *heartburn*, which may help you remember the term *cardiac sphincter*.) The dome-shaped area next to the cardiac region is called the *fundus*; it expands superiorly (upward) with really big meals. The middle part, or *body*, of the stomach forms a large curve on the left side called the *greater curvature*. The right, much shorter edge is the *lesser curvature*. The lower part of the stomach, shaped like the letter J, is the *pylorus*. The far end of the stomach remains closed off by the *pyloric sphincter* until its contents have been digested sufficiently to pass into the *duodenum* of the small intestine.

The wave of peristalsis carrying the bolus down continues into the stomach, triggering it to release *gastrin,* the hormone that kick-starts all the stomach's activities. It uses its additional layer of smooth muscle to *churn* the food, tossing it around inside to mix it with the *gastric juice.* This process continues the mechanical breakdown, creating the pastelike chyme that travels through the intestines. The stomach's work begins with the first bite — or sometimes before; we've all felt and heard our stomachs growl in anticipation, as though our smooth muscles are doing their warmup. As we continue to put more food in, the wall expands. Eventually, stretch receptors in the muscularis layer send a message of fullness to the brain.

REMEMBER

Though we associate gastric juice with stomach acid, it's far more than that. Further, we tend to give stomach acid all the credit for the breakdown of the food itself. This isn't accurate, however. Although it does *denature* proteins (causing them to unfold), it's not what breaks apart a protein's components. In fact, it's not doing chemical digestion at all. What it does do is create the environment (a low pH) that our enzymes need so they can perform chemical digestion — with the added bonus of killing most of the germs we inadvertently eat with our food.

The enzymes and other secretions of the stomach are released by several exocrine glands embedded in the stomach's mucosa. Three main cell types contribute to the formation of gastric juice:

>> **Mucous cells:** These cells secrete mucus to protect the lining from the high acidity of the gastric juices. The main component of mucus is the protein *mucin.*

>> **Chief cells:** These cells secrete *pepsinogen,* a precursor to the enzyme pepsin that helps break down proteins into peptides. Chief cells also secrete *lipase* to break fats down into lipids and *chymosin* (also referred to as *rennin*) to break down milk proteins, though production of the latter is often lost by adulthood.

>> **Parietal cells:** These cells secrete hydrochloric acid (HCl), which then combines with pepsinogen to form pepsin and begin protein digestion.

REMEMBER

The main goal of the stomach is mechanical and chemical digestion, not absorption. Alcohol, caffeine, and many other drugs are absorbed here, which is convenient but outside the scope of our normal physiology. Water can move through the lining here, however. If any small molecules happen to be dissolved in it, they might be able to come along for the ride, but that's an exception, not a rule.

Breaking Down Digestive Enzymes

Now that you've broken everything down into small-enough physical pieces, it's time to get things into small-enough molecular pieces. And that means you're ready for the next stop in the alimentary canal: the duodenum of the small intestine.

Entry into the duodenum is controlled by the *pyloric sphincter.* As the chyme enters, the walls of the small intestine stretch, triggering the *enterogastric reflex,* which prevents the small intestine from being overloaded by squeezing the pyloric sphincter shut. Now is when chemical digestion really gets going. The presence of lipids and proteins stimulates cells in the mucosa of the duodenum to release *cholecystokinin* (CCK), the hormone that prompts the release of digestive enzymes into the lumen. The acidity activates other cells to release *secretin*. This hormone triggers the release of a buffer into the lumen, as most enzymes need a more neutral pH to function.

So I've been throwing around this word *enzyme* quite a bit in this chapter. Perhaps it's time to explain what it means. *Enzymes* are proteins that act as catalysts, meaning that they initiate and accelerate chemical reactions without themselves being permanently changed in the reaction. In this case, the reaction is the breakdown of biomolecules. Enzymes are very picky proteins indeed. They're effective only in their own pH range; they catalyze only a single chemical reaction; they act on a specific substance, called a *substrate*; and they function best at 98.6 degrees Fahrenheit, which just happens to be normal body temperature. Three organs contribute enzymes to the small intestine for chemical breakdown: the liver, the pancreas, and the duodenum itself.

The liver delivers

The *liver* is the largest internal organ, weighing in at about 3 to 3.5 pounds (1.4 to 1.6 kilograms). It's located below your diaphragm and takes up most of the upper-right quadrant of the abdomen. (See Figure 14-3.) The liver is soft, pinkish-brown, and triangular, with four lobes of unequal size and shape: the *right lobe, left lobe, quadrate lobe,* and *caudate lobe.* The liver is covered by a connective tissue capsule that branches and extends throughout its insides, providing a scaffolding of support for the blood vessels, lymphatic vessels, and ducts that weave throughout the organ.

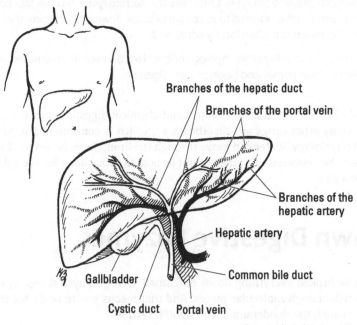

Branches of the hepatic duct

Branches of the portal vein

Branches of the hepatic artery

Hepatic artery

Common bile duct

Gallbladder

Cystic duct   Portal vein

FIGURE 14-3: A closer look at the liver.

Illustration by Kathryn Born, MA

The role of the liver in digestion is the production of *bile*, which aids in both the digestion and absorption of fats and lipids. Bile is a watery fluid made of many molecules, including cholesterol, electrolytes, and a variety of *bile salts*. Bile can be released directly into the duodenum from the liver via the common bile duct, but most often, it branches off through the *cystic duct* to the *gallbladder* to be concentrated and stored. Your gallbladder is a pear-shaped sac tucked into the curve of your liver whose only function is to store bile and deliver it on demand to the small intestine when it releases CCK. Bile both enters and leaves the gallbladder through the *cystic duct*. The cystic duct joins with the common hepatic duct to form the *common bile duct*, which releases bile into the duodenum.

Pancreas party

The *pancreas* sits in the abdominal cavity next to the duodenum and behind the stomach. Although it's a major player in the endocrine system (which I cover in Chapter 9), it also produces *pancreatic juice*, which is full of digestive enzymes. In response to CCK, the acinar cells that create tiny tubes secrete their enzymes into the pancreatic duct; Table 14-1 details which enzymes I'm talking about. Other cells along the pancreas's internal ducts secrete bicarbonate (which is alkaline) in response to secretin's being added to the mix. (The bicarbonate works to neutralize the acidic chyme.) The pancreatic duct extends from the head to the tail, receiving the ducts of various lobules (sections) that make up the gland. It generally joins the *common bile duct*, but some 40 percent of humans have a pancreatic duct and a common bile duct that open separately into the duodenum.

Table 14-1   Pancreatic Enzymes

Enzyme	Targeted Nutrient	Result of Breakdown
Trypsin, chymotrypsin	Proteins	Peptides (chains of amino acids)
Peptidases (various)	Peptides	Individual amino acids
Lipase	Fats	Fatty acids and glycerol
Nuclease	Nucleic acids (DNA, RNA)	Nucleotides
Amylase	Carbohydrates	Glucose and fructose

Down in the duodenum

The role of the duodenum is to complete chemical digestion. Lined with large and numerous *villi* (fingerlike projections), the vast surface area of the duodenum is well suited to its digestive role. Many cells can be packed in a short distance to release enzymes, and the projections jut into the chyme, giving maximum exposure. The main enzymes of the small intestine are

>> **Enterokinase:** It has no enzyme action by itself, but when added to pancreatic juices, it combines with *trypsinogen* to form *trypsin*, which can break down proteins.

>> **Erepsins, or proteolytic enzymes:** These enzymes don't digest proteins directly, but complete the protein digestion started elsewhere. They split polypeptide bonds, thus separating amino acids; one example would be *peptidase*.

>> **Lipases:** These target fats and lipids, splitting them into fatty acids and glycerol.

>> **Inverting enzymes:** They split disaccharides into monosaccharides in line with the following minitable.

Enzyme	Disaccharide	Monosaccharides
Maltase	Maltose	Glucose + Glucose
Lactase	Lactose	Glucose + Galactose
Sucrase	Sucrose	Glucose + Fructose

The duodenum also has *Brunner's glands*, which secrete a clear, alkaline mucus. The glands are most numerous near the stomach and decrease in number toward the opposite, or *jejunum*, end. Nearly all chemical digestion is complete by the time the chyme exits the duodenum.

That was a lot to digest! See if you're full of new knowledge.

PRACTICE   For questions 23–30, use the terms that follow to identify the anatomy of the stomach, as shown in Figure 14-4.

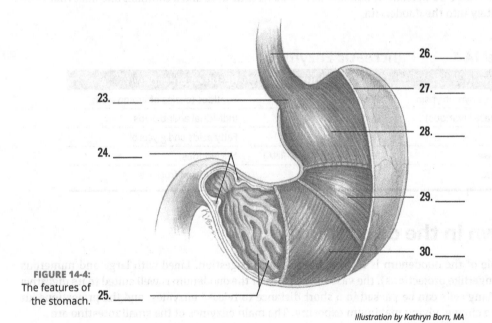

26. _____

27. _____

28. _____

29. _____

30. _____

23. _____

24. _____

FIGURE 14-4:
The features of
the stomach.   25. _____

Illustration by Kathryn Born, MA

(a) Circular muscle layer

(b) Esophagus

(c) Rugae of the mucosa

(d) Cardiac (lower esophageal) sphincter

(e) Serosa

(f) Oblique muscle layer

(g) Pyloric sphincter

(h) Longitudinal muscle layer

31 Explain how peristalsis moves food through the alimentary canal.

_____

_____

_____

_____

32 Which part of the digestive tract isn't doing its job when you have heartburn?

(a) The esophagus

(b) The pyloric sphincter

(c) The peritoneal fold

(d) The cardiac sphincter

(e) The rugae

33 Which of the following terms doesn't belong with the others?

(a) Enterokinase

(b) Maltose

(c) Amylase

(d) Peptidase

(e) Sucrase

34 What is the role of enterokinase in digestion?

(a) It activates a pancreatic enzyme that can break down protein.

(b) It acts as a proteolytic enzyme.

(c) It prompts the release of erepsin.

(d) It triggers enzyme release from the pancreas and liver.

(e) It inverts the functions of other enzymes.

35 Secretin is released by the small intestine to do what? Choose all that apply.

(a) Begin chemical digestion

(b) Trigger the liver to release its chemicals

(c) Stimulate the gallbladder to secrete its fluid

(d) Prompt the pancreas to release its enzymes

(e) Signal the need for buffers

36   When the small intestine receives the acidic chyme, the enterogastric reflex occurs, causing

   (a) the pancreas to release alkaline fluid.

   (b) the pyloric sphincter to close.

   (c) the ileocecal sphincter to open.

   (d) the jejunum to begin peristalsis.

   (e) the pancreas and gallbladder to release enzymes and bile.

For questions 37–42, fill in the blanks.

The hormone 37 _____ is released from the small intestine, specifically the 38 _____, when food has entered. This hormone travels through the bloodstream, targeting two organs. First is the 39 _____, triggering it to release its numerous digestive enzymes. It also targets the 40 _____, prompting it to release its fluid, called 41 _____, which helps break down fats. This organ, however, only stores the fluid; it's actually made by the 42 _____.

Double Duty: The Small Intestine and the Liver

The *intestine* is a long muscular tube (up to about 20 feet, or 6 meters) that extends from the pyloric sphincter to the anal sphincter. Up to this point in the chapter, we've made it only about 1 foot in. How does 20 feet of tubing fit into a relatively small space that's also crowded with other organs? The tube becomes narrow and convoluted. The intestines are classified as small and large based on their width, not their length. The lumen of the small intestine is about 1 inch (2.5 centimeters) in diameter; the large intestine is about 2.5 inches (6.4 centimeters).

The extensive curves of the small intestine are held in place by the *mesentery*, which also routes the blood and lymphatic vessels as well as nerves. Recent research has shown that the mesentery is in fact an organ and may play a role in digestive disorders for which researchers haven't yet pinpointed a cause. The *greater omentum* hangs down from the stomach, covering nearly all the intestines. It is laden with fat and contains numerous macrophages to battle pathogens. The greater omentum has been known to migrate to areas of damage, walling the area off to prevent the spread of infection. Both the mesentery and greater omentum are folded extensions of the visceral peritoneum.

The jejunum and ileum make up the bulk of the small intestine and are the sites of nutrient absorption. Like the duodenum, the mucosa is extensively folded into villi, and the cells have microvilli to further increase the surface area for absorption. The jejunum also has circular folds called *plicae circularis,* which are permanent folds that don't smooth when the intestine is distended. Each villus contains a central lymph vessel, or *lacteal,* which absorbs fatty acids, forming a milk-white substance called *chyle.* The chyle is carried through lymphatic vessels to the subclavian vein. The remaining nutrients are pulled into the capillary network in the submucosa of each villus.

All the capillaries in the submucosa of the jejunum and ileum merge into the veins that run through the mesentery to the *hepatic portal vein*. Everything absorbed from the small intestine and entering the bloodstream is routed to the liver for processing. Oxygenated blood enters the liver area via the *hepatic artery*, which branches off the aorta. Thus, the liver plays an important role in detoxifying our bodies. Following is a brief overview of the liver's numerous functions:

>> **It processes and eliminates toxins.** Toxic byproducts of some drugs, including alcohol, and other substances arrive from the digestive organs through the portal vein.

>> **It processes and helps eliminate metabolic waste.** The liver removes dying red blood cells from the blood and converts the hemoglobin to *bilirubin* and other byproducts. These substances are used to form bile. (The iron is recycled.)

>> **It stores glucose in the form of glycogen and reconverts it when blood glucose levels get low.** This function is mediated by insulin and glucagon. (Turn to Chapter 9 for more on the endocrine system.)

>> **It stores vitamins and minerals.** Vitamins A, B12, D, E, and K are all stored in the liver, as well as minerals such as iron and copper.

>> **It produces many kinds of protein, including protein hormones, the plasma proteins, and the proteins of the clotting cascade (see Chapter 10) and the complement system (see Chapter 12).**

By the time peristalsis pushes the chyme through all three parts of your small intestine, the nutrients that your body needs have been absorbed into the blood and sent to the liver. At the ileum, the indigestible matter passes through the *ileocecal sphincter* and into the large intestine.

One Last Look Before Leaving

Chyme oozes from the small intestine to the large intestine (also called the *colon*), passing out of the ileum through the *ileocecal sphincter* into the *cecum*, the first portion of the large intestine. The material is now called *feces*.

The unappreciated appendix

Just on the other side of the sphincter is an often-overlooked structure. Once considered to be a *vestigial organ* (one that has no known purpose), the *appendix* is found where the small and large intestine meet. It's true that this tiny, wormlike tube has no digestive function. However, it's comprised of lymphatic tissue and is thought to house our (good) bacteria. This way, we can always repopulate the *intestinal flora* after a bout of diarrhea.

The big bowel

The large intestine is about 6 feet (almost 2 meters) long and is positioned anatomically like a frame around the small intestine. Beyond the cecum, the large intestine moves upward as the *ascending colon*, across as the *transverse colon*, and downward as the *descending colon* and finally becomes the *sigmoid colon*.

In the large intestine, water is reabsorbed from the feces by diffusion across the intestinal wall into the capillaries within the submucosa. Because electrolytes are dissolved in water, the large intestine functions to absorb them too (but not nutrients). The removal of water compacts the indigestible material in the colon and, with the addition of mucus, forms the characteristic texture of the feces.

In addition to undigested food, the feces contain the remnants of digestive secretions such as bile. The brown color of feces comes from the combination of greenish-yellow bile pigments, bilirubin (also from bile), and bacteria.

Your intestines are home to unimaginably large numbers of bacteria, including hundreds of species. Though all the functions of our gut flora are an active area of research, we do know that some of the bacteria make vitamins, and if these vitamins are soluble in water, they can easily find their way into our bloodstream. Vitamin K is the most notable example, and it seems increasingly likely that some B vitamins are obtained this way too. The normal resident bacteria also take up space that could otherwise be occupied by pathogenic bacteria that could make us sick.

So what do our bacteria gain by helping us with our digestion? Why, a consistent meal from our undigestables such as cellulose (plant starch), of course! Unfortunately for us, they create gas as a byproduct. The gas, or *flatus*, builds up and is released via *flatulence*, which can lead to some awkward social situations. But that situation is a trade-off. Perhaps when we pass gas, instead of saying "Excuse me," we should say "Thank you!"

Mass movements

Peristalsis moves the chyme through the large intestine slowly, segmenting it for maximum water absorption. Following a meal, a *mass movement* occurs; a strong wave of peristalsis pushes the newly formed feces toward the rectum. Once filled, stretch in the walls of the rectum initiates the *defecation reflex*. This reflex leads to the relaxation of the internal anal sphincter in preparation to dispose of the remnants of previous meals. Fortunately, we can block this reflex voluntarily by contracting the *external anal sphincter*. When we've prepared ourselves accordingly, we stop contracting the sphincter and allow the reflex to continue. This reflex triggers one final peristaltic wave through the descending colon, and feces are forced out of the rectum through the anus.

And that's the end of the line.

  Which of the following is *not* one of the liver's functions?

PRACTICE

(a) Secretion of chemicals that neutralize acids

(b) Production of blood proteins such as clotting factors

(c) Storage of vitamins and minerals

(d) Metabolism of absorbed nutrients

(e) Removal of old blood cells and toxins

44 What is the role of mucus in the large intestine?

_____

_____

 45 This structure covers the organs of the alimentary canal. It's made of adipose tissue as well as lymphatic tissue, and contains white blood cells that help prevent the spread of any infection.

(a) Villi

(b) Appendix

(c) Omentum

(d) Mesentery

(e) Cecum

 46 When the defecation reflex has been initiated, we can pause it voluntarily by controlling which of the following structure(s)? Choose all that apply.

(a) Rectum

(b) Sigmoid sphincter

(c) Internal anal sphincter

(d) Colon

(e) External anal sphincter

 47 Which of the following nutrients are absorbed into lymphatic flow rather than blood flow?

(a) Nucleotides

(b) Simple sugars

(c) Amino acids

(d) Fatty acids

(e) Water

48 How does the gas we release, or flatus, get created?

_____

_____

Practice Questions Answers and Explanations

(1) **Both chemical and mechanical digestion are breakdown processes; the difference is scale. Mechanical digestion is physical; it uses force to break something into smaller pieces. Chewing is the perfect example; your bite of food is broken into bits, but the molecules that make it up remain unchanged. That's where chemical digestion comes in — to break the bonds between atoms.** The biomolecules that make up our food are far too large to be absorbed by the small intestine.

(2) **5, 4, 6, 2, 1, 3** Lumen; Mucosa; Submucosa; Muscularis, circular; Muscularis, longitudinal; Serosa

(3) **Ingestion, Mastication, Deglutition, Digestion (enzymatic), Absorption, Defecation.** Eat, chew, swallow, break down, get nutrients, release waste

(4) **Hard palate**

(5) **Tongue**

(6) **Mandible**

(7) **Hyoid bone**

(8) **Trachea**

(9) **Soft palate**

(10) **Oral cavity**

(11) **Pharynx**

(12) **Epiglottis**

(13) **Larynx**

(14) **Esophagus**

(15) **(d) It prevents food and liquid from getting into the nasal cavity.** So as irritating as snoring may be, you need that soft palate to stay put.

(16) **(c) The lingual frenulum.** The gingiva are the gums, and a labial frenulum anchors the lips. If you're bilingual, you can speak two languages, which definitely require the use of your tongue.

(17) **(e) Dentin.** Enamel does this too, but it isn't a choice.

(18) **(g) Pulp**

(19) **(b) Apical foramen.** The term *foramen* always refers to a hole, usually in a bone.

(20) **(c) Cementum**

(21) **(a) Alveoli.** *Alveolus* is Latin for *small cavity.*

(22) **All of them.** Saliva is multifunctional stuff; it facilitates swallowing, initiates the digestion of certain carbohydrates, and moistens and lubricates the mouth and lips. Molecules that stimulate our taste receptors can be carried deeper into the papillae for a stronger taste sensation. And as a bonus, it contains nifty lysozymes to destroy bacteria (the ones that have cell walls, anyway).

(23) **(d) Cardiac (lower esophageal) sphincter**

(24) **(g) Pyloric sphincter**

(25) **(c) Rugae of the mucosa**

(26) **(b) Esophagus**

(27) **(e) Serosa**

(28) **(h) Longitudinal muscle layer**

(29) **(a) Circular muscle layer**

(30) **(f) Oblique muscle layer**

(31) **As the food is pushed through a tube, the receiving section is stretched as a result, which is called** *receptive relaxation*. **In response to the stretch, the ring of smooth muscle reflexively contracts. But because the previous section just did the same thing, the contraction ends up occurring behind the food rather than around it. The longitudinal muscles also contract to continuously squeeze the food through and keep it from going backward.** The result is the rhythmic wave of contractions that is the definition of peristalsis.

(32) **(d) The cardiac sphincter.** Biggest clue? Cardiac = heart.

(33) **(b) Maltose.** Maltose is a sugar, whereas the other terms are enzymes. (Note the ending –*ase*.)

(34) **(a) It activates a pancreatic enzyme that can break down protein.**

(35) **(e) Signal the need for buffers.** Secretin is released in response to the acidic pH of the chyme entering from the stomach. Answers (b), (c), and (d) are responses to CCK.

(36) **(b) the pyloric sphincter to close.** Although answer (a) seems to be reasonable, the release of alkaline fluid is triggered by the hormone secretin and isn't part of the reflex; the reflex merely controls how much chyme is allowed in at one time.

(37) **CCK**

(38) **duodenum**

(39) **pancreas**

(40) **gallbladder**

(41) **bile**

42. liver

43. **(a) Secretion of chemicals that neutralize acids.** The pancreas does this job.

44. **Mucus lines the large intestine to allow the feces to slide through with ease.** It also helps hold together the remnants of food, which is especially important because we're absorbing water; the longer the feces sit in the large intestine, the more water gets absorbed and the more difficult they become to release.

45. **(c) Omentum**

46. **(e) External anal sphincter.** None of the other structures contains skeletal muscle; as such, they're under autonomic control.

47. **(d) Fatty acids.** These substances are absorbed directly into the lacteal and pass through the lymphatic vessels as chyle.

48. **Gas is created mostly by the bacteria that inhabit our intestines. When they make energy, they produce a gas byproduct, just as we do. This gas builds up and needs to be released, and the only way out is the anus.** Eating too fast can also contribute to flatulence because you swallow more air when you eat quickly, though that air often comes out as eructation *(burps)*.

If you're ready to test your skills a bit more, take the following chapter quiz, which incorporates all the chapter topics.

Whaddya Know? Chapter 14 Quiz

Did you digest and absorb all this information? It's time to find out. Answers and explanations are in the following section.

1 Identify the correct sequence of the movement of food through the body.

(a) Mouth ⇨ Pharynx ⇨ Esophagus ⇨ Stomach ⇨ Small intestine ⇨ Large intestine

(b) Mouth ⇨ Esophagus ⇨ Pharynx ⇨ Stomach ⇨ Small intestine ⇨ Large intestine

(c) Mouth ⇨ Pharynx ⇨ Esophagus ⇨ Stomach ⇨ Large intestine ⇨ Small intestine

(d) Mouth ⇨ Pharynx ⇨ Stomach ⇨ Esophagus ⇨ Small intestine ⇨ Large intestine

(e) Mouth ⇨ Esophagus ⇨ Stomach ⇨ Small intestine ⇨ Pharynx ⇨ Large intestine

2 The hydrochloric acid produced in the stomach

(a) converts gastrin into pepsinogen.

(b) increases release of enzyme from the chief cells.

(c) chemically breaks down food (breaks bonds).

(d) inhibits the production of mucus.

(e) converts pepsinogen into pepsin.

3 Which structure controls the release of chyme into the large intestine?

(a) Cardiac sphincter

(b) Esophageal sphincter

(c) Ileocecal sphincter

(d) Pyloric sphincter

(e) Internal anal sphincter

4 What is the purpose of villi in the alimentary canal, and why does the pattern vary in different organs?

_____

_____

_____

5 Which digestive organ must process all absorbed nutrients before they can be used?

(a) Liver

(b) Pancreas

(c) Large intestine

(d) Small intestine

(e) Kidney

6 What biomolecules do inverting enzymes target for breakdown?

(a) Carbohydrates

(b) Lipids

(c) Proteins

(d) Nucleic acids

7 Which salivary gland produces the bulk of your saliva?

(a) Parotid

(b) Sublingual

(c) Submandibular

(d) Palatine

(e) Buccinator

8 Cholecystokinin is released from where in response to the presence of fats and proteins?

(a) Stomach

(b) Small intestine

(c) Liver

(d) Pancreas

(e) Large intestine

9 Where can chyme first be found along the digestive tract?

(a) Pylorus

(b) Peritoneum

(c) Jejunum

(d) Esophagus

(e) Large intestine

For questions 10–14, use Figure 14-5 to identify the organ that completes the function.

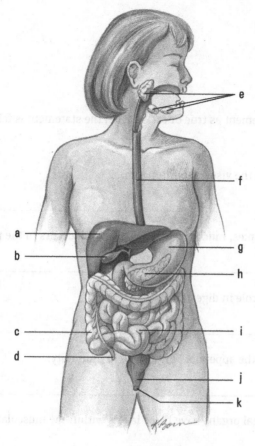

FIGURE 14-5:
The organs and glands of the digestive system.

Illustration by Kathryn Born, MA

10 _____ Secretes enzymes destined for the duodenum

11 _____ Area where most water reabsorption occurs

12 _____ Creates the enzymes that food is first exposed to

13 _____ Concentrates fluid for breakdown of fats

14 _____ Controls defecation reflex

15 What does peristalsis do to food?

(a) These contractions of the small intestine wring nutrients from partially digested food.

(b) This reflexive upward muscular motion mixes food with stomach acids to create vomit.

(c) This grinding action of cartilaginous tissue in the esophagus breaks down food before it enters the stomach.

(d) This sequential contraction of circular muscles helps move food through the esophagus.

(e) These contractions occur in the colon, holding feces there until defecation.

16 The enzyme nuclease is produced in which digestive organ?

(a) Stomach

(b) Esophagus

(c) Small intestine

(d) Pancreas

(e) Salivary glands

For questions 17–22, identify the statement as true or false. When the statement is false, identify the error.

17 The serosa is the same thing as the visceral peritoneum.

_____

18 Because it makes so many enzymes, much of chemical digestion occurs in the pancreas.

_____

19 The large intestine performs a role in digestion.

_____

20 Though we can live without it, the appendix plays a role in immunity.

_____

21 The walls of the alimentary canal organs have three layers within the muscularis: longitudinal, circular, and oblique.

_____

22 The role of the uvula is to close off the nasal cavity when you swallow.

_____

23 The function of the mouth is to

(a) mix solid foods with saliva.

(b) break down milk protein via the enzyme rennin.

(c) masticate or break down food into small particles.

(d) (a) and (c)

(e) (a), (b), and (c)

24 What role does the rectum play in the digestive system?

_____

_____

25 When the small intestine receives the acidic chyme, the enterogastric reflex occurs, causing

(a) the pancreas to release alkaline fluid.

(b) the pyloric sphincter to close.

(c) the ileocecal sphincter to open.

(d) the jejunum to begin peristalsis.

(e) the pancreas and gallbladder to release enzymes and bile.

26 Which of the following sphincters are under voluntary control?

I. External anal sphincter

II. Internal anal sphincter

III. Upper esophageal sphincter

(a) I

(b) III

(c) I and III

(d) II and III

(e) I, II, and III

27 What role do villi play throughout the small intestine?

(a) They neutralize acids.

(b) They propel chyme along the intestinal route.

(c) They increase the surface area available to absorb nutrients.

(d) They perform mixing movements to increase access to nutrients.

(e) They further masticate the chyme.

28 Which of the following statements about teeth is _not_ true?

(a) The permanent teeth in each human jaw are four incisors, two canines, four premolars, and six molars.

(b) Each tooth has a single cuspid anchoring it.

(c) Teeth are covered in both dentin and cementum.

(d) The tooth cavity contains the tooth pulp.

(e) The enamel consists of 94 percent calcium phosphate and calcium carbonate.

29 Where are most nutrients absorbed?

(a) Pylorus

(b) Jejunum

(c) Ileum

(d) Fundus

(e) Duodenum

30 Which structure connects the curves of the small intestine and routes veins from the submucosa to the liver?

(a) Hepatic vein

(b) Portal artery

(c) Greater omentum

(d) Mesentery

(e) Cystic duct

31 What cells in the stomach are responsible for secreting gastric enzymes?

(a) Mucin

(b) Chief

(c) Parietal

(d) Rennin

(e) Bolus

32 What is the role of the masseter?

(a) Deglutition

(b) Churning

(c) Defecation

(d) Mastication

(e) Peristalsis

Answers to Chapter 14 Quiz

1 (a) **Mouth ⇨ Pharynx ⇨ Esophagus ⇨ Stomach ⇨ Small intestine ⇨ Large intestine**

2 (e) **converts pepsinogen into pepsin.**

3 (c) **Ileocecal sphincter.** It's where the ileum, the last part of the small intestine, meets the cecum, the first part of the large intestine.

4 **Villi are tiny folds in the pattern of the mucosa. These fingerlike projections serve to increase the surface area of a tissue. The benefit of increasing surface area is creating more opportunities for absorption and secretion; thus, the organs with these functions have the most prolific villi.** Cells can also show this pattern, called microvilli, to further increase surface area.

5 (a) **Liver.** Think about all the molecules in our food that we'd rather not absorb into our bodies. Yet we can't always prevent this absorption, so it's a good idea to send the blood through our detox center before adding it to the full supply.

6 (a) **Carbohydrates.** When I see the word "invert," I always think "flip," and when considering these choices, all but carbohydrates are complicated molecules — much too difficult to simply flip part of them.

7 (c) **Submandibular**

8 (b) **Small intestine.** It calls for enzymes, and the small intestine (the duodenum in particular) is where the bulk of chemical digestion occurs, so it makes sense that CCK would need to come from there.

9 (a) **Pylorus.** Chyme is the thick, semiliquid mass of food that's ready to leave the stomach. Here's a silly but effective memory tool for this term: When food is ready to leave the stomach, it rings a chime.

10 (h) **(Pancreas).** The gallbladder and liver also secrete fluid into the duodenum, but bile isn't an enzyme. (Technically, it's an emulsifier.)

11 (c) **(Large intestine)**

12 (e) **(Salivary glands).** Don't forget that the first enzyme exposure occurs in the mouth!

13 (b) **(gallbladder).** Because it's storing bile, it makes sense that it would be concentrated in the process.

14 (k) **(Anus).** More specifically, it's the external anal sphincter that we can exercise control of. That sphincter combined with the internal one comprises the anus.

15 (d) **This sequential contraction of circular muscles helps move food through the esophagus.**

16 (d) **Pancreas**

17 **True.** When we're looking at everything in the cavity, we're concerned with the visceral (and parietal) peritoneum, but when we're looking at an organ by itself, it creates the outer layer, which is called the serosa.

18 **False.** Although it does make a large number of enzymes, the food doesn't enter the pancreas, so chemical digestion can't occur there.

19. **False.** The large intestine's job is to deal with what remains of the food we consumed when we're done getting nutrients out of it; the time for mechanical and chemical digestion has long passed. The large intestine absorbs water, and if anything happens to be dissolved in it, that's just an added bonus. I suppose you could argue that the bacteria digest molecules such as cellulose, but that's not part of our chemical digestion; we didn't make the bacteria.

20. **True.**

21. **False.** That's the correct order of the muscularis layers (from the outside in), but only the stomach has the third, oblique layer.

22. **True.**

23. **(d) (a and c).** The mouth does lots of things, including mixing saliva into the food to add the enzyme amylase, but that's not rennin. With answer options like these, it's best to stick to the basics.

24. **The rectum is the portion of the large intestine that connects to the anus. Its job is to hold feces until enough have gathered for defecation.** As it fills, the walls stretch, and the rectum sends messages to the brain, communicating how dire the need for release is.

25. **(b) the pyloric sphincter to close.** This reflex prevents overfilling of the duodenum.

26. **(a) I.** Fortunately, the sphincter that controls the exit of feces (the external anal sphincter) is under our control. The upper esophageal sphincter is made of skeletal muscle, leading you to believe that it could be controlled voluntarily, but its relaxation is part of the swallowing reflex and can't be overridden.

27. **(c). They increase the surface area available to absorb nutrients.** They do other things along the way, but none of those functions is included in the other answers.

28. **(b) Each tooth has a single cuspid anchoring it.** You can rule out this answer as false because a cuspid is a type of tooth, so it makes no sense that each tooth would have another type of tooth anchoring it.

29. **(b) Jejunum.** The duodenum does digestion (both words start with *d*), and the ileum is our last-ditch effort to get nutrients out; thus, the jejunum does the bulk of the absorption. Also, the *ille*um keeps you from getting ill!

30. **(d) Mesentery**

31. **(b) Chief**

32. **(d) Mastication.** That's the proper word for *chewing*, which is what the masseter, or mass *eater*, does.

Chapter **15**

Cleaning Up Your Act: The Urinary System

I f you just happened to have read Chapter 14 (on the digestive system), you may be chewing on the idea that undigested food is the body's primary waste product. But it's not; that title belongs to urine. We make more of it than we do feces — in fact, our bodies are making small amounts of urine all the time — and we release it more often throughout the day. Most important, urine captures all the leftovers from our cells' metabolic activities and jettisons them before they can build up and become toxic. In addition, urine helps maintain the *homeostasis* of body fluids: electrolytes that control the acid–base ratio and the mixture of salt and water that regulates blood pressure.

Meet & Greet the Body's Recycling and Waste Management System

Simply put, the function of the *urinary system* is to filter blood. Unlike with some other organ systems, you can identify the point where *the urinary system* begins and another point, not too far away, where it ends. See the "Urinary System" color plate in the center of the book.

Like the alimentary canal, the urinary system is essentially a system of tubes through which a substance passes, undergoing a series of physiological processes as it does so. The familiar tissue layers — an outer fibrous covering, a muscular layer, and a mucous layer lining the interior surface — appear throughout the urinary system, beginning at the ureter.

Underneath all the organs in the abdominal cavity lies the *abdominal aorta.* Blood is diverted from here by the right and left *renal arteries,* which then pass through the *parietal peritoneum.* (See Chapter 1 if you need a refresher on what that is). Thus, the kidneys are *retroperitoneal.* The average adult kidney is about 4.5 inches (11 cm) long and 2 inches (5 cm) in thickness and width, though the right one is often slightly smaller due to the liver. The outer connective tissue layer of the kidneys, the *renal capsule,* is full of collagen fibers that extend outward to anchor them into the surrounding tissue, including the back of the parietal peritoneum and the muscles of the lower back.

The renal arteries split into four or five branches that enter the kidneys at their *hilum,* the middle of their inner curve. After entering the kidney, these arteries branch extensively, carrying the blood to the outer band of kidney tissue, the *renal cortex.* You can see the internal kidney structures in the color plate titled "Kidney and Nephron." Here, the blood runs through the *nephrons,* the kidney's filters. Then the blood retraces its steps to exit the kidney through the *renal veins,* which empty directly into the *inferior vena cava.*

Beneath the cortex is the *medulla,* a series of fan-shaped structures called *renal pyramids.* The spaces between each pyramid, aptly named *renal columns,* allow blood vessels through to reach the cortex. The pyramids are comprised of microscopic tubes that drip urine into saclike structures. The innermost layer, known as the *renal pelvis,* serves as a container for these urine-collecting sacs and works to channel the urine into the *ureter.* The right and left ureters carry the urine to the *bladder,* where it's stored until there's enough to trigger urination. At that point, muscle contractions force the urine through the *urethra* and out of the body.

Focus on Filtration

Together, the kidneys receive about 20 percent of all the blood pumped by the heart each minute, so they need a lot of tiny filters. At the microscopic level, each kidney contains more than 1 million tiny sets of tubes known as *nephrons* — a fancy word considering their rather unfancy task of filtering. Nephrons are the primary functional units of the urinary system. At one end, each nephron is closed off and folded into a small, double-cupped structure called a *Bowman's capsule* or *glomerular capsule,* where the actual process of filtration occurs. Leading away from the capsule, the nephron forms into the *proximal convoluted tubule* (PCT), which is lined with cuboidal epithelial cells with microvilli that increase the area of absorption. This tube straightens to form a structure called the *descending loop of Henle* and then bends back in a hairpin turn into the *ascending loop of Henle.* After that, the tube becomes convoluted again, forming the *distal convoluted tubule* (DCT), which is made of the same types of cells as the first, or proximal, convoluted tubule but without any microvilli. This tubule connects to a *collecting duct* that it shares with the output ends of many other nephrons. The collecting ducts lie within the pyramids and drain into the renal pelvis. (You can see all these structures in the "Kidney and Nephron" color plate.)

Blood is carried to the cortex through the renal columns in *interlobular arteries,* which branch to route the blood into a nephron by *afferent arterioles.* Inside the glomerular (Bowman's) capsule, the afferent arteriole branches into a bundle of five to eight capillaries called a *glomerulus.* In the glomerulus, the pressure of the blood on the capillaries' walls forces plasma out between the cells. The glomerular capsule catches the fluid, which is carrying metabolic wastes with it; we now refer to the fluid as *filtrate*. Figure 15-1 shows this process, called *glomerular filtration*. The glomerular capillaries come back together to form *efferent arterioles*. These arterioles branch to form the *peritubular capillaries*, which surround the convoluted tubules, and the *vasa recta*, which surround the loop of Henle. The capillaries come together once again to form a *venule* (a small vein) that merges with others into the *interlobular veins* that run between the pyramids. Then these veins merge until blood leaves the kidney via the renal vein.

TIP

Each glomerulus and its surrounding glomerular capsule make up a single renal corpuscle where glomerular filtration takes place. To understand how the renal corpuscles work, think of an espresso machine: Water is forced under pressure through a sieve containing ground coffee beans, and a filtrate called brewed coffee trickles out the other end. Something similar takes place in the renal corpuscles.

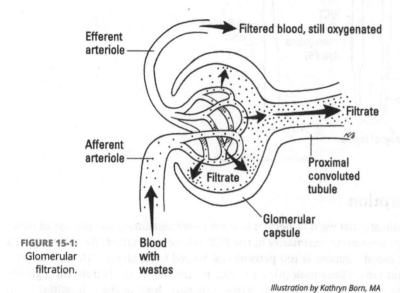

FIGURE 15-1:
Glomerular
filtration.

Illustration by Kathryn Born, MA

WARNING

Like all capillaries, glomeruli have thin, membranous walls, but unlike their capillary cousins elsewhere, these vessels have unusually large pores called *fenestrations* or *fenestrae* (from the Latin word *fenestra*, which means *window*). Plasma is pushed out here, but gas exchange does not take place, which is why the glomerular capillaries merge into an efferent arteriole (as opposed to a vein). Normal capillary exchange takes place in the peritubular capillaries and vasa recta.

Can't we just take out the trash?

The process of *glomerular filtration* creates filtrate to the tune of about 125 mL/min. After filtrate is collected by the glomerular capsule, it's ushered into the PCT. (Refer to Figure 15-1 earlier in this chapter.) So we have a rather effective filter, but we can't afford to lose that much water; we couldn't possibly drink fast enough to replace it. Also, we'd need to urinate constantly. Thus, we have to get most of that water back into the plasma (blood flow), and we do, at 124 mL/minute on average. Furthermore, there are molecules in the filtrate that we'd rather not get rid of such, as ions and glucose. Holding on to those molecules is the job of the nephron tubules, as they work to get back everything in the filtrate that may be useful to us.

Resource recovery

The work of the nephron takes place in three phases, each corresponding to a different section, which you can see in Figure 15-2.

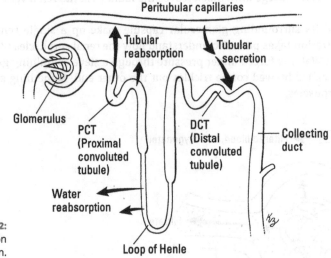

Peritubular capillaries

Tubular reabsorption

Tubular secretion

Glomerulus

PCT (Proximal convoluted tubule)

DCT (Distal convoluted tubule)

Collecting duct

Water reabsorption

Loop of Henle

FIGURE 15-2: Nephron action.

Illustration by Kathryn Born, MA

Tubular reabsorption

The molecules in the filtrate that we'd rather not lose are recovered during the process of *tubular reabsorption*. This process occurs primarily in the PCT, where cells target the lost resources to return them to the blood. Glucose is 100 percent reabsorbed here, along with various proteins, amino acids, and ions. These molecules are transported out of the filtrate through the cells lining the PCT by facilitated diffusion or active transport. Now in the interstitial fluid, they can enter the peritubular capillaries via capillary exchange and thus back into blood flow. Wastes such as *urea* and *uric acid* remain in the filtrate.

Water reabsorption

The function of the loop of Henle is *water reabsorption*. Unfortunately, we don't have a mechanism to transport water directly. But we can pump ions, which can lead to the movement of water by osmosis. Along the ascending loop are sodium pumps that transport Na$^+$ out of the filtrate. Anions such as Cl$^-$ follow the sodium ions out, attracted to the opposing charge. In the interstitial fluid, these ions bond, forming salts. Consequently, the interstitial fluid becomes a *hypertonic solution*. That is, the solute concentration has increased, which means that the water concentration decreased. Because now the concentration of water is lower outside the nephron, it moves out between the cells of the descending loop of Henle. Because so much water was pushed out during glomerular filtration, the concentration of water is even lower inside the capillaries of the vasa recta, so water flows back into the bloodstream. This process, called *countercurrent exchange*, ensures that we're always getting water out of the filtrate by controlling the solute concentration of the interstitial fluid.

WARNING

Though each filtration event corresponds primarily to a specific nephron section, it's important to realize that many of these events — especially water reabsorption — can occur anywhere.

Tubular secretion

After leaving the loop of Henle, the filtrate enters the DCT, where the goal is *tubular secretion*. Some of the metabolic wastes (hydrogen ions, for example) are too numerous to be filtered out at the glomerulus; others, such as proteins like histamine, are too large. These wastes are still in our blood even after glomerular filtration. As a result, pumps along the peritubular capillaries target these wastes, secreting them from blood flow and into the interstitial fluid. Cells along the DCT use active transport to pull the wastes into the filtrate, which will soon be urine. It is through tubular secretion that creatinine (a byproduct of protein breakdown) and ammonia exit the blood for excretion; the same is true for drug compounds that aren't processed by the liver.

Last, the filtrate enters the collecting duct. When we are well hydrated, not a lot happens here. The remaining filtrate joins tiny drops of filtrate from other nephrons that share this collecting duct. Then these drops roll to the bottom of the pyramid, where the filtrate is officially considered to be urine. If we're not well hydrated, however, this process is our last chance to get water out of the filtrate and back into our blood. The amount of water reabsorbed from the filtrate back into the blood at any part of the nephron depends on the hydration situation in the body. We can reabsorb upward of 99 percent of the filtrate we produce each day. But even in cases of extreme dehydration, the kidneys need to produce around 16 ounces (about half a liter) of urine per day just to excrete the toxic waste substances.

REMEMBER

When reading about the actions of the nephron, keep your focus on the big picture. In the end, you don't care about what's in the urine; you care about what's in (or isn't in) the blood. So when I say "reabsorption," I'm talking about reabsorbing *back into* the bloodstream. When I say "secretion," I'm talking about secreting *from the* bloodstream.

PRACTICE

You've absorbed a lot in the preceding paragraphs. See how much of it is getting caught in your filters.

 Create a flow chart to show the pathway of blood into, through, and out of a kidney.

For questions 2–9, use Figure 15-3 to match nephron structures with events.

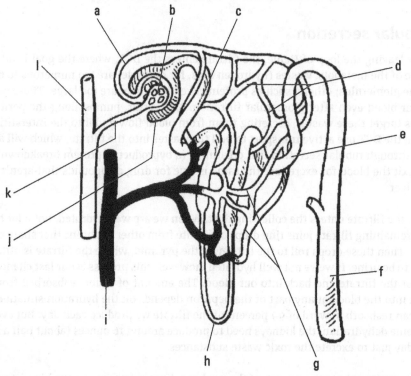

FIGURE 15-3:
Nephron
structures.

Illustration by Kathryn Born, MA

2. _____ Actively transports positive ions out so salts will form in the interstitial fluid

3. _____ Collects the fluid forced out of the fenestrated capillaries

4. _____ Transports filtrate toward the renal pelvis

5. _____ Selectively reabsorbs molecules such as glucose

6. _____ Site of exchange with interstitial fluid

7. _____ Carries filtered, oxygenated blood

8. _____ Collects blood to transport to the kidney's exit

9. _____ Site of tubular secretion

10. Explain how filtrate is created in the renal corpuscle.

_____

_____

_____

(11) Which part of a nephron has microvilli? What is their purpose?

_____

_____

_____

(12) Which of the following is *most likely* to be transported during tubular secretion?

(a) Urea

(b) Water

(c) Creatinine

(d) Glucose

(e) Salt

The Yellow River

We run our blood through the filters, pulling out plasma to create filtrate. Then we target molecules to either pull back into blood flow or pull out of that same blood flow, and put those molecules into the filtrate. What's left when the filtrate reaches the end of the collecting duct is urine. In this section, you follow its journey out of the body.

The collecting ducts are bundled in the renal medulla, creating the *renal pyramids*. The tips of the pyramids, the *renal papillae*, empty their contents into a collecting area called the *minor calyx* — one of several saclike structures referred to as the *minor* and *major calyces*, which form the start of the urinary tract's "plumbing" system and collect urine dripped through the papillae of the pyramids. Although the number varies from person to person, a single minor calyx surrounding the papilla of one pyramid combines into four or five minor calyces, which merge into two or three major calyces. Urine passes through the minor calyx into its major calyx, through the pelvis, and then into the ureter for the trip to the bladder.

Surfing the ureters

The two *ureters* are narrow tubes that transport the urine from each kidney to the bladder. The ureter emerges from the renal pelvis at the kidney's hilum, behind the renal artery and vein. About 10 inches (25 cm) long, each ureter descends from a kidney and joins the bladder on its posterior side. Like the kidneys themselves, the ureters are *behind* the peritoneum and *outside* the abdominal cavity, so the term *retroperitoneal* applies to them too. (They pass through the parietal peritoneum just before reaching the bladder.) The inner wall of the ureter is a simple mucous membrane — one that's necessarily protective, because urine is slightly acidic. The middle layer of the ureter wall is smooth muscle tissue that propels the urine by peristalsis — the same process that moves food through the digestive system. So rather than trickling into the bladder, urine arrives in small spurts as the muscular contractions force it down. The tube is surrounded by an outer layer of fibrous connective tissue that supports it during peristalsis.

Ballooning the bladder

The urinary *bladder* is a large, muscular bag that lies in the pelvis behind the pubis bones, anterior to (in front of) the rectum. In females, the bladder sits between the uterus and the vagina (see the "Female Reproductive System" color plate). The bladder has three openings: two on the back, where the ureters enter, and one on the bottom for the *urethra*, the tube that carries urine outside the body.

Like other organs in the urinary and digestive systems, the bladder is made up of an outer protective membrane, several layers of muscles arranged in opposing directions, and an inner mucosal layer. The mucosa is made up of a special kind of epithelial tissue called *transitional epithelium* in which the cells can change shape from cuboidal to squamous to accommodate larger volumes of urine. Stretch receptors in the muscle layer send impulses to the brain when the bladder is becoming full. A full bladder can hold about 2 cups (475 mL) of urine.

The *urethra* is the tube that carries urine from the bladder to the outside world. The male and female urethras are adapted to interact with the respective reproductive systems and therefore differ in some aspects of their anatomy.

In both males and females, a sphincter at the proximal end of the urethra (between the bladder and the urethra) keeps urine in the bladder. This ring of smooth muscle, called the *internal urethral sphincter*, is under the control of the autonomic nervous system. It opens to release urine into the urethra for urination. Where the urethra passes through the pelvic floor, you find another sphincter made of skeletal muscle, called the *external urethral sphincter*, which is under voluntary control.

Differences in the male and female urethra include:

>> **Female urethra:** In females, the urethra is about 1.5 inches (3.8 cm) long. It runs along the anterior wall of the vagina and opens between the clitoris and the vaginal orifice (opening). The external sphincter is located just inside the exit point.

>> **Male urethra:** In males, the urethra is about 8 inches (20 cm) long from the bladder to its opening at the tip of the penis, called the *urethral meatus*. The male urethra is divided into three named sections, based on anatomical structures:

- The *prostatic urethra* contains the internal sphincter and passes through the prostate. Openings in this region allow sperm and fluid from the prostate gland to enter the urethra during ejaculation.

- The *membranous urethra* contains the external sphincter. It's only about 1 inch (2.5 cm) long, spanning from where the ejaculatory ducts join in to the base of the penis. (These structures can be seen in the "Male Reproductive System" color plate.)

- The *cavernous urethra,* or *spongy urethra,* runs the length of the penis, ending at the urethral meatus.

Sweet relief

Urination, known by the proper term *micturition*, occurs when the bladder is emptied through the urethra. Although urine is created continuously, it's stored in the bladder until the person finds a convenient time to release it.

The first message about the impending need to urinate is sent when the bladder is about half full — around 6 to 8 ounces (175 to 235 mL). At about 10 ounces (300 mL), the *micturition reflex* is initiated, causing the internal sphincter to relax and forcing urine into the urethra. Because the external sphincter is composed of skeletal muscle tissue, no urine is released yet. Additionally, we can contract our *pelvic floor muscles* to continue blocking the reflex.

As the bladder approaches its maximum capacity, the messages become stronger, and it becomes difficult to control the external urethral sphincter. When we've deemed it to be time for urination, we relax our external urethral sphincter, and the micturition reflex restarts. We contract the *detrusor muscles* surrounding the bladder, and urine flows through the urethra and out of the body.

Are you getting the flow of all this? Try answering a couple of questions.

PRACTICE For questions 13–17, match the descriptions with their anatomical terms. Not all choices will be used.

13 _____ Site of glomerular filtration

14 _____ Transports urine from the nephrons

15 _____ Saclike structure for collecting urine in the renal pelvis

16 _____ Performs peristalsis to transport urine

17 _____ Structure that releases urine from the pyramids

(a) Calyx

(b) Collecting duct

(c) Column

(d) Cortex

(e) Pyramid

(f) Renal papillae

(g) Renal pelvis

(h) Ureter

(i) Urethra

18 The separation of the reproductive and urinary systems is complete in

(a) males.

(b) females.

(c) both males and females.

(d) neither males nor females.

19 What is the function of the internal sphincter at the junction of the bladder neck and the urethra?

(a) To stimulate the expulsion of urine from the bladder

(b) To keep foreign substances and infections from entering the bladder

(c) To prevent urine leakage from the bladder

(d) To prevent the return of urine from the urethra to the bladder

(e) To relax when voluntarily triggered to expel urine

The Homeostasis Machines

The kidneys are nonstop filters that sift through 5 cups (1.2 liters) of blood per minute, which makes them well suited to do far more than dispose of our metabolic wastes. In fact, they're quite the little homeostasis machines. Responding to hormones and releasing their own, they play a critical role in the stunningly complex and precise control of our life's chemistry.

In addition to disposing of our body's metabolic wastes, the kidneys

>> Respond to *hypocalcemia* (low blood calcium) by releasing *calcitriol,* which stimulates the digestive tract to absorb more of it

>> Counteract *hypoxemia* (low blood oxygen) by releasing *erythropoietin,* which travels to bone marrow to stimulate the production of *erythrocytes* (red blood cells)

>> Maintain proper *osmotic pressure* (fluid balance) by eliminating excess water

>> Stabilize blood pH and, thus, the pH of other body fluids

>> Regulate blood volume and, thus, blood pressure by initiating the *renin-angiotensin system,* ultimately leading to increased water reabsorption by the nephrons

TECHNICAL STUFF

The adult body contains about 3 gallons (11 liters) of extracellular fluid, constituting about 16 percent of body weight, as well as about 12 cups (almost 3 liters) of plasma, constituting about 4 percent of body weight. Blood plasma and extracellular fluid are very similar chemically, and in conjunction with *intracellular fluid* (fluid inside the cells), they help control the movement of water and electrolytes throughout the body. Some of the important ions found in interstitial fluid (fluid outside the cells but within the tissues) are Na^+, K^+, Cl^-, and Ca^{2+}.

The kidneys can control how much filtrate is created by adjusting the *glomerular filtration rate* (GFR), which is especially useful when the concentration of toxins (metabolic or otherwise) in the blood is particularly high. Because the process uses blood pressure as the driving force behind the filtration, increasing it within the glomerular capillaries causes more filtrate to be pushed out, resulting in a higher filtration rate. The need to process more blood over a shorter period of time, however, doesn't warrant a *systemic* (full-body) increase in blood pressure. In other words, just because we need our kidneys to work more efficiently doesn't mean that we need to increase our blood pressure everywhere in the body.

Because the afferent arteriole branches into the glomerular capillaries, its diameter directly affects the blood pressure and the resulting amount of filtrate. Dilating it (*vasodilation*) would cause more blood to enter the glomerulus; constricting it would do the reverse. Increased volume within the glomerulus causes the blood to push harder on the walls, forcing out more plasma. So with vasodilation of the afferent arteriole, more filtrate is created as the blood moves through, resulting in a higher GFR. Vasoconstriction of the efferent arteriole has the same effect; less blood exits, creating a backup in the capillaries that increases local blood pressure.

Regulating blood pH

The homeostatic range for blood pH is narrow; the optimum value is around 7.4. *Alkalosis* (increase in alkalinity) is life-threatening at 7.8, whereas *acidosis* (increased acidity) is life-threatening at 7.0. To maintain the blood pH within its homeostatic range (7.3 to 7.4), the kidney can produce urine with a pH as low as 4.5 or as high as 8.5.

Although the body has buffers circulating in the blood to resist pH changes, acids and to a lesser extent *alkalis* (bases) are byproducts of metabolic processes. Digesting fats, for example, produces fatty acids; the carbon dioxide produced by cellular respiration can form carbonic acid if it reacts with water; and muscle activity produces lactic acid. Then there are the acids we ingest through food and drink. The kidneys respond to changes in blood pH by excreting acidic (H^+) or basic (OH^-) ions into the urine.

TECHNICAL STUFF

The *pH scale* measures the concentration of hydrogen ions (H^+) in a solution. The scale ranges from 0 (extreme acidity, high concentration of hydrogen ions) to 14 (extreme alkalinity, low concentration of H^+) and consequently, high concentration of hydroxide (OH^-). Solutions with balanced amounts of these two components have a pH in the neutral range. We use our kidneys to maintain ion balances, and these two — H^+ and OH^- — are no exceptions, which accounts for their roles in pH control.

Controlling blood volume

While all this filtering and absorption is going on, the kidneys play a role in maintaining blood pressure by influencing blood volume — specifically, the amount of water in the blood plasma. When blood pressure drops, not as much filtrate is created. Cells in the nephron notice this situation and secrete a hormone called *renin*, which kicks off the *renin-angiotensin system* (RAS) — a cascade of reactions that ultimately leads to an increase in blood pressure. To prevent a local change from initiating a global response, the RAS uses multiple hormones as a means of checks and balances; renin by itself does nothing. Other organs — the liver and the lungs — must "agree" that blood pressure is too low and then initiate the changes that increase it.

The liver, in response to low blood pressure, releases *angiotensinogen*, which reacts with renin to create *angiotensin I*, a substance that by itself also does nothing. The lungs release *angiotensin converting enzyme* (ACE), which coverts angiotensin I into *angiotensin II*. Now we can finally address the problem. Angiotensin II causes *vasoconstriction*, or narrowing of the blood vessels, which serves to increase blood pressure but is only a temporary solution. What we really need is to increase the blood volume, which we accomplish by retaining more water in the kidneys.

Angiotensin II also triggers the posterior pituitary to release *antidiuretic hormone* (ADH) and stimulates the adrenal cortex to release *aldosterone*, both of which head to the kidneys. Aldosterone triggers sodium reabsorption in the DCT; ADH does so in the collecting duct (which is our last chance to get anything out of the filtrate). Water follows salt, and voilà! We have more water in our bloodstream, and the problem of low blood pressure is solved.

REMEMBER

The release of ADH also triggers the sensation of thirst, prompting you to help the cause by drinking more water.

PRACTICE

Did that section quench your thirst for knowledge of the urinary system? Check your understanding before moving on to the chapter quiz.

20 To increase the rate of pressure filtration, the kidney

(a) Constricts the afferent arteriole

(b) Dilates the efferent arteriole

(c) Constricts the efferent arteriole

(d) Dilates the proximal convoluted tubule

(e) Constricts the proximal convoluted tubule

21 Circle all the following hormones that have a direct effect on blood pressure.

Renin   Calcitriol   Angiotensinogen

Angiotensin II   Angiotensin I   Aldosterone

ADH   ACE

22 What is the function of calcitriol and how would it affect bone health?

_____

_____

23 When the blood pH drops below 7.3, what do the kidneys do?

(a) Increase excretion of hydrogen ions

(b) Increase excretion of hydroxide ions

(c) Increase excretion of both hydroxide and hydrogen ions

(d) Increase conservation of both hydroxide and hydrogen ions

24. and 25. The kidneys affect blood oxygen concentration by releasing (24) _____
so that (25)_____.

24 choices:

(a) aldosterone

(b) renin

(c) calcitriol

(d) erythropoietin

(e) angiotensin

25 choices:

(a) an increase in red blood cell production will occur

(b) a decrease in hemoglobin excretion will occur

(c) oxygen levels drop through the formation of water

(d) oxygen is made available by breaking apart water

(e) more oxygen absorption occurs in the lungs

Practice Questions Answers and Explanations

(1) The following figure shows one possible answer.

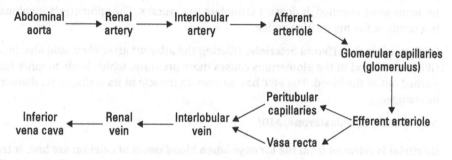

(2) **(h)** (loop of Henle)

(3) **(j)** (glomerular capsule)

(4) **(f)** (collecting duct)

(5) **(c)** (proximal convoluted tubule)

(6) **(g)** (vasa recta) and **(d)** (peritubular capillaries)

(7) **(a)** (efferent arteriole)

(8) **(i)** (interlobular vein)

(9) **(e)** (distal convoluted tubule)

(10) **Blood enters the glomerulus, which are specialized capillaries with holes in them (fenestrations). The walls are only one layer of cells thick, and as the blood enters, it pushes on them. Thus, plasma is forced out between the cells, carrying anything dissolved in it and anything that can fit through the space.** As the plasma is filtered out, it's caught by the glomerular (Bowman's) capsule to be run through the rest of the nephron.

(11) **The proximal convoluted tubule is made of cells with microvilli. Microvilli serve to increase surface area, providing more opportunities for the cells to interact with the filtrate and reabsorbing the useful resources that left with the plasma during glomerular filtration.**

(12) **(c) Creatinine.** The only molecule in the list that wouldn't exit with the plasma during glomerular filtration is creatinine; thus, it must be targeted and pulled into the filtrate from the blood during tubular secretion.

(13) **(d) Cortex.** Most of a nephron's structure lies in the renal cortex, including the glomerulus.

(14) **(b) Collecting duct**

(15) **(a) Calyx.** The renal pelvis is the larger area that contains all the minor and major calyces.

(16) **(h) Ureter**

17 (f) **Renal papillae**

18 (b) **females.** Only the female body keeps these pathways separate. I imagine that otherwise, urinating while pregnant would be a logistical nightmare!

19 (c) **To prevent urine leakage from the bladder.** Although the internal sphincter must relax for urine to be expelled, it doesn't stimulate the process. The sphincter's involuntary control is a result of the micturition reflex.

20 (c) **Constricts the efferent arteriole.** Dilating the afferent arteriole would also increase the GFR; more blood in the glomerulus causes more pressure, which leads to more fluid being pushed out of the blood. The PCT has no smooth muscle in its walls, so its diameter cannot be changed.

21 **Angiotensin II, Aldosterone, ADH**

22 **Calcitriol is released from the kidneys when blood levels of calcium are low. It travels to the lining of the small intestine to trigger the absorption of calcium. An influx of calcium balances the blood calcium but also provides the necessary supplies for osteoblasts (bone-building cells).**

23 (a) **Increase excretion of hydrogen ions.** A low pH indicates acidity, which means that there is an excess of H⁺.

24 (d) **erythropoietin**

25 (a) **an increase in red blood cell production will occur**

If you're ready to test your skills a bit more, take the following chapter quiz, which incorporates all the chapter topics.

Whaddya Know? Chapter 15 Quiz

I hope that you were able to filter through all that information and reabsorb it into your memory, because it's time for the chapter quiz. Answers and explanations are in the following section.

1 Which of the following functions are considered to be functions of the kidneys? Choose all that apply.

(a) Regulation of body fluid concentration

(b) Control of body fluid volume

(c) Removal of waste products from the body

(d) Adjusting the pH of the blood

(e) Balancing electrolyte concentrations

2 What is the primary functional unit of a kidney?

(a) Medulla

(b) Loop of Henle

(c) Renal papillae

(d) Nephron

(e) Renal pelvis

For questions 3–7, match each hormone with its origin. Choices can be used more than once or not at all.

3 _____ Renin

4 _____ Angiotensinogen

5 _____ Aldosterone

6 _____ ADH

7 _____ ACE

(a) Adrenal cortex

(b) Adrenal medulla

(c) Anterior pituitary

(d) Kidney

(e) Liver

(f) Lungs

(g) Posterior pituitary

(h) Thyroid

8 What is the main function of the proximal convoluted tubule?

(a) Peristalsis

(b) Pressure filtration

(c) Tubular reabsorption

(d) Tubular secretion

(e) Micturition

9 The external sphincter of the urinary tract is made up of

   (a) circular skeletal muscle.

   (b) rugae.

   (c) simple mucous membrane.

   (d) the membranous urethra.

   (e) smooth muscle.

10 Although the kidney has many functional parts, where does filtration primarily occur?

   (a) Inside the distal convoluted tubules

   (b) Outside the loop of Henle

   (c) Just beyond the renal cortex

   (d) In conjunction with the renal pelvis

   (e) Within the glomerular capsule

11 Because some of the wastes we need to excrete are large and/or don't dissolve in water, this process gets them out of the bloodstream and into the pathway that creates urine.

   (a) Tubular secretion

   (b) Tubular absorption

   (c) Tubular reabsorption

   (d) Glomerular filtration

12 How can the interior of the bladder expand as much as it does?

   (a) Its ciliated columnar epithelial cells can shift away from each other.

   (b) It's lined with transitional epithelium.

   (c) Its cuboidal epithelium can realign its structure as needed.

   (d) Its white fibrous connective tissue gives way under pressure.

For questions 13–18, identify the statement as true or false. When the statement is false, identify the error.

13 When everything in the kidney is working properly, glucose, oxygen, and water are completely reabsorbed.

_____

14 The kidney lies outside the abdominal cavity, which makes it retroperitoneal.

_____

15 Countercurrent exchange occurs when water is actively transported out of the nephron and into the interstitial fluid, where it can passively enter the peritubular capillaries.

_____

16 Urine production occurs continuously throughout the day.

_____

17 During the sympathetic stress response, the afferent arteriole constricts in order to increase the GFR.

_____

18 In times of extreme dehydration, we can pull water out of the urine being stored in the bladder.

_____

19 Which process does the ureter use to transport urine?

(a) Micturition

(b) Excretion

(c) Urination

(d) Peristalsis

(e) None (it just rolls through)

20 Which nephritic (nephron) structure initially receives the filtrate that is pulled from the blood?

(a) Glomerular capsule

(b) Convoluted tubules

(c) Efferent arteriole

(d) Loop of Henle

(e) Collecting duct

21 Which hormone is responsible for water retention in the kidney from the last place possible before the filtrate becomes urine?

(a) Renin

(b) Angiotensin

(c) ACE

(d) Aldosterone

(e) ADH

22 Compare and contrast aldosterone and ADH.

_____

_____

_____

For questions 23–26, match the event with the location in the nephron where it's most likely to occur.

23 _____ Sodium is actively transported.

24 _____ Toxins are secreted.

25 _____ Plasma is pushed out.

26 _____ Glucose is reabsorbed.

(a) Collecting duct

(b) DCT

(c) Loop of Henle

(d) PCT

(e) Renal corpuscle

Answers to Chapter 15 Quiz

1. **(a), (b), (c), (d),** and **(e).** All these functions are functions of the kidneys.

2. **(d) Nephron.** Each nephron contains a series of the parts needed to do the kidney's filtering job.

3. **(d) Kidney**

4. **(e) Liver**

5. **(a) Adrenal cortex**

6. **(g) Posterior pituitary**

7. **(f) Lungs**

8. **(c) Tubular reabsorption**

9. **(a) circular skeletal muscle.** The external sphincter is under a person's control. The only tissue in this list of choices that features voluntary control is skeletal muscle.

10. **(e) Within the glomerular capsule**

11. **(a) Tubular secretion**

12. **(b) It's lined with transitional epithelium.** That tissue type is able to stretch without the cells separating as a result. This way, the bladder can easily stretch and collapse as needed without damage.

13. **False.** If water were completely reabsorbed, the filtrate would turn solid, and we wouldn't be able to expel it in urine.

14. **True**

15. **False.** We can't control water directly; we can only persuade it to move (so to speak) by actively transporting solutes. (See Chapter 2.)

16. **True**

17. **False.** Although the sympathetic stress response would cause vasoconstriction of the afferent arteriole, this process would lead to less blood entering the glomerulus, thus decreasing the GFR.

18. **False.** The lining of the bladder is not equipped for absorption; after urine drips out of the collecting duct, we can't get anything out or put anything in.

19. **(d) Peristalsis**

20. **(a) Glomerular capsule**

21. **(e) ADH**

22 Both hormones are released in response to low blood pressure and prompt water reabsorption in the kidney. They do so by stimulating the nephron to pump positive ions (such as sodium) out of the filtrate. Aldosterone, which comes from the adrenal cortex, targets the DCT; ADH from the posterior pituitary targets the collecting duct.

23 **(c) Loop of Henle.** Yes, sodium pumps can be activated in other points of the nephron, but the loop of Henle is still the *most likely* location where this event would occur.

24 **(b) DCT**

25 **(e) Renal corpuscle**

26 **(d) PCT**

6

Making a Mini-Me: Behind the Scenes

In This Unit . . .

IN THIS CHAPTER

» Getting to know gamete production

» Making sense of the male reproductive system

» Studying semen and its release

» Figuring out the female reproductive system

» Sorting out the menstrual cycle

Chapter **16**

Why Ask Y? Gamete Production

The reproductive system is different from all the other body systems. The other systems focus entirely on our survival, but the reproductive system risks it all in an effort to contribute genes to future generations. It does nothing to enhance physiological well-being; in fact, it can pose severe threats to the organism's survival.

Here's an overview of what the reproductive system is responsible for:

» **Making gametes:** The *gametes,* also called *sex cells,* are made within the organs of the female and male reproductive systems. There are two kinds of gametes. The *ova* (singular, *ovum)* are the female gametes, and the *sperm* (singular, *sperm)* are the male gametes.

» **Moving gametes into place:** If the reproductive system is to succeed, one ovum and one sperm cell must make their way to the same place at the same time under the right conditions for them to fuse. Many of the reproductive system's tissues and organs chaperone the gametes from the place and time of their production to another place, where they're most likely to encounter their destiny.

» **Gestating and giving birth:** Only the female reproductive system has organs for gestating a fetus and giving birth. (See Chapter 17 for a detailed discussion of pregnancy.)

» **Nurturing the newborn:** The female body has specialized tissues and organs for nourishing a newborn in the first few months of life, until the older baby is capable of digesting other food. (This topic is covered in Chapter 18.)

New life starts with the creation of a single cell: a *zygote*. From this one cell come all the others — a combination of the two parents, carrying two copies of every chromosome, one from each. Thus, all our *somatic* (body) cells are *diploid*. This means, though, that each parent cell must be *haploid*, carrying only one copy of each chromosome. Our usual process of cell division *(mitosis)*, then, isn't going to work. Instead, stem cells in the testes and ovaries called *germ cells* undergo meiosis.

Producing Gametes

The process of meiosis is a sequence of cell-level events that result in the formation of sex cells (gametes) from germ cells. At the cellular level, the processes are essentially identical in female and male bodies. At the tissue, organ, system, and organism levels, the processes are very different. (I discuss the various processes throughout this chapter.)

Meiosis is the only cellular process in the human life cycle that produces *haploid* cells. It is similar in its mechanics to the process of mitosis but has several key distinctions.

TIP

The terminology of meiosis can get disorienting quickly because some general terms also have sex-specific variants. Table 16-1 breaks down these terms for your reference. You might also find it helpful to review my discussion of cell division in Chapter 3 before moving on in this section.

Table 16-1   Reproduction Terms to Know

Terms That Vary by Gender	General Term	Male Term	Female Term
Sex organs	Gonads	Testes	Ovaries
Germ cells	Gametogonia	Spermatogonia	Oogonia
Original cell	Gametocyte	Spermatocyte	Oocyte
Meiosis	Gametogenesis	Spermatogenesis	Oogenesis
Sex cell	Gamete	Sperm or Spermatozoa	Ovum

All somatic cells spend most of their time in *interphase*. Although it's closely associated with both mitosis and meiosis (and often mistakenly listed as the first step), this phase precedes the division process. In interphase, cells go about their daily business and slowly proceed through steps that prepare them to split in two — namely, growing in size and copying all the DNA. When the cell — in this case, a gametocyte — completes interphase and passes all the checkpoints, meiosis begins.

Meiosis

The most obvious difference from mitosis is that meiosis has two parts, called *meiosis I* and *meiosis II*. Each part proceeds in a sequence of events similar to that of mitosis (prophase, metaphase, anaphase, and telophase). In mitosis, the mother cell is diploid, and both daughter cells are also diploid, each having one complete and identical copy of the mother cell's genome. By contrast, meiosis results in four haploid daughter cells. What's more, the four haploid genomes are different.

TECHNICAL STUFF

During the early stages of meiosis, the two copies of a chromosome can exchange genes by swapping pieces (like doing a trade-trade-lemonade, but with your arms). This process, called *crossing-over* or *recombination*, increases variety in the human genome. The result is gametes that are completely unique from the mother cell's chromosomes and from one another. Like a lot of processes in cell biology, the complexity of this process is beyond the scope of this book.

The events of prophase occur to prepare for the nucleus to divide, so in regard to the structures of the cells, the two versions (mitosis and meiosis I) include the same events. With meiosis, however, cells spend far more time in prophase I than in the mitotic counterpart because the two copies of each chromosome must be linked when they're condensing; this process is called *synapsis*. Perhaps you're thinking, "If the whole point is to separate them, why are we linking them?" The answer lies in the events of metaphase I.

During metaphase I, instead of lining up single-file down the middle, the chromosomes align in *homologous pairs*. (You can see the result in Figure 16-1.) When the first division is complete, each resulting cell contains either Mom's copy or Dad's copy for each chromosome. Remember, though, that each butterfly-shaped chromosome is actually two linked *sister chromatids*: two identical copies of a single piece of DNA. (Well, they would be identical if it weren't for crossing-over.) So at this point, we still have double the DNA we need for the gamete, even though the cells are haploid.

Between meiosis I and II, *interkinesis* occurs. The cytoplasm fully separates, creating two haploid daughter cells. Though each cell contains one of every chromosome, the cells aren't genetically identical; they contain different *alleles* (versions of the genetic instructions). During this brief period of rest, the chromosomes don't unwind, and unlike in interphase, the DNA won't be replicated again.

Meiosis II follows the same process as mitosis, which is why meiosis I is sometimes referred to as the *reduction division* and meiosis II as the *mitotic division*. The chromosomes align along the equator and are pulled apart into sister chromatids in anaphase II. When telophase II occurs, the chromatids unwind into chromatin, and the nuclear membrane reforms, creating the pronucleus. When cytokinesis is complete, we have our new, haploid gametes.

TECHNICAL STUFF

Meiosis includes several mechanisms intended to ensure that each gamete has exactly one complete and correct copy of each gene. Any omission, duplication, or error is very likely to be fatal to the gamete (or, later, to the embryo).

Male versus female

The difference between gametogenesis in males and females begins during interkinesis, as you can see in Figure 16-1. In males, the cells are split evenly, and again during cytokinesis following telophase II. The result is four *spermatids* that will mature into sperm cells, which are capable of fertilizing an egg. In *oogenesis*, though, the division during interkinesis is uneven. One of the cells gets all the cytoplasm and organelles; the other cell is merely a packet of chromosomes called a *polar body*. The polar body disintegrates, and the cell, called the *secondary oocyte*, completes meiosis and cytokinesis, again dividing unevenly, creating an ovum (egg cell) and a second polar body that also disintegrates.

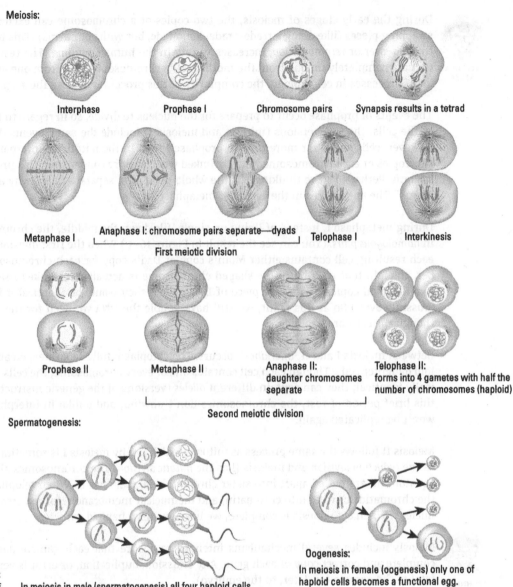

Meiosis:

Interphase

Prophase I

Chromosome pairs

Synapsis results in a tetrad

Metaphase I

Anaphase I: chromosome pairs separate—dyads

Telophase I

Interkinesis

First meiotic division

Prophase II

Metaphase II

Anaphase II:
daughter chromosomes
separate

Telophase II:
forms into 4 gametes with half the
number of chromosomes (haploid)

Second meiotic division

Spermatogenesis:

FIGURE 16-1:
The process of
meiosis.

In meiosis in male (spermatogenesis) all four haploid cells
become functional sperms.

Oogenesis:
In meiosis in female (oogenesis) only one of
haploid cells becomes a functional egg.

Illustration by Imagineering Media Services, Inc.

Female gametes: Ova

A mature ovum (see Figure 16-2) is the only cell in the human body that is visible without magnification. The ovum contains a haploid pronucleus, ample cytoplasm, and all the types of organelles usually found in a somatic cell. The cell membrane is enclosed within a glycoprotein membrane called the *zona pellucida*, which will protect the zygote and pre-embryo until implantation. Around that membrane is the *corona radiata*, a layer of cells from the ovary that came along when it was released to protect and provide nutrients to the egg.

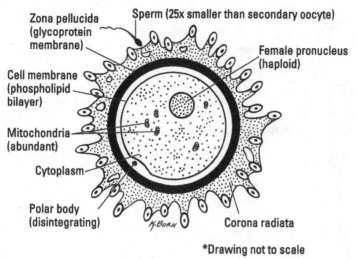

Zona pellucida
(glycoprotein
membrane)

Sperm (25x smaller than secondary oocyte)

Female pronucleus
(haploid)

Cell membrane
(phospholipid
bilayer)

Mitochondria
(abundant)

Cytoplasm

Polar body
(disintegrating)

K.Born

Corona radiata

FIGURE 16-2:
The human
ovum.

*Drawing not to scale

Illustration by Kathryn Born, MA

Oogenesis actually begins way back in embryonic development, when specialized somatic cells called *oogonia* undergo mitosis. Millions of new cells called *primary oocytes* are created. As the fetus develops, the primary oocytes begin meiosis, but they're stopped during prophase I until the female reaches puberty. Many of these cells don't survive. By birth, the female has only about 700,000 primary oocytes left, and half of those will die before puberty.

After the onset of puberty, the primary oocyte resumes meiosis I, producing two cells, called a *secondary oocyte* and the *first polar body*. The secondary oocyte continues with meiosis II but is suspended again, this time during metaphase II. The polar body degenerates.

The cells released from an ovary at ovulation are actually secondary oocytes, not technically ova. If the cell isn't fertilized, it degenerates without completing meiosis II. When (or if) a sperm initiates fertilization, the secondary oocyte immediately resumes meiosis II, producing the ovum (plus a second polar body, which degenerates). Following fertilization, the ovum contains the sperm pronucleus, and after approximately 12 hours, the two haploid nuclei fuse, producing the *zygote*.

Male gametes: Spermatozoa

The cells produced in *spermatogenesis* are spermatids, not yet mature sperm cells. Mature *spermatozoa* have three parts: a head, containing a haploid pronucleus; a short middle section; and a long flagellum. The sperm is adapted to travel light; it has very little cytoplasm. (See Figure 16-3.) The head is covered by a structure called the *acrosome*, containing enzymes that break down the ovum's membranes to allow entry. The middle section contains mitochondria and little else. Mitochondria produce the energy that fuels the sperm's highly active flagellum, which propels the sperm through the female reproductive tract.

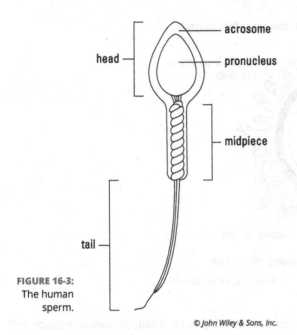

- acrosome
- head
- pronucleus
- midpiece
- tail

FIGURE 16-3:
The human sperm.

© John Wiley & Sons, Inc.

Males are born with *spermatogonia*, which remain dormant until puberty. During puberty, hormonal mechanisms activate them. Throughout the male's life, the population of spermatogonia is maintained by mitosis; one cell remains, and the other becomes the *primary spermatocyte*, which will undergo meiosis.

Whereas oogenesis is cyclic, spermatogenesis is continuous, beginning at puberty and occurring lifelong in most men. By contrast with the one-per-month gametogenesis in females, males produce astronomical numbers of sperm. Each ejaculation produces about 1 teaspoon of semen, which contains about 400 million sperm in a matrix of seminal fluid. Mature sperm can live in the epididymis and vas deferens for up to six weeks. An unfertilized egg lives only 24 hours.

Is all that information about meiosis swimming around in your head? It's time for some review questions:

PRACTICE

1. A man has 46 chromosomes in a spermatocyte. How many chromosomes are in each sperm?

   (a) 23 pairs

   (b) 23

   (c) 184

   (d) 46

   (e) 46 pairs

2. Synapsis of homologous chromosomes

   (a) occurs in mitosis.

   (b) completes fertilization.

   (c) occurs in meiosis.

   (d) signifies the end of meiosis I.

   (e) signifies the end of prophase of the second meiotic division.

3. During oogenesis, the nonfunctional cells produced are called

   (a) cross-over gametes.

   (b) spermatozoa.

   (c) polar bodies.

   (d) oogonia.

   (e) somatic cells.

4. Crossing over is a key process in increasing our genetic variability. It occurs when two chromatids (of the same chromosome but with different alleles) basically swap ends, mixing up the varieties of the genes they're carrying. During which phase is this process most likely to occur?

   (a) Prophase II

   (b) Metaphase II

   (c) Anaphase I

   (d) Anaphase II

   (e) Prophase I

5 Complete the following worksheet on the stages of meiosis. Draw the stages of meiosis and describe the changes in each stage.

1.

Late Prophase I

2.

Metaphase I

3.

Anaphase I

4.

Telophase I

5.

Interkinesis

6.

Prophase II

7.

Metaphase II

8.

Anaphase II

9.

Telophase II

10.

Sperm

11.

Ovum and Polar Bodies

© John Wiley & Sons, Inc.

Meet & Greet the Male Reproductive System

The function of the male reproductive system is to make sperm and deliver them into the female body. The anatomical structures are built to do just that, which is why they're located on the periphery of the body rather than protected inside. These structures are illustrated in the "Male Reproductive Organs" color plate, which you'll want to refer to as you make your way through the production and pathway of sperm.

Sperm development

The reproductive process in males begins with meiosis of cells located in the *testes* (singular, *testis*). These paired organs are contained in a pouch of skin called the *scrotum*, which serves to hold the testes away from the body. Beneath the skin is an inner smooth muscle layer that contracts when the area is cold and elongates when it's warm. Why? As it turns out, the production of sperm is most effective at a temperature that's a bit cooler than the body. If the testes become too cold, the scrotum contracts and draws the testes toward the body for warmth. When the testes are overly warm, the scrotum elongates to allow the testes to hang farther from the heat of the body.

A fibrous capsule called the *tunica albuginea* encases each testis and extends into the gland, forming incomplete *septa* (partitions), which divide the testis into about 200 *lobules* (shown in Figure 16-4). These compartments contain small, coiled *seminiferous tubules*, which are the site of spermatogenesis. The walls of the seminiferous tubules contain numerous cell types:

>> **Spermatogonia:** Stem cells that undergo mitosis, creating a *type A* cell that will remain a stem cell as well as a *type B* cell that will differentiate

>> **Primary spermatocytes:** The differentiated type B spermatogonia that will undergo meiosis

>> **Spermatids:** The immature gametes

>> **Sertoli cells:** Cells that nourish maturing spermatids and regulate spermatogonia development; stimulated by follicle-stimulating hormone (FSH)

>> **Leydig cells:** Cells that produce testosterone, which is necessary for meiosis and maturation of the sperm; stimulated by luteinizing hormone (LH)

After about a month, the gametes are released into the lumen (space) of the seminiferous tubules. Now called spermatozoa, they are not yet fully developed, still having a rounded head and a short tail.

The seminiferous tubules come together in a network called the *rete testis*. Ciliated cells in these tubules shuttle the spermatozoa into the *epididymis*. This extremely long (about 20 feet or 6 meters), tightly coiled tube is the site of sperm maturation. In the two weeks that spermatozoa spend here, they develop their flagella and an *acrosome*, the outer casing around the head that contains enzymes to aid in fertilization. (Refer to Figure 16-3 for an image of a mature sperm cell.)

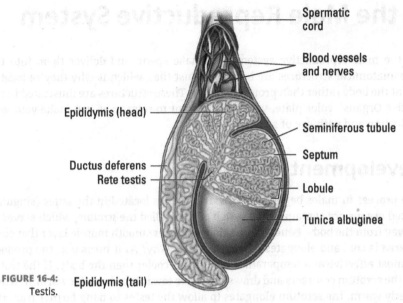

Spermatic cord

Blood vessels and nerves

Epididymis (head)

Seminiferous tubule

Septum

Ductus deferens

Rete testis

Lobule

Tunica albuginea

FIGURE 16-4: Epididymis (tail)
Testis.

Illustration by Imagineering Media Services Inc.

The epididymis leads into the *ductus deferens,* or *vas deferens,* which carries sperm up into the abdominal cavity through the *spermatic cord,* which also encases the testicular artery and vein, lymphatic vessels, and nerves. Sperm are stored in the ductus deferens until they're released from the body or they die (and their parts are recycled).

Semen production

The ductus deferens carries sperm up and around the bladder to the *ejaculatory duct,* which lies within the *prostate gland.* Convoluted pouches called *seminal vesicles* lie behind the base of the bladder on either side of the ductus deferens. These pouches produce a thick fluid to nourish the sperm. The prostate gland adds its own fluid to the ejaculatory duct as the sperm enter. The prostate's secretions contain mainly citric acid and a variety of enzymes that thin the fluid, aiding in sperm *motility* (ability to move).

The components of the *semen* (the fluid that's now in the ejaculatory duct) have several characteristics, including the following:

>> They're slightly basic, with a pH of 7.5, just the way sperm like their environment to be.

>> They nourish the sperm by providing the sugar fructose so that a sperm's mitochondria can make enough energy to move its tail and travel all the way to the egg.

>> They contain *prostaglandins,* which are chemicals that cause smooth-muscle contractions. The lining of the uterus contracts on exposure, and the sperm are pulled upward into the female's reproductive tract.

It's within the prostate gland that the semen enters the urethra. The ejaculatory duct enters into the *prostatic urethra,* which is the first portion as it exits the bladder. (See Chapter 15 for more about the male urethra.) Then the semen, carrying millions of sperm, is forced through the *cavernous urethra* of the penis during *ejaculation.*

Arousal and ejaculation

The function of the penis is to deliver the sperm into the female reproductive tract. It consists of a tube-shaped shaft that encases the urethra and the *glans penis* (the head or tip), which contains the *urethral orifice* (opening of the urethra). The *prepuce,* or foreskin, forms a protective covering around the glans penis, leaving the urethral orifice exposed so that sperm may exit easily.

REMEMBER

The prepuce is commonly removed in certain parts of the world during the circumcision procedure performed on newborns.

Because the urethra is also the exit tube for urine, though, the tissue of the penis needs to remain flexible, or *flaccid,* until it's time to fulfill its reproductive role. The internal structures of the penis allow for the transition of the penis to and from its flaccid and rigid forms.

TECHNICAL STUFF

Though urine and semen exit through the same tube, semen doesn't contain urine. At the point of ejaculation, a sphincter closes off the bladder to keep urine, which is acidic, from mixing with the sperm, which prefer a more basic environment.

Since the female reproductive anatomy is internal, a male must undergo *arousal* for sperm delivery to be possible. This process begins when physical and/or mental stimulation causes an autonomic reflex that triggers *erection.* As a result of arousal, arterioles dilate in three columns of erectile tissue in the penis: the *corpus spongiosum,* surrounding the urethra, and the two *corpora cavernosa,* on the dorsal surface of the erect penis. As the arterioles dilate, blood flow increases, while other vessels constrict, causing the entire penis to swell with blood and creating a rigid erection that makes the penis capable of entering the female's vagina. Most of the rigidity of the erect penis results from blood pressure in the corpora cavernosa; high blood pressure in the corpus spongiosum could close the urethra, preventing passage of the semen.

For ejaculation to occur, the male must achieve *orgasm,* which requires repeated sexual stimulation. The tissue of the shaft and glans penis has a high concentration of sensory receptors for just that purpose. The mechanical actions of intercourse provide the repeated stimulation for the male to approach orgasm. The two small *bulbourethral glands,* also called *Cowper's glands,* sit at the base of the penis on either side of the urethra. During erection, these glands are compressed, causing them to release their mucuslike fluid into the urethra, clearing the pathway of any acidity from urine and providing lubrication for intercourse. Most of the lubrication, however, comes from the *vestibular glands* of the female, located near the vaginal opening.

When a sufficient level of stimulation has been received, the *sympathetic nervous system* triggers the two-part process of release. First, rhythmic contractions along the ductus deferens force sperm into the ejaculatory duct. Simultaneously, the seminal vesicles and the prostate gland are triggered to release their secretions into the duct. Then muscles in the prostate propel the semen into the urethra during the process of *emission.* Next, through continuing sympathetic impulse, muscles along the shaft of the penis contract to propel the semen out of the urethra. When the fluid exits the urethral orifice, *ejaculation* has occurred.

See how familiar you are with male reproduction by tackling these practice questions.

PRACTICE For questions 6–15, label the structures in Figure 16-5.

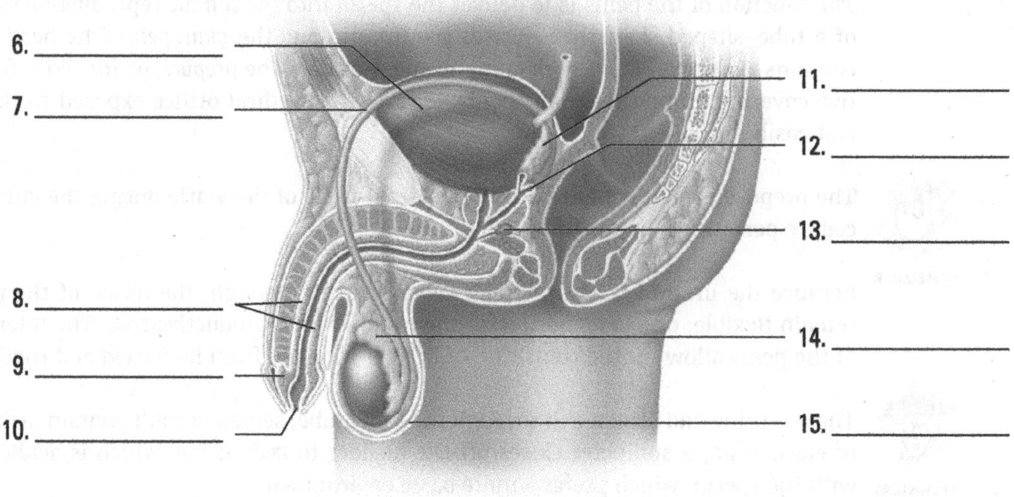

6. _____

7. _____

8. _____

9. _____

10. _____

11. _____

12. _____

13. _____

14. _____

15. _____

FIGURE 16-5:
The male
reproductive
system.

Illustration by Imagineering Media Services, Inc.

16 How does the scrotum contribute to the efficiency of sperm production?

_____

_____

_____

For questions 17–22, fill in the blanks in the following sentences.

Sperm-forming cells, or 17 _____, divide throughout the reproductive

lifetime of the male by a process called 18 _____. But after puberty, these

cells begin to produce two types of 19 _____ cells. Type A cells remain

in the wall of the 20 _____, producing more of themselves. Type B cells

produce 21 _____ through the process of 22 _____.

23 An average ejaculation contains about _____ sperm.

(a) 40 to 50 million

(b) 400 to 500 million

(c) 400 to 500

(d) 400,000 to 500,000

(e) 4 to 5 million

24　During sexual arousal, the penis becomes hard and erect. What is happening to cause that reaction?

(a) Muscles in the base of the penis are contracting, while other muscles toward the end of the penis are expanding.

(b) The corpus spongiosum and the corpora cavernosa draw slowly away from each other.

(c) Blood vessels are dilating to allow free flow of blood into the penis, while vascular shunts constrict drainage and cause blood pressure to rise in the corpora cavernosa.

(d) The Cowper's glands are squeezing additional fluids into the shaft of the penis.

(e) A sympathetic impulse triggers the blood vessels to close, trapping blood in the corpus spongiosum.

25　Explain the difference between emission and ejaculation.

_____

_____

_____

Meet & Greet the Female Reproductive System

Men may have quite a few hard-working parts in their reproductive systems, but women are the ones who are truly responsible for survival of the species (biologically speaking, anyway). The female body prepares for reproduction every month for most of a woman's adult life, producing a secondary oocyte and then measuring out delicate levels of hormones to prepare for nurturing a developing embryo. When a fertilized ovum fails to show up, the body hits the biological reset button and sloughs off the uterine lining before building it up all over again for the next month's reproductive roulette.

Mapping the female organs

Most of the organs of the female reproductive system are concentrated in the pelvic cavity; you can see them in the color plate titled "Female Reproductive System." These organs are attached to the *broad ligament,* a sheet of tissue that supports the organs and anchors them to the walls and floor of the cavity. It also contains the blood and lymphatic vessels that supply the organs, as well as the nerves.

Ovaries

The *ovaries* are two almond-shaped structures approximately 2 inches (5 centimeters) wide, one on each side of the pelvic cavity. Each ovary has two additional ligaments: a *suspensory ligament* attaching it to the pelvis and an *ovarian ligament* holding it in position relative to the uterus and uterine tubes.

The ovaries are the primary sex organs because they're the site of *oogenesis,* the process of oocyte maturation. Each oocyte is encased in a primordial follicle that undergoes changes along with the developing egg. This process contributes to the ovaries' major role in endocrine signaling, especially the production and control of hormones related to sex and reproduction: estrogen and progesterone. (The details of ovarian function are covered in the next section.)

Beginning at the female's puberty, the process of ovulation, or releasing eggs, begins. The primary oocytes that have been dormant in her ovaries since early in her fetal development are hormonally activated, and secondary oocytes are released at a rate of approximately one per month from *menarche* (the first menstrual period) to *menopause* (the last) — that is, from her early teen years to her late 40s or early 50s. The human female ovulates about 400 times during her lifetime.

Uterine tubes

The *uterine tubes,* also called the *fallopian tubes,* run from the ovary to the uterus. These tubes aren't literally connected to the ovaries; they just kind of hang over them. At the ovary end, the uterine tube expands into a funnel called the *infundibulum.* This funnel branches into finger-like structures called *fimbriae,* which guide the egg into the uterine tube. The ciliated cells that line the tube gently sweep the egg toward the uterus over the course of three or four days. The process of fertilization usually occurs in the uterine tube.

Uterus

The *uterus* or *womb* nourishes and shelters the developing fetus during gestation. It's a muscular organ about the size and shape of an upside-down pear. The walls of the uterus are thick and capable of stretching as a fetus grows.

The lining of the uterus, called the *endometrium,* is built up under the control of *estrogen* and *progesterone* from the ovaries, preparing it for an embryo. A portion of the endometrium becomes part of the placenta during pregnancy. If the egg isn't fertilized, the woman undergoes *menstruation,* shedding the extra tissue and starting over.

The *cervix* is a cylindrical muscular structure about 1 inch (2.5 centimeters) long that rests at the bottom of the uterus. It controls the movement of biological fluids and other material (not to mention occasionally a baby) into and out of the uterus. Normally, the cervix is open ever so slightly to allow sperm to pass into the uterus. During childbirth, the cervix opens wide to allow the fetus to move out of the uterus.

Vagina

The *vagina* is the part of the female body that receives the male penis during sexual intercourse and serves as a passageway for sperm to enter the uterus and uterine tubes. The vagina is about

3 to 4 inches (8 to 10 centimeters) long. The cervix marks the top of the vagina. Two pea-size glands are located just inside the vagina, within the lateral walls. Called the *vestibular* or *Bartholin's glands*, these glands secrete a mucuslike fluid to provide lubrication for intercourse.

During childbirth, the vagina must accommodate the passage of a fetus weighing on average about 7 pounds (3 kilograms), so the vagina's walls are made of stretchy tissues — some fibrous, some muscular, and some erectile. In their normal state, the vagina's walls have many folds; when the vagina needs to stretch, the folds flatten, providing more volume.

Vulva

In females, the external genitalia, collectively termed the *vulva*, include the following:

>> **Mons pubis:** Fatty tissue that lies atop the pubic symphysis, secretes pheromones, and grows hair to trap particles, preventing them from entering the reproductive tract

>> **Labia majora:** The outer folds of tissue that protect the area, containing sweat and sebaceous (oil) glands to aid in lubrication; homologous to the scrotum

>> **Labia minora:** Surround the openings to the urethra and vagina, are highly vascularized, and become engorged during arousal to increase sensitivity to pleasurable sensation

>> **Clitoris:** A small mound of erectile tissue on the anterior end of where there the labia minora meet and the most sensitive area of the female genitalia; repeated stimulation can lead to orgasm; homologous to the penis

Breasts

Although they're not technically reproductive organs, the breasts do provide an important reproductive function: nourishment for offspring. Actually, all humans have *mammary glands*, but only females produce a substance we call *milk* for the nutrition of relatively helpless infants with high calorie requirements. Besides providing nutrition, breast milk boosts the infant's immune system.

Each breast contains about two dozen lobules that are filled with *alveoli* (hollow sacs) that make and store milk. The milk is released into *lactiferous ducts*, which merge at the *nipple*. (See Figure 16-6). During puberty, the lobules and ducts develop, and adipose tissue is deposited under the skin to protect the lobules and ducts and to give shape to the breast. During pregnancy, hormones increase the number of milk-producing cells and increase the sizes of the lobules and ducts.

After the infant is born, the mother's pituitary gland secretes the hormone *prolactin*, which causes the milk-producing cells to create milk, and *lactation* begins. When the infant latches onto the nipple and begins to suckle, the mother's posterior pituitary is stimulated to release *oxytocin* as part of the *let-down reflex*. This reflex causes the lobules to contract, releasing the milk into the *lactiferous sinus*. As long as the child nurses regularly, lactation will continue.

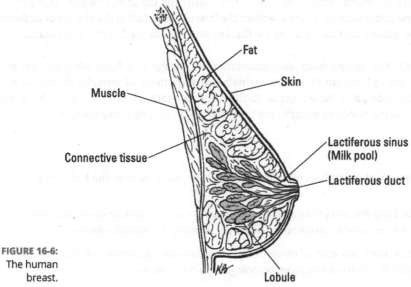

Fat

Skin

Muscle

Connective tissue

Lactiferous sinus (Milk pool)

Lactiferous duct

FIGURE 16-6:
The human breast.

Lobule

Illustration by Kathryn Born, MA

The vicious cycle

The *menstrual cycle* (monthly cycle) consists of both the *ovarian cycle* and the *uterine cycle*, both of which are approximately 28 days in duration. These cycles run concurrently to prepare the ovum and the uterus, respectively, for pregnancy.

By convention, the first day of menstrual bleeding is counted as Day 1 of the menstrual cycle. Menstrual bleeding begins at a point in the cycle when the levels of estrogen and progesterone are at their lowest. But the entire menstrual cycle is directed by several hormones, not just estrogen and progesterone.

TIP

The term *estrogen* actually refers to a group of hormones primarily produced in the ovaries. These hormones are responsible for developing the reproductive organs as well as providing the female secondary sex characteristics (as testosterone does in males). The primary estrogen hormone is *estradiol*, so when *estrogen* is used generically, that term refers to this hormone.

The ovarian cycle

The 28-day *ovarian cycle* is the most important part of the menstrual cycle because it's responsible for producing the hormones that control the uterine cycle. The structures involved in this process are shown in Figure 16-7. From Day 1 to Day 13, triggered by low estrogen level, *follicle-stimulating hormone* (FSH) stimulates the development of a follicle, and *luteinizing hormone* (LH) stimulates the maturation of an oocyte in one of the ovaries.

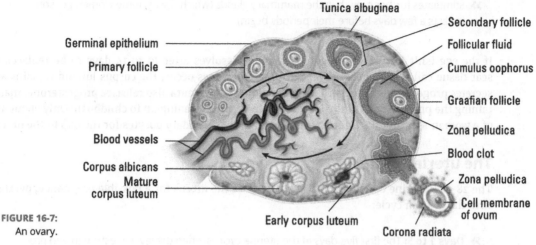

Germinal epithelium

Primary follicle

Tunica albuginea

Secondary follicle

Follicular fluid

Cumulus oophorus

Graafian follicle

Zona pelludica

Blood vessels

Blood clot

Corpus albicans

Mature
corpus luteum

Zona pelludica

Cell membrane
of ovum

FIGURE 16-7:
An ovary.

Early corpus luteum

Corona radiata

Illustration by Imagineering Media Services, Inc.

The ovarian cycle occurs as follows:

1. At the onset of menstruation, the anterior pituitary gland secretes FSH, which prompts about 1,000 of the primordial follicles to resume meiosis.

2. While the oocyte continues meiosis, follicular cells divide and surround the oocyte with several layers of cells called the *cumulus oophorus*, creating the *secondary follicle*.

   Usually, one follicle outgrows the rest, maturing into a *Graafian follicle.* Nonidentical, or *fraternal*, twins, triplets, or even more embryos result if more than one follicle matures and more than one secondary oocyte is released (and fertilized).

3. As maturation continues, the cell layers separate, leaving a fluid-filled space between them. The cells that are in contact with the oocyte form the *zona pellucida*; the others secrete estrogen to signal the uterus to prepare by thickening the endometrium.

4. As blood levels of estrogen begin to rise, the pituitary stops releasing FSH and begins releasing LH, which prompts the Graafian follicle now at the surface of the ovary to rupture, triggering *ovulation* — the release of the secondary oocyte, more commonly referred to as an egg cell.

5. The oocyte, wrapped in the zona pellucida, takes some follicle cells with it when released. The cells form the *corona radiata*, which serves to provide nourishment for the journey to the uterus.

Meanwhile, back at the ovary, a clot has formed inside the ruptured follicle, and the remaining follicle cells form the *corpus luteum* (literally, *yellow body)*. This new but temporary endocrine gland does the following things:

>> Secretes progesterone, signaling the uterine lining to prepare for possible implantation of a fertilized egg

>> Inhibits the maturing of follicles, ovulation, and the production of estrogen to prevent menstruation

>> Stimulates further growth in the mammary glands (which is why some women get sore breasts a few days before their periods begin)

If the egg isn't fertilized, the corpus luteum dissolves after 10 to 14 days to be replaced by scar tissue called the *corpus albicans*. If pregnancy does occur, the corpus luteum remains and secretes progesterone to prevent menstruation. The placenta also releases progesterone, maintaining the pregnancy until the levels start to decrease leading up to childbirth. Only about 400 of a woman's primordial follicles ever develop into secondary oocytes for the trip to the uterus.

The uterine cycle

The 28-day uterine cycle, which aims to prepare the uterus for a possible pregnancy, overlaps with the ovarian cycle:

>> **Days 1 to 5:** The first five days of the uterine cycle is when the level of estrogen and progesterone is lowest — the period of menstruation. The low level of sex hormones fails to prevent the tissues lining the uterus (the *endometrium)* from disintegrating and shedding. As the hormone levels drop, blood vessels spasm; cells undergo *autolysis* (self-destruction); tissues tear apart from the uterus wall; and blood vessels rupture, causing the bleeding that occurs during a period. The blood and tissue (menstrual flow) passes out of the uterus through the cervix and then out of the body through the vagina.

>> **Days 6 to 14:** During this *proliferative phase,* the developed follicle secretes high levels of estrogen, which makes the endometrium regenerate fresh tissue. The tissues lining the uterus and the glands in the uterine wall grow and develop an increased supply of blood. All these changes are preparation for nourishing an embryo and supporting a pregnancy, should the oocyte become fertilized and implant in the wall of the uterus.

>> **Days 15 to 28:** During this *secretory phase,* the corpus luteum secretes an increasing level of progesterone (which further thickens the endometrium), and the glands of the uterus secrete thick mucus as well. If the egg becomes fertilized, the thickened endometrium and mucus help trap the fertilized egg so that it implants properly in the uterus. If the egg doesn't become fertilized, no signal is sent back to the ovary, and the corpus luteum begins to shrink. As the corpus luteum shrinks, the progesterone and estrogen levels decline, causing the endometrium to shred and shed just before menstruation.

When there are no more primordial follicles to be stimulated by FSH, there is no more production of estrogen and progesterone by the ovaries. The ovarian and uterine cycles are stopped in their tracks, and the woman reaches *menopause.*

PRACTICE

It's time to pause for a review of the female reproductive system.

For questions 26–41, use the terms that follow to identify the anatomy of the female reproductive system shown in Figure 16-8.

(a) Cervix

(b) Fimbriae

(c) Urinary bladder

(d) Clitoris

(e) Labium major

(f) Ovary

(g) Rectum

(h) Vestibular (Bartholin's) glands

(i) Vaginal orifice

(j) Anus

(k) Mons pubis

(l) Uterus

(m) Labium minor

(n) Uterine (fallopian) tube

(o) Vagina

(p) Urethra

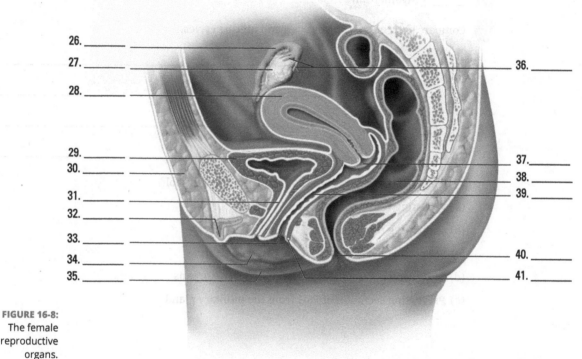

26. _____
27. _____
28. _____
29. _____
30. _____
31. _____
32. _____
33. _____
34. _____
35. _____
36. _____
37. _____
38. _____
39. _____
40. _____
41. _____

FIGURE 16-8: The female reproductive organs.

Illustration by Imagineering Media Services, Inc.

For questions 42–46, match each term with its description.

   (a) Fingerlike projections at the end of a uterine tube

   (b) Protective layer directly surrounding the secondary oocyte

   (c) Granulosa cells forming the outer layer of the Graafian follicle

   (d) Layer of smooth muscle in uterine wall

   (e) Inner lining of the uterus

42 _____ Corona radiata

43 _____ Endometrium

44 _____ Fimbriae

45 _____ Zona pelludica

46 _____ Myometrium

47 Compare and contrast the ovarian and uterine cycles.

_____

_____

_____

_____

48 Which one of the following is *not* a function of estradiol?

   (a) Preparing the endometrium

   (b) Supporting development of the secondary oocyte

   (c) Supporting development of the corpus luteum

   (d) Triggering development of the reproductive organs

   (e) Preventing secretion of FSH from the pituitary gland

Practice Questions Answers and Explanations

1. **(b) 23.** Because the spermatocyte is the original cell that undergoes division, and because a human has a total of 46 chromosomes only *after* sperm meets egg, you must divide the number 46 in half.

2. **(c) occurs in meiosis.** Specifically, synapsis occurs during prophase I, which is the first stage of meiosis.

3. **(c) polar bodies.** These cells eventually disintegrate.

4. **(e) Prophase I**

5. Following is a summary of what should appear in your drawings and descriptions of the stages of meiosis. For further reference, check out Figure 16-1.

   In the drawing for late prophase I, at least two pairs of homologous chromosomes should be grouped into tetrads. (In truth, there are 23 pairs, but simplified illustrations tend to show just two.) The description for prophase I should include reference to the tetrad formation. The drawing for metaphase I should show the equatorial plane (a center horizontal line) with the tetrads aligned along it and an X on each side. The illustration also should show spindles radiating from each pole, with the tetrads attached to them by their centromeres. The description should include reference to the equatorial plane, the poles, and the spindles.

   The drawing for anaphase I should show the tetrads moving to the top and bottom of the cell along the spindles and the cytoplasm slowly beginning to divide. In telophase I, the division becomes more pronounced, and two new nuclei form. The chromosomes remain in the X shape. As the process enters interkinesis, the cytoplasm pinches off into two cells.

   During prophase II, which is also the start of the second meiotic division, the chromosomes migrate toward a new equatorial plane. The drawing of metaphase II should show all chromosomes aligned on the equatorial plane, with the centromere along the (invisible) center line. For anaphase II, you should show the chromosomes pulling apart into chromatids and moving toward the poles. In the final stage, telophase II, you should draw new nuclei forming around the chromatin as it uncoils.

6. **Bladder**

7. **Ductus deferens** or vas deferens

8. **Corpus spongiosum**

9. **Glans penis**

10. **Urethral orifice**

11. **Seminal vesicle**

12. **Ejaculatory duct**

13. **Bulbourethral gland**

(14) **Epididymis**

(15) **Scrotum**

(16) **Sperm are sensitive to temperature, and production is most effective when the temperature is just below that of the body. The scrotum holds the testes away from the body to account for that fact. If the testes get too cold, the smooth muscle of the scrotum contracts and pulls them closer to the body to warm them up. If the testes get too warm, the scrotum lengthens to hold them away from the body's heat.**

(17) **spermatogonia**

(18) **mitosis**

(19) **daughter**

(20) **seminiferous tubules**

(21) **four sperm**

(22) **meiosis**

(23) **(b) 400 to 500 million sperm.** Keep in mind that sperm are microscopically small, so quite a few can fit in a tiny amount of semen.

(24) **(c) Blood vessels are dilating to allow free flow of blood into the penis, while vascular shunts constrict drainage and cause blood pressure to rise in the corpora cavernosa.**

(25) **Although both processes are under sympathetic control and seem to happen simultaneously, there's an important distinction: Erection is triggered via a parasympathetic pathway, so it's important to note that emission doesn't begin until repeated stimulation initiates it.** Here's a simple way to think about the process: Emission is preparation, and ejaculation is release.

(26) **(n) Uterine (fallopian) tube**

(27) **(f) Ovary**

(28) **(l) Uterus**

(29) **(c) Urinary bladder**

(30) **(k) Mons pubis**

(31) **(p) Urethra**

(32) **(d) Clitoris**

(33) **(i) Vaginal orifice**

(34) **(m) Labium minor**

(35) **(e) Labium major**

(36) **(b) Fimbriae**

(37) **(a) Cervix**

(38) **(g) Rectum**

(39) **(o) Vagina**

(40) **(j) Anus**

(41) **(h) Vestibular (Bartholin's) glands**

(42) **(c) Granulosa cells forming the outer layer of the Graafian follicle**

(43) **(e) Inner lining of the uterus**

(44) **(a) Fingerlike projections at the end of a uterine tube**

(45) **(b) Protective layer directly surrounding the secondary oocyte**

(46) **(d) Layer of smooth muscle in the uterine wall**

(47) **Both cycles take about 28 days and restart with menstruation (so Day 1 is the first day of a period). The ovarian cycle serves to prepare the follicle, restart meiosis in the oocyte, and then release it via ovulation. From there, the remainder of the follicle produces progesterone, which influences the uterine cycle. The purpose of the uterine cycle is to prepare the uterus for implantation. Before ovulation, estrogen triggers cell division in the endometrium. Following ovulation, progesterone prompts the tissue to take on more fluid and for glands in the uterine lining to secrete mucus.** If fertilization doesn't occur, the production of progesterone halts, and the cycles start over.

(48) **(c) Supporting development of the corpus luteum.** By the time the corpus luteum starts to develop, estrogen already has begun to bow out of the reproductive equation, allowing progesterone to take the lead.

If you're ready to test your skills a bit more, take the following chapter quiz, which incorporates all the chapter topics.

Whaddya Know? Chapter 16 Quiz

Quiz time! Complete each problem to test your knowledge of the various topics covered in this chapter. You can find the solutions and explanations in the next section.

For questions 1–9, fill in the blanks to complete the following sentences:

Meiosis produces sperm and ova, which when combined make a (1) _____ (fertilized egg) with its full complement of chromosomes. Normally, humans have (2) _____, or cells containing 23 pairs of homologous chromosomes for a total of 46 chromosomes each. A pair of chromosomes containing the same type of genetic information are (3) _____ chromosomes, though they may have different (4) _____ or varieties of a gene. Ova and sperm are called sex cells or (5) _____. The number of chromosomes is cut in half during the first meiotic division, producing gametes that are (6) _____. The second meiotic division produces (7) _____ sperm in the male but only (8) _____ secondary oocyte in the female. Then the zygote proceeds through (9) _____ to produce the body's cells.

(10) Select the correct sequence for the movement of sperm.

  (a) Rete testis → Seminiferous tubules → Epididymis → Ductus deferens → Urethra

  (b) Seminiferous tubules → Rete testis → Ejaculatory duct → Epididymis → Ductus deferens → Urethra

  (c) Epididymis → Ejaculatory duct → Rete testis → Seminiferous tubules → Ductus deferens → Urethra

  (d) Seminiferous tubules → Ejaculatory duct → Rete testis → Epididymis → Ductus deferens → Urethra

  (e) Seminiferous tubules → Rete testis → Epididymis → Ductus deferens → Ejaculatory duct → Urethra

(11) Why does the corpus luteum produce progesterone?

  (a) To stimulate development of the cumulus oophorus

  (b) To trigger ovulation

  (c) To prepare a woman's uterus for pregnancy and prevent menstruation

  (d) To trigger the resumption of meiosis

  (e) To prepare the infundibulum to fragment into fimbriae

12   After all the proper glands have secreted fluids to nourish and protect the departing sperm, the substance ejaculated is called

(a) stroma.

(b) semen.

(c) prepuce.

(d) spermatozoa.

(e) inguinal.

13   Anaphase I of meiosis is characterized by which of the following?

(a) Synapsed chromosomes move away from the poles.

(b) DNA duplicates itself and condenses into chromosomes.

(c) Synapsis of homologous chromosomes occurs.

(d) Homologous chromosomes separate and move poleward with centromeres intact.

(e) Chromosomes are split at the centromere and pulled away from the poles.

For questions 14–20, identify the statement as true or false. When the statement is false, identify the error.

14   Meiosis I is the reduction division because it creates haploid cells.

_____

15   When a sperm penetrates the cell membrane of the ovum, the zona pelludica immediately hardens.

_____

16   The corpus albicans continues making progesterone if the egg is fertilized.

_____

17   The clitoris is akin to the glans penis, containing numerous receptors for sexual stimulation.

_____

18   Meiosis begins before birth and resumes during puberty.

_____

19 The prostate gland contributes fructose to the semen to power cellular respiration in the sperm.

_____

20 A lack of primordial follicles leads to the onset of menopause in females.

_____

21 What structure releases its secretion near the onset of erection?

(a) Leydig cells

(b) Bulbourethral glands

(c) Seminal vesicle

(d) Prostate gland

(e) Sertoli cells

22 Where is testosterone produced?

(a) Seminal vesicles

(b) Leydig cells

(c) Prostate gland

(d) Sertoli cells

(e) Epididymis

For questions 23–32, fill in the blanks to complete the following sentences.

The 23 _____ in a developing female fetus begin meiosis. Upward of 2 million 24 _____ are produced, but they're suspended in 25 _____. By the onset of 26 _____, only about 300,000 to 400,000 remain. Each month, about 1,000 primary follicles are triggered to resume meiosis by 27 _____. One usually outgrows the rest, forming the 28 _____. Meiosis is again suspended, this time at 29 _____. Release of the secondary oocyte, or 30 _____, is triggered by the release of 31 _____ from the anterior pituitary. Meiosis II isn't complete until 32 _____ has begun.

33 Spermatid maturation takes place in the

(a) seminiferous tubule.

(b) rete testis.

(c) spermatic cord.

(d) scrotum.

(e) epididymis.

34 Why are both prolactin and oxytocin necessary for nursing an infant?

_____

_____

_____

35 What is the function of the prostaglandins released from the seminal vesicles?

(a) To control the pH of the semen, keeping it from getting too acidic

(b) To provide lubrication to the penis to reduce friction during intercourse

(c) To provide an energy source for the sperm

(d) To trigger arousal in the female

(e) To trigger muscle contractions in the uterus to help the sperm travel

Answers to Chapter 16 Quiz

1. **zygote**

2. **diploid**

3. **homologous**

4. **alleles**

5. **gametes**

6. **haploid**

7. **four haploid**

8. **one functional haploid**

9. **mitosis**

10. **(e) Seminiferous tubules → Rete testis → Epididymis → Ductus deferens → Ejaculatory duct → Urethra.** The sperm develop in the coiled tubules, move through the straighter tubes (tubuli recti), continue across the network of the testis (rete testis), through the efferent ducts, into the epididymis (the really long tube), and past the ductus (or vas) deferens and the ejaculatory duct into the urethra.

11. **(c) To prepare a woman's uterus for pregnancy and prevent menstruation.** It stands to reason that if the corpus luteum forms after the Graafian follicle ruptures, the progesterone it's producing is signaling the uterus that a fertilized egg may be on its way.

12. **(b) semen**

13. **(d) Homologous chromosomes separate and move poleward with centromeres intact.** Think anaphase = apart. Then you just have to take note of whether it's meiosis I or II to determine what is being pulled apart.

14. **True.** Even though they contain two copies of it, the new cells contain only one of the original chromosomes, so they're considered to be haploid cells.

15. **True.** This process keeps other sperm from contributing a pronucleus, leading to a zygote that isn't viable.

16. **False.** The corpus luteum is what secretes progesterone. The corpus albicans is what is left behind as the corpus luteum degrades because the egg wasn't fertilized.

17. **True**

18. **False.** This statement is true for women but not for males; they don't begin performing meiosis until puberty.

19. **True**

(20) True

(21) **(b) Bulbourethral glands.** Their mucuslike fluid clears the urethra of any urine and helps with lubrication, so this secretion must occur early in the process.

(22) **(b) Leydig cells**

(23) oogonia

(24) primordial follicles

(25) prophase I

(26) puberty

(27) FSH

(28) Graafian follicle

(29) metaphase II

(30) ovulation

(31) LH

(32) fertilization

(33) (e) epididymis

(34) The prolactin triggers the secretion of milk by the mammary glands, but these glands are located back from the exit in the nipple. With the baby's limited ability to create suction, this location is problematic. Oxytocin triggers the lobules housing the glands to contract, forcing the milk into the lactiferous ducts so that when the baby suckles, it's able to pull in milk. This process is called let-down.

(35) (e) To trigger muscle contractions in the uterus to help the sperm travel

Chapter **17**

Baking the Bundle of Joy

Pregnancy is divided into three periods called *trimesters* (although many new parents bemoan a postnatal fourth trimester until the baby sleeps through the night). By convention, this periodization begins before fertilization, as the counting begins on the first day of the woman's last period. Actually, the soon-to-be bundle of joy has undergone quite a bit of development before even the earliest store-bought pregnancy test could produce those telling two lines (or smiley face, words, or whatever clever stuff marketing teams have devised).

The first 12 weeks of development mark the first trimester, during which *organogenesis* (organ formation) is established. Most of this trimester is spent during the embryonic period, with the fetal period beginning at week 10. During the second trimester (typically considered to be weeks 13 to 27), all fetal systems continue to develop, and rapid growth triples the fetus's length. By the third trimester (typically considered to be weeks 28 to 40), all organ systems are functional, and the fetus usually is considered to be *viable* (capable of surviving outside the womb) even if it's born prematurely. The overall growth rate slows in the third trimester, but the fetus gains weight rapidly.

Fertilization: A Battle for the Ages!

Fertilization happens in the uterine tube and has only a small window of time to occur. Following ovulation, which occurs around day 12, an oocyte can only survive for about 24 hours. Fortunately, sperm can survive for up to a week; they may be there anxiously awaiting the egg's arrival.

The cells of the *corona radiata* (refer to Figure 16-2 in Chapter 16) release chemicals that attract the sperm. After the sperm arrive, they burrow through the corona radiata by releasing enzymes from their *acrosome* (refer to Figure 16-3 in Chapter 16). These sperm, however, won't be the ones to make it to the oocyte; they work to clear the way for others to make it through without having to use their enzymes. When the corona radiata is breached, sperm are able to bind with the receptors in the *zona pellucida*. Again, enzymes are used to digest the egg's protective layer. When a sperm has finally reached the cell membrane of the oocyte, and with its acrosome gone, it attaches to receptors there. This event immediately triggers two other important events:

>> **The zona pellucida hardens,** preventing another sperm from fusing with the egg's actual cell membrane.

>> **The oocyte restarts meiosis II.** At this point, the sperm's pronucleus is allowed into the cell.

When the oocyte completes meiosis II, its pronucleus will fuse with that of the sperm. The cell now contains two of each of our 23 different chromosomes, including the sex chromosomes. Since Mom only has X chromosomes to contribute, it is the fertilizing sperm that determines the sex of the zygote; if it brings an X, then the sex is female, if it brings a Y, then the sex is male. Fertilization is now complete, and the zygote is ready to begin its journey to the uterus.

And So It Begins

The zygote undergoes *cleavage* (mitotic division) immediately. Over the next few days, the daughter cells (called *blastomeres*) divide twice more, for a total of 16 blastomeres, all within the rigid wall of the zona pellucida with no increase in overall size. You can see the progression of this early prenatal development in Figure 17-1.

The mass, now called a *morula* (which means *mulberry-shaped*), leaves the uterine tube and enters the uterine cavity. Cell division continues, still confined within the zona pellucida, and a cavity known as the *blastocoel* begins to form in the morula's center.

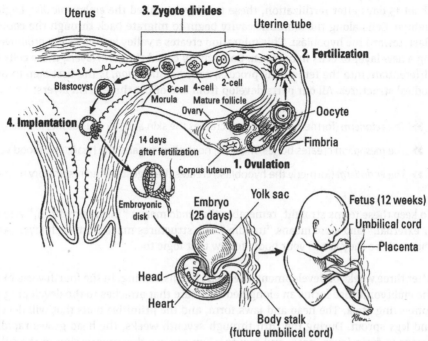

Uterus · **3. Zygote divides** · **Uterine tube**

2. Fertilization

Blastocyst · 8-cell · 4-cell · 2-cell

Morula · Mature follicle · **Oocyte**

Ovary

4. Implantation

Fimbria

14 days after fertilization · Corpus luteum · **1. Ovulation**

Embryonic disk · Embryo (25 days) · Yolk sac · Fetus (12 weeks)

Umbilical cord

Placenta

Head

Heart

Body stalk (future umbilical cord)

FIGURE 17-1: Early prenatal development.

Illustration by Kathryn Born, MA

Implantation

Around the sixth day after fertilization, the structure, now called a *blastocyst*, has hollowed out except for a pile of cells called the *inner cell mass*. The blastocyst hatches from the eroded zona pellucida within the uterine cavity (no, really, it's actually called *hatching*). The cells of the outer layer of blastocyst — the *trophoblast* — secrete an enzyme that facilitates *implantation* in the endometrium. That is, the pre-embryo attaches to wall of the uterus. Now diffusion between cells of the mother and the blastocyst can occur. When this diffusion is established, implantation is complete, and the pregnancy is established.

Embryonic development

Weeks three through eight after fertilization, or weeks five through ten of pregnancy, are known as the *embryonic stage*. During these weeks, the embryo's cells begin to differentiate and specialize. The inner cell mass splits off into two layers:

>> **Epiblast:** The cells deep in the endometrium, which begin to create the *amniotic cavity*. This fluid-filled space grows with the baby, providing structural support throughout the pregnancy.

>> **Hypoblast:** The cells along the interior space of the blastocyst (the blastocoel), which begin to form the *yolk sac*. The cells of the developing embryo are nourished by the yolk sac until the placenta is fully formed.

About 15 days after fertilization, these two layers, called the *embryonic disc*, begin to form the embryo. Cells along the amniotic cavity begin to migrate back through the center of the epiblast, toward the hypoblast. This migration creates a valley known as the *primitive streak*, forming a new layer of cells in between. Collectively called the *germ layers*, these cells duplicate and differentiate into the fetus. This process, called *gastrulation*, is the first step in organizing our bodies' structures. All our parts develop from one of the three germ layers:

>> The *ectoderm* (formerly the epiblast) creates the skin and nervous tissue, including the brain.

>> The *mesoderm* creates bones, muscles, nonhollow organs, the heart, and blood vessels.

>> The *endoderm* (formerly the hypoblast) creates the digestive and respiratory tracts.

TIP

To keep these terms straight, remember that *endo* means "inside or within," *ecto* means "outer or external," and *meso* means "middle." The structures made from each layer, when you think about where they are in your body, follow that logic too.

After three weeks of development, the heart begins beating. In the fourth week of development, the embryonic disc forms an elongated structure that attaches to the developing placenta via a connecting stalk. The head and jaws form, and the primitive buds that will develop into arms and legs sprout. During the fifth through seventh weeks, the head grows rapidly, and a face begins to form (eyes, nose, and mouth). You can see this progression in the color plate titled "Fetal Development." Fingers and toes grow at the ends of the elongating limb buds. All internal organs have started to form. After eight weeks of development, the embryo begins to have a more human appearance and is referred to as a *fetus.*

PRACTICE

Just like embryonic development, this section covered a lot in very little time! Pause now to check your understanding.

For questions 1–5, match the terms with their descriptions. Not all the answer choices will be used.

(a) Acrosome   (f) Morula

(b) Blastocyst   (g) Oocyte

(c) Blastomeres   (h) Trophoblast

(d) Epiblast   (i) Zygote

(e) Hypoblast

1 _____ The outer layer of a blastocyst

2 _____ The solid ball of cells created in the first three days after fertilization

3 _____ The cells created by mitosis of a zygote

4 _____ The layer of the inner cell mass that is farthest from the uterine cavity

5 _____ The pre-embryo that is a ball of cells with a hollow middle

For questions 6–11, identify the statement as true or false. When the statement is false, identify the error.

6 The mesoderm forms nervous tissue and skin.

_____

7 The embryonic stage is complete at the end of the eighth week.

_____

8 Outer cells of the embryonic disc form the germ layers.

_____

9 During the fifth through seventh weeks, the arm and leg buds elongate, and fingers and toes begin to form.

_____

10 As the zygote moves through the uterine tube, it undergoes meiosis.

_____

11 After five days of cleavage, the cells form into a hollow ball called the morula.

_____

12 What is gastrulation?

(a) The creation of a hollow space within the developing embryo

(b) The beginning of mitosis following fertilization

(c) The attachment of the pre-embryo to the wall of the uterus

(d) The removal of the zona pellucida

(e) The creation of the three germ layers

Finishing Out the Trimesters

Following week eight and lasting until birth is the *fetal stage*. The placenta attaches the fetus to the uterine wall, exchanges gases and waste between the maternal and fetal bloodstreams, and secretes hormones to sustain the pregnancy. Growth and development occur rapidly during this time. The *primary germ layers* continue their development as the fetus becomes more recognizable as a baby. The ectoderm develops into the skin and nervous tissues, while the endoderm forms the inner tubes: the alimentary canal and the respiratory tract. The mesoderm develops into everything in between, including the bones and muscles.

The placenta

Immediately after implantation, formation of the *placenta*, a special organ that exists only during pregnancy, begins. The cells of the trophoblast (the outer layer of the blastocyst) and cells of the uterus begin to interact, building the blood vessels and other tissues that will take over for the yolk sac about 10 weeks into pregnancy.

The placenta is a dark red disc of tissue about 9 inches (23 centimeters) in diameter and 1 inch (2.5 centimeters) thick in the center; and it weighs about a pound (roughly half a kilogram). It connects to the fetus via an *umbilical cord* of approximately 22 to 24 inches (56 to 61 centimeters) long that contains two arteries and one vein. The placenta grows along with the fetus.

The placenta serves to support the sharing of physiological functions between the mother and the fetus: nourishment (provision of energy and nutrients), gas exchange (a fetus must take in oxygen and eliminate carbon dioxide before birth), and the elimination of metabolic waste. The placenta allows some substances to enter the fetal body and blocks others. It does a good job of delivering nutrients and maintaining fluid balance, but it's permeable to alcohol, many drugs, and some toxic substances.

Nutrients and oxygen diffuse through the placenta; then the fetal blood picks them up and carries them through the umbilical cord. The wastes generated by the fetus are carried back out through the umbilical cord and diffused into the placenta. The mother's blood picks up the wastes from the placenta, and her body excretes them. (Geez, moms start cleaning up after their kids before they're even born!)

Both the fetus and the placenta are enclosed within the *amniotic sac*, a double-membrane structure filled with a fluid matrix called *amniotic fluid*. The fluid keeps the temperature constant for the developing fetus, allows for movement, and absorbs the shock from the mother's movements.

Fetal development

The fetus, with all its systems in place, continues programmed development in the second trimester. *Ultrasound imaging* can reveal the skeleton, head details, and external genitalia. Bone begins to replace the cartilage that formed during the embryonic stage. At the end of the second trimester, the fetus is about 12 to 14 inches (30 to 36 centimeters) long and weighs about 3 pounds (1.4 kilograms). The fetal development program speeds up in the third trimester. The fetus grows dramatically in height and weight, as you can see in the fetal development color plate. Subcutaneous fat is deposited, serving as a critical energy reserve for brain and nervous-system development. The kidneys and lungs continue their fine-tuning. And the baby works out its muscles by moving around, kicking and punching Mom and her bladder at inopportune times.

The milestones of fetal development are marked monthly:

REMEMBER

>> **End of the second month:** The terminology changes from *embryo* to *fetus*. The head remains overly large compared with the developing body, and the limbs are still short. All major regions of the brain have formed.

>> **Third month:** Body growth accelerates, and head growth slows. The arms reach the length they will maintain during fetal development. The bones begin to ossify, and all body systems have begun to form. The cardiovascular (circulatory) system supplies blood to all the developing extremities, and even the lungs begin to practice "breathing" amniotic fluid. By the end of the third month, the external genitalia are visible in the male. The fetus is a bit less than 4 inches (10 cm) long and weighs about 1 ounce (28 g).

>> **Fourth month:** The body grows rapidly during the fourth month as legs lengthen, and the skeleton continues to ossify as joints begin to form. The face looks more human. The fetus is about 7 inches (18 cm) long and weighs 4 ounces (113 g).

>> **Fifth month:** Growth slows during the fifth month, and the legs reach their final fetal proportions. Skeletal muscles become so active that the mother can feel fetal movement. Hair grows on the head, and *lanugo,* a covering of fine, soft hair, covers the skin. The fetus is about 10 inches (25 cm) long and weighs ½ to 1 pound (227–454 g).

>> **Sixth month:** The fetus gains weight during the sixth month, and eyebrows and eyelashes form. The skin is wrinkled, translucent, and reddish because of the blood vessels in the dermis. The fetus is between 11 and 14 inches (28–35 cm) long and weighs a bit less than 1½ pounds (680 g).

>> **Seventh month:** During the seventh month, skin becomes smoother as the fetus gains subcutaneous fat tissue. The eyelids, which are fused during the sixth month, open. Usually, the fetus turns to an upside-down position. It's between 13 and 17 inches (33–43 cm) long and weighs 2½ to 3 pounds (1–1⅓ kg).

>> **Eighth month:** During the eighth month, subcutaneous fat increases, and the fetus shows more babylike proportions. The testes of a male fetus descend into the scrotum. The fetus is 16 to 18 inches (41–46 cm) long and has grown to just under 5 pounds (2¼ kg).

>> **Ninth month:** During the ninth month, the fetus plumps up considerably with additional subcutaneous fat. Much of the lanugo is shed, and fingernails extend all the way to the tips of the fingers. The average newborn at the end of the ninth month is 20 inches (51 cm) long and weighs about 7½ pounds (3½ kg).

Near the end of the third trimester, the fetus positions itself for birth, turns its head down, and aims for the exit. When the fetus's head reaches the *ischial spines* of the pelvic bones (see Chapter 6), the fetus is said to be *engaged* for birth. (See Figure 17-2.)

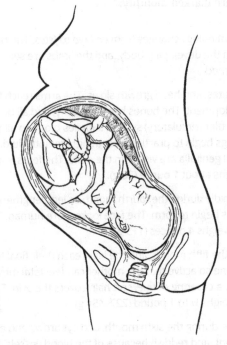

FIGURE 17-2:
A fetus late in
the third
trimester.

Illustration by Kathryn Born, MA

PRACTICE

The following practice questions deal with the development of the fetus during its 40 weeks in the womb.

13 What needs to happen before a fetus is considered to be viable?

(a) The heart rate must be easily detected.

(b) All organ systems must be functional.

(c) Subcutaneous fat must be built up to at least an inch.

(d) The brain is fully formed.

(e) The fetus must have turned to an upside-down position.

14 Describe one new fetal development for each month.

Third month: _____

Fourth month: _____

Fifth month: _____

Sixth month: _____

Seventh month: _____

Eighth month: _____

Ninth month: _____

 15 When does the placenta take over for the yolk sac to provide nourishment for the baby?

(a) Immediately after implantation

(b) After about a month of development

(c) Near the end of the first trimester

(d) When the heart starts circulating blood

The Fun Part: Labor and Delivery

After 9½ months of mood swings, hunger, and countless sudden urges to pee, Mom is ready to send her little bundle of joy packing. And 40 weeks after the egg left the ovary, the baby is ready to meet and greet the world. All that remains is labor and delivery. Like throughout the pregnancy, both mother and baby contribute to this process. Much of the work here, though, is left to the mother.

Near the end of the fetus's stay in the womb, the woman's body begins preparations for *parturition,* or childbirth. Progesterone drops steadily as the big day nears, leading to mild contractions that most women don't even feel. The cervix begins to dilate to allow the fetus into the birth canal (vagina). But the trigger that sets off the cascade of hormones that maintain the process of *labor* remains a mystery. It is widely believed that the process is initiated by the fetus. Regardless of the trigger, parturition is a classic example of a *positive feedback loop.*

Prostaglandins cause early uterine contractions, which lead to stretching of the uterus and cervix. Receptors communicate the stretch to the hypothalamus, causing it to release *oxytocin* from the posterior pituitary. This release triggers uterine contractions, causing more stretch, which causes more oxytocin and more contractions, and so on, until the baby is delivered.

TIP

It doesn't seem to make sense. I mean, really. An organ's contractions cause that very same organ to stretch? But think of the uterus as a water-filled balloon. You squeeze around it, and one end bulges as the water is pushed in that direction. That's exactly what happens to the uterus. Eventually, the uterus contracts with enough force that the water balloon pops. That is, the amniotic sac ruptures, and the fluid spills out, which we call the *water breaking.* This event usually happens during early, or *latent,* labor.

Parturition occurs in three stages. The first stage is labor, when the cervix fully dilates to allow the baby to pass through. Stage two is the actual delivery of the baby. This stage is followed by a final round of contractions to deliver the placenta.

A labor of love

Stage 1, *labor*, is the long process that a woman's body undergoes to get ready to deliver the baby; to physically push it out. It actually begins before any noticeable contractions occur, even a week or so before. *Lightening* occurs, when the baby's head drops down into the pelvis (often referred to as *dropping*). The mucus plug that sealed off the cervix also passes through the vagina.

Labor progresses through three phases, the first of which is the longest:

>> **Latent labor:** Short contractions every 5–20 minutes, cervix dilates up to 5–6 cm (1–2 inches), typically lasts 6–12 hours

>> **Active labor:** Increasingly stronger contractions every 3–5 minutes that last about 60 seconds, cervix dilates up to 8 cm (3 in.), typically lasts 3–5 hours

>> **Transition:** Intense contractions every 2–3 minutes for 60–90 seconds, cervix dilates fully, typically lasts 30–60 minutes

The goal of labor is to cause changes to the cervix that will allow the baby's head to pass through it. In addition to fully dilating, which is when the opening is about 10 cm (4 in.) across, the cervix must undergo *effacement*: the thinning of the wall of the cervix itself, as you can see in Figure 17-3.

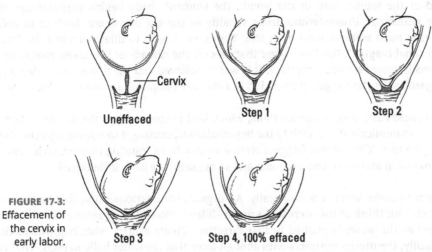

FIGURE 17-3: Effacement of the cervix in early labor.

Uneffaced — Cervix

Step 1

Step 2

Step 3

Step 4, 100% effaced

Illustration by Kathryn Born, MA

For a first-time mom, labor can take 12 to 24 hours; fortunately, the process is often shorter every time after. Though half of that time is spent in latent labor, which doesn't sound so bad, the remainder is a misery that leads many women to opt for an *epidural*, an injection in the spine that blocks the pain signals from making it to the brain. Active labor begins, and the contractions become more painful, squeezing inward from all directions. The contractions continually become stronger, last longer (45 to 60 seconds), and occur more frequently (every 3 to 5 minutes). Active labor generally lasts around 3 to 5 hours, until the cervix reaches about 8 cm (3 in.).

During *transition,* the contractions shift from squeezing in to pushing down. This phase lasts only 30 minutes to 2 hours but is the most intense. The powerful contractions last 60 to 90 seconds, with only 30 seconds to 2 minutes between. Many women who don't opt for the epidural state that transition is the most painful part of the entire parturition process. On completion of this phase, the cervix is dilated to 10 cm (4 in.), and the second stage of parturition begins.

Special delivery

As the contractions continue at a now-regular pace, the mother has a natural urge to push. As she pushes with each contraction, the baby makes specific motions to ease its passage through the birth canal. (See Figure 17-4.) It flexes its head out and then turns to aid passage of the shoulders. This stage lasts about 30 minutes to 2 hours. Just after delivery, the infant's umbilical cord is cut and tied off. Now the infant is totally separated from the mother and will soon have a stylish belly button.

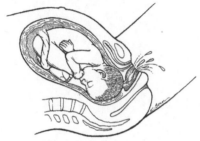

a. Dilation of the cervix and breaking of amniotic sac

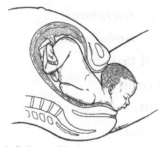

b. Delivery of the head

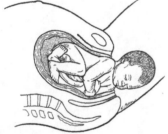

c. Delivery of the body

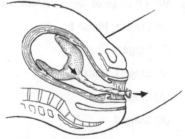

d. Delivery of the placenta

FIGURE 17-4: An overview of delivery.

Illustration by Kathryn Born, MA

Following a brief reprieve, the final stage of parturition begins. About 15 minutes after the baby has been delivered, uterine contractions resume to deliver the placenta. The contractions serve to separate the placenta from the wall of the uterus, squeezing the blood vessels shut in the process. This stage usually takes about 30 minutes.

Over the next week or so, uterine contractions continue, though they are much milder and sporadic. These contractions help the uterus return to its pre-pregnancy size, a process that takes up to six weeks.

The hours following parturition are critical for the bond between mother and child. *Prolactin* begins triggering milk production in the mammary glands, and the nursing infant stimulates the release of oxytocin to release the milk, called *let-down*. Oxytocin is known as the "love hormone"; it's the neurotransmitter that provides the love sensation we feel for family members.

REMEMBER

Are you feeling the love? Find out by answering some questions.

PRACTICE

16 Explain how childbirth is an example of a positive feedback loop.

_____

_____

_____

17 Which hormones play an active role in the progression of labor?

    I. Progesterone

    II. Prostaglandins

    III. Oxytocin

(a) I only

(b) III only

(c) I and II

(d) II and III

(e) I, II, and III

18 Which stage of labor usually lasts longest?

(a) Stage 1, Phase 1

(b) Stage 1, Phase 2

(c) Stage 1, Phase 3

(d) Stage 2

(e) Stage 3

Practice Questions Answers and Explanations

1. **(h) Trophoblast**

2. **(f) Morula**

3. **(c) Blastomeres**

4. **(d) Epiblast.** The hypoblast is the layer on the other side of the inner cell mass.

5. **(b) Blastocyst**

6. **False.** Skin is the outside of your body, so it forms from the outer germ layer: ectoderm.

7. **True**

8. **False.** The inner cells do this work; the outer cells form the amniotic sac.

9. **True**

10. **False.** Most of meiosis was completed before ovulation.

11. **False.** *Morula* is the term for the ball of cells that exists before the inner cavity is created; now it's a blastocyst. If you remember that a cyst is a fluid-filled sac, remembering this term is easier.

12. **(e) The creation of the three germ layers.** This process happens when cells become mobile, moving between the two layers of the embryonic disc and creating the primitive streak. The result is the formation of the mesoderm: the third germ layer.

13. **(b) All organ systems must be functional.** This stage of development generally occurs by the third trimester, but even then, premature birth can have serious health consequences for the newborn.

14. **Third month:** Bones begin to ossify; body growth accelerates and head growth slows; lungs begin to practice breathing amniotic fluid; external genitalia is visible on the male; the fetus is 4 inches (10 cm) long and weighs about 1 ounce (28 g).

    Fourth month: Body grows rapidly; legs lengthen; joints begin to form; face looks more human; the fetus is roughly 7 inches (18 cm) long and weighs about 4 ounces (113 g).

    Fifth month: The mother can feel fetal movement; hair grows on the head; lanugo covers the skin; the fetus is about 10 inches (25 cm) long and weighs ½ to 1 pound (227-454 g).

    Sixth month: Eyebrows and eyelashes form; weight gain accelerates; the fetus is roughly 11 to 14 inches (28-35 cm) long and weighs just under 1½ pounds (680 g).

    Seventh month: Subcutaneous fat begins to form; eyelids open; fetus usually turns to upside-down position; the fetus is 13 to 17 inches (33-43 cm) long and weighs 2½ to 3 (1-1⅓ kg) pounds.

Eighth month: Subcutaneous fat increases; fetus appears more babylike; testes of male descend into scrotum; the fetus is 16 to 18 inches (41–46 cm) long and weighs roughly 5 pounds (2¼ kg).

Ninth month: Substantial plumping occurs due to subcutaneous fat; lanugo is shed; fingernails extend to fingertips; the average newborn is 20 inches (51 cm) long and weighs 7½ pounds (3½ kg).

15 **(c) Near the end of the first trimester**

16 In negative feedback, the body's response to a stimulus is to push in the opposite direction. For example, the body sweats in response to a temperature increase to bring the body temperature down. **In positive feedback, the body's response is to push further in the same direction. There are numerous players in this loop, but the bottom line is: contractions create contractions, which in turn create more contractions.**

17 **(d) II and III (prostaglandins and oxytocin).** Progesterone blocks uterine contractions. Its decline at the end of pregnancy allows labor, which is stimulated by oxytocin.

18 **(a) Stage 1, Phase 1.** Although it's certainly not the most painful, early labor can last more than a day.

If you're ready to test your skills a bit more, take the following chapter quiz, which incorporates all the chapter topics.

Whaddya Know? Chapter 17 Quiz

Are you ready to deliver on your knowledge of the creation of new life?

1 Which statement is the best description of how the placenta supports the fetus?

(a) It's a sac of nutrients from which the fetus can draw energy, much like the yolk sac.

(b) It's a common tissue in which maternal and fetal capillaries can exchange materials.

(c) It provides a centralized location for oxygenated maternal blood to enter the fetal blood flow, and vice versa.

(d) It surrounds the baby in a protective fluid from which it can draw nutrients.

2 What is the term for the successive mitotic divisions of the embryonic cells into smaller cells?

(a) Hatching

(b) Gastrulation

(c) Cleavage

(d) Implantation

(e) Transition

3 Which cells of the embryonic disc create the amniotic cavity?

(a) Blastocoels

(b) Blastomeres

(c) Epiblast

(d) Hypoblast

(e) Trophoblast

For questions 4–8, identify the month of fetal development in which each event occurs.

4 _____ Legs are proportional.

5 _____ External genitalia are visible (in males).

6 _____ Joints form.

7 _____ Eyes open.

8 _____ Cartilage is replaced by osseous tissue.

9 Which structure(s) are derived from the endoderm? Choose all that apply.

(a) Stomach

(b) Liver

(c) Lungs

(d) Heart

(e) Hair and nails

(f) Duodenum

(g) Pharynx

(h) Bones

10 What role does progesterone play during pregnancy?

(a) Its release stimulates parturition.

(b) It attracts sperm to the oocyte for fertilization.

(c) It prevents contractions to maintain the pregnancy.

(d) Its decline allows for implantation.

11 If only one sperm managed to reach the oocyte, fertilization wouldn't occur. Why not?

_____

_____

12 When is the sex of the fetus determined?

(a) During fertilization

(b) During gastrulation

(c) During implantation

(d) During embryonic development

(e) During fetal development

13 During labor, the cervix must dilate and efface. What does this mean?

_____

_____

For questions 14–18, answer true or false. When the statement is false, identify the error.

14 The placenta exchanges gases and waste between maternal and fetal blood.

_____

15 Fertilization occurs before an ovum is created.

_____

16 Sexual intercourse five days before ovulation cannot lead to pregnancy.

_____

17 The zona pellucida secretes chemicals that attract sperm to the oocyte.

_____

18 The fetus responds to contractions during delivery and shifts position to help the process along.

_____

Answers to Chapter 17 Quiz

1. **(b) It's a common tissue in which maternal and fetal capillaries can exchange materials.** It's a common misconception that mother and fetus share circulation — that is, the mother's blood enters the fetus's body, and vice versa.

2. **(b) Gastrulation**

3. **(c) Epiblast**

4. **Fifth**

5. **Third**

6. **Fourth**

7. **Seventh**

8. **Third**

9. **(a) Stomach, (c) Lungs, (f) Duodenum, and (g) Pharynx.** The endoderm creates the internal tubes: the respiratory and digestive tracts. The liver is a digestive organ but not part of the digestive tract.

10. **(c) It prevents contractions to maintain the pregnancy.** Its decline allows parturition to occur.

11. **The sperm must use the enzymes in its acrosome to breach the corona radiata. By the time it reached the zona pellucida, there wouldn't be any left to burrow in. The sperm that arrive first clear paths through the outer protective layer so that other sperm can get through and then use their enzymes to get through the zona and bind to the membrane of the oocyte.**

12. **(a) During fertilization.** Though no one else will know before the end of the first trimester, sex is determined at fertilization by the sex chromosome the sperm is carrying.

13. *Dilation* **refers to an increase in the size of the opening of the cervix. Normally, the opening is effectively closed (no more than a slit), but to allow the baby to pass through, it must dilate to 10 cm (4 in.) across. If that process were all that occurred, the cervix would still take up space.** *Effacement* **is the thinning of the cervix wall, creating the maximum amount of space for the baby to pass through.**

14. **True**

15. **True.** The secondary oocyte pauses meiosis during metaphase II. It doesn't actually finish until a sperm binds to the cell membrane, which is the beginning of fertilization.

16. **False.** Pregnancy at that time certainly can happen — and does. As long as viable sperm are present when the egg arrives, conception can occur.

17. **False.** The cells of the corona radiata secrete the attractant chemicals.

18. **True**

Chapter **18**

From Cradle to Grave

I n the context of anatomy and physiology, *development* means the pattern of change through an organism's existence. Development is closely related to the specialized branch of biology called *ontogeny*, which studies an organism's history within its lifetime. Human development has been a subject of much study for thousands of years, especially for parents and grandparents of young human children. It's a good thing for babies that we find them so fascinating because they take a great deal of effort to maintain and a long time to mature.

In this chapter, we take a look at human development, from the creation of the zygote through old age. Although you've already experienced some of this development, you get a glimpse of some of the changes your body will go through as you age.

Programming Development

One way to think of development is as "the unfolding in real time and space of a program for generating a unique biological organism." The program is launched when a new *zygote* comes into existence (and, just to be clear, all zygotes are created the same way). It follows the program encoded in its DNA to develop, creating its various human characteristics but with its new and unique varieties.

The totality of the DNA of a zygote — that is, its *genome* — comes into existence at the time of fertilization. DNA is the master code, divided into sections called *genes* that give instructions for specific characteristics. There are genes for all our variable traits, such as hair and eye color, as well as those that aren't quite so flexible, such as having a diaphragm or a four-chambered heart. With a full set of directions from Mom and a full set from Dad, each new offspring inherits every trait they have from their parents (unless mutation creates a new one). But the combination of varieties is new and unique, never to be repeated.

TECHNICAL STUFF

The exception here seems to be identical twins; but their identical combination of possibilities isn't repeated; it's created at the same time. Identical twins occur when the zygote splits in two sometime during cleavage. That is, it began as a single zygote, started doing mitosis as usual, and then split into two cells before implantation. That's why these twins are referred to as *monozygotic twins. Dizygotic* (fraternal) *twins* result when two eggs are released and fertilized by different sperm.

Just as development is encoded into the program, aging and death are built into the genome as well. It seems to be counterproductive, but just as there are portions of the DNA for our development, there are parts that are responsible for our predictable pattern of decline. Everyone dies sooner or later. A few people survive until their program has fully unfolded and reached its end; they die of old age, so to speak.

Stages of development

The process of development begins in the zygote and continues until death. There's no universally agreed-on definition of the development stages (although birth and puberty are universally acknowledged), and the age range at which a person passes from one stage to the next is wide. Change is more or less continuous through life, and different organ systems undergo significant changes on their own development timetable. Conventionally in human biology, however, the development milestones that mark the stages are based on developments in the nervous and reproductive systems.

Dimensions of development

The structural and physiological changes that happen during human development include an increase in size, the acquisition of some specialized abilities, and the loss of some other specialized abilities continuously throughout life.

TIP

In this chapter, we assume that we're dealing with an organism for whom all is going well, biologically speaking. In other words, the person has no fatal errors in the genome itself and adequate resources to sustain nutrition, homeostasis, and all the rest of the life-maintaining physiological reactions.

Growth

Part of human development involves an increase in size. Increased size is accomplished primarily through the growth of organs that exist in some form in the embryo. The heart grows larger, the brain grows larger, and the bones get longer and heavier. The organs grow by building more of their own tissues, and tissues get bigger by adding cells or increasing cell size. Everything grows in tandem, mostly by adding cells.

Not everything grows equally, however. Different stages of development are characterized by different proportions of tissue types. Both the brain and skeletal muscle increase in size from infancy to adulthood, for example, but the proportion of muscle tissue to brain tissue is much higher in adulthood.

TECHNICAL STUFF

When a three-dimensional object such as a living body increases in size, the *surface-to-volume ratio* decreases. (To put it another way, the *volume-to-surface ratio* increases; more of your inside parts are dependent on fewer of your outside parts to interact directly with your environment.) The size of a human body strongly influences thermoregulation, fluid balance, and other key aspects of homeostasis, which is likely why the average life span decreases as height increases.

Differentiation

New physiological abilities usually come about because of cell *differentiation* — the process that cells undergo to take on their specialized functions. A newborn has some version of (more or less) all the cell and tissue types, but having them doesn't mean that they're functional. Cells must be fully differentiated into their corresponding tissues to gain function. An example is the development of a long bone via cell growth and differentiation, which I cover in Chapter 6.

REMEMBER

Lots of human-body functions aren't learned; they're developed. The ability to digest starch, for example, is acquired during the first year of life, when the body starts producing the necessary enzymes — not when someone teaches a baby how. Toilet training is more about the maturity of the nervous and muscular systems than the diligence of the parents.

Gaining a new skill, structure, or process is sometimes accompanied by the loss of existing abilities. A young adult is better at planning than a teenager but has most likely lost some stamina for all-nighters, parties, and road trips. The stages of human development can be characterized by abilities that are gained and lost.

Decline

According to recent studies, age-related decline in specialized, and even basic physiological functions are built into new genomes at the start. Structures at the ends of the chromosomes called *telomeres* get shorter as a genome ages. As a result, the telomeres effectively control the number of times that the genome can replicate. As a result, cells gradually lose the ability to divide. The aged and dying cells in a tissue eventually outnumber the new cells being made to replace them, and the tissue loses its ability to function, which impairs the organism's survival.

The aging processes are an active area of research in anatomy and physiology. In recent decades, therapies and devices to counter aging's effects have dominated the medical-products marketplace worldwide.

Human Life Span

The development that happens before birth gets covered in Chapter 17, so here, I start with the first moment a new human exists outside its mother's protective womb.

TECHNICAL STUFF

The timing of birth is a compromise between the anatomical needs of a large brain and those of a bipedal gait. The fetus is born a little earlier than is probably ideal from the point of view of its development, but the pelvis of the adult female has grown narrower and less flexible to support a redistribution of weight and bipedal mobility. Evolution has favored a compromise: a period of a few weeks when the baby, though fully separate from the mother, is still developing in ways that other mammals have completed before birth. Evolution continues to support this compromise relentlessly.

Changes at birth

From birth to 4 weeks of age, the newborn is called a *neonate.* Faced with survival after its physical separation from the mother, a neonate must abruptly begin to adjust to its cold, bright, dry, and loud environment. The immediate adjustments include

>> **The first breath:** The fetus exchanges oxygen and carbon dioxide through the placenta. At birth, the newborn's lungs are not inflated and contain some amniotic fluid. Within about 10 seconds after delivery, the newborn's central nervous system reacts to the sudden change in temperature and environment by stimulating the first breath, which usually precedes the crying. The lungs inflate and begin working on their own as the fluid is absorbed (by the lung tissue) or coughed out.

>> **Thermoregulation:** Newborns' muscles aren't developed enough to play a large role in thermoregulation. Fortunately, they spent time in the womb building up *brown fat* (called that because unlike regular fat tissue, the cells of brown fat have numerous mitochondria, which are brown because they contain iron). This specialized adipose tissue not only provides insulation to keep them warm, but also uses glucose and lipids to generate heat. Because newborns also have a high surface-to-volume ratio, they lose heat quickly; as a result, they continue to build up their fat stores until childhood.

>> **Digestive system:** The newborn's digestive system starts to work in a limited way immediately after birth. Although it isn't yet equipped to process food, it can digest breast milk, absorb its nutrients, and pass the remnants. Even so, the digestive system can take several weeks to settle down to efficient functioning.

>> **Urinary system:** A newborn usually urinates within the first 24 hours. The capabilities of the kidneys increase sharply through the first two weeks after birth. The kidneys gradually become able to maintain the body's fluid and electrolyte balance.

>> **Immunity:** The immune system begins to develop in the fetus and continues to mature through the child's first few years of life. The development of immunity is an active area of research; there's still much we don't understand. Mothers pass their antibodies on through fetal circulation and when nursing. Breast milk also contains components that have been shown to promote the development of an infant's own immune system.

Infancy and childhood

From 4 weeks to 2 years of age, the baby is called an *infant.* Growth during this period is explosive under the stimulation of *growth hormone* from the pituitary gland. All organ systems grow and develop in infancy almost as rapidly as they do during fetal development— well, all except the reproductive system, which is effectively put on hold until puberty. The physical development milestones in the first year alone would take a full book this size to describe. Just looking at a year-old infant and then at the infant's photos at birth paints a telling picture.

Most infants double or triple their birth weight in the first year. Besides adding size to all tissues and organs, the new cells differentiate elaborately and add functionality in accordance with the individual's genomic scheme. The baby's skeleton changes size, proportion, and composition, and the muscles strengthen to move and support it. Teeth begin to erupt through the gums, and the mouth acquires the subtle muscle control to form words and kisses. The baby starts applying the characteristically human opposable thumb to everyday tasks (such as picking up toys and throwing them down again) in the second half of the first year. The brain grows in size, and just as important, the number and complexity of connections grow astronomically.

From 2 years to puberty, you're looking at a *child*. Influenced by growth hormones, growth continues its rapid pace. Muscle coordination, language skills, and intellectual skills also develop rapidly. A toddler (between ages 1 and 3) develops sphincter control and begins learning to control urination and defecation. Social life begins. Vigorous play coordinates the development of the musculoskeletal and nervous systems. The mechanisms of homeostasis gradually strengthen.

Overall, from infancy to adolescence, the human child becomes bigger, stronger, and smarter every day. The child steadily gains control of the body on the conscious and physiological levels. The child is typically fluent in at least one spoken language by age 6. They start caring for the world around them and the people in it, as well as developing a sense of autonomy.

The physical and mental development of children is devoted to mastering the unique aspect of human life called culture, which takes these brilliant human children about 20 years of intense study. During the period of human evolution, the survival of any individual depended primarily on the survival of the individual's kin group. Participating effectively in the culture has always been the best way for humans to increase the likelihood of their own survival and the survival of those who carry their genes.

Adolescence

From *puberty*, which starts between the ages of 11 and 14, to adulthood, the child is called an *adolescent*. Although the main purpose of this phase is approaching reproductive maturity, every body system responds to the increase in *gonadotropic hormones*. Though the trigger for it is unclear, the hypothalamus releases large amounts of *gonadotropin-releasing hormone (GnRH)* during puberty. GnRH in turn stimulates the anterior pituitary to release *luteinizing hormone (LH)* and *follicle stimulating hormone (FSH)*, the two hormones responsible for initiating meiosis and prompting the gonads (testes and ovaries) to produce the sex hormones (testosterone and estrogen, respectively). Flip back to Chapter 16 for a refresher on the gonadotropic hormones.

Female puberty

In females, the growth and development of the reproductive organs are mostly unnoticeable, though the vulva and pubic hair begin to grow. As the ovaries become active, the ovarian and uterine cycles begin (see Chapter 16), leading to menstruation. It can take several years for the cycles to settle into a consistent pattern, but after a female starts ovulating, pregnancy is possible.

The breasts develop as fat is redistributed to provide the needed resources for lactation in the mammary glands. The hips widen as the pelvis changes shape to accommodate the baby's head during *parturition* (birth).

Male puberty

In males, the testes and penis grow, and sperm production begins. Though the process isn't cyclical, there's still an adjustment period during which males experience random erections and nocturnal emissions (semen released during sleep) as the reproductive structures learn to work together and respond to both hormone shifts and external stimuli.

Because the hormone that drives male puberty is a steroid (testosterone, to be specific; see Chapter 9), its effects are far-reaching. Hair grows on the face, chest, armpits, and groin, and becomes thicker and coarser on the arms and legs. The shoulders and chest broaden. The larynx (and other cartilaginous structures) enlarge, deepening the voice.

Ending adolescence

The release of sex hormones in adolescence coincides with a massive release of growth hormone, thyroid hormone, and steroid hormones from the adrenal cortex. This onslaught of hormones causes the last phase of major developmental changes within a relatively short period.

Most noticeably during adolescence, growth occurs in spurts, and muscle mass increases to support it. Girls achieve their maximum growth rate between the ages of 10 and 13, whereas boys experience their fastest growth between the ages of 12 and 15. Primary and secondary sex characteristics begin to appear. Growth terminates when the epiphyseal plates of the long bones ossify (see Chapter 6), sometime between the ages of 18 and 21.

Motor skills continue to develop, becoming much more fine-tuned. The growth of body hair increases, as does the production of sweat and sebaceous (oil) glands in the skin (causing the bane of teenage existence: acne). Lung capacity increases, and energy levels can seem to be infinite. By the end of adolescence, the organs are fully mature in size and shape, and so is the body as a whole. There is, of course, still some fine-tuning to come, but physically, development is complete — emphasis on *physically*. As anyone who interacts with adolescents on a daily basis will tell you, I'm leaving out the most volatile part of adolescence: brain development.

Intellectual abilities increase dramatically, but those critical *executive functions* (organization, judgment, and impulse control) do not. The frontal lobe of the brain is beginning its development but is nearly a decade away from completion. As the brain begins more advanced processing, adolescents just don't have the life experiences to draw on to make the most intelligent choices, so this stage is often a time of poor decision-making. Teens' sleep–wake cycles are disrupted, and their surging hormones can lead to mood swings. Parents and adolescents alike need reassurance during this developmental phase, which is in fact just a phase.

Young adulthood

The young-adult stage covers roughly 20 years, from age 20 to about 40. Physical development reaches its peak, and the person assumes adult responsibilities, often including embarking on a career, marriage, and a family. By age 25, brain development is complete, marked by the maturation of the *prefrontal cortex* — the part of the brain responsible for our most advanced functions, such as planning, decision-making, problem-solving, and critical thinking.

Typically, young adulthood is a time of good health and resilience. Many people complete their parents' bid for evolutionary success during this stage. (That is, they become parents themselves.) Energy levels can remain high throughout the 30s, 40s, and 50s for some people. After growth is complete, however, metabolic requirements decrease, and adults must decrease their caloric intake to avoid accumulating body fat that can put long-term stress on several organ systems.

The gradual physical decline of *senescence* actually begins during these years. Influenced by genetic and environmental factors, slightly more bone is lost than is made, and the same thing happens with the structural proteins of the skin. Muscle mass declines slowly. Damage accumulates from repetitive injuries, bad habits, and bad genes.

Pausing for pregnancy

Many females go through an additional, albeit temporary, developmental phase in adulthood. The maternal body responds to pregnancy with many anatomical and physiological changes to accommodate the growth and development of the fetus. Most structures and processes revert to the nonpregnant form (more or less) after the end of the pregnancy.

Uterus

During pregnancy, the uterus grows to about five times its nonpregnant size and weight to accommodate not only the fetus, but also the placenta, the umbilical cord, about a quart (about 950 mL) of amniotic fluid, and the fetal membranes. The uterus usually reaches its peak size at about 38 weeks gestation. During the last few weeks of pregnancy, the uterus has expanded to fill the abdominal cavity all the way up to the ribs. The size of the expanded uterus and the pressure of the full-grown fetus make many things difficult for the mother, particularly getting up and down.

Though the placenta is built from fetal cells, maternal blood vessels must incorporate into it for the exchange of materials with the fetal blood vessels. In addition to providing resources to the growing fetus, the placenta acts as a temporary endocrine gland during pregnancy, producing large amounts of estrogen and progesterone by 10 to 12 weeks. It serves to maintain the growth of the uterus, helps control uterine activity, and is responsible for many of the changes in the maternal body.

Throughout the pregnancy, enlarged and active mucus glands in the cervix produce the *operculum*, a mucus plug that protects the fetus and fetal membranes from infection. In the final month, the cervix begins to soften to prepare for parturition, and the mucus plug may be expelled (though sometimes, it doesn't come out until labor). Finally, dilation and effacement of the cervix occur at the onset of labor.

Ovaries

Hormonal mechanisms prevent follicle development and ovulation in the ovaries.

Breasts

The breasts usually increase in size as pregnancy progresses and may feel inflamed or tender. The areolas of the nipples enlarge and darken. The areola's sebaceous (oil) glands enlarge and tend to protrude. By the 16th week (second trimester), the breasts begin to produce *colostrum*, the precursor of breast milk.

Other organ systems

Pregnancy affects all organ systems as they support the growth and development of the fetus and maintain homeostasis in the female. Here are a few important physiological consequences of pregnancy:

>> The other abdominal organs are displaced to the sides as the uterus grows.

>> Decreased tone and mobility of smooth muscles slows peristalsis and enhances the absorption of nutrients. An increase in water uptake from the large intestines increases the risk of constipation. Relaxation of the cardiac sphincter may increase regurgitation and heartburn. Nausea and other gastric discomforts are common.

>> Increases occur in blood volume, cardiac output, body core temperature, respiration rate, urine volume, and output from sweat glands.

>> Immunity is partially suppressed.

>> Spinal curvature is realigned to counterbalance the growing uterus. Slight relaxation and increased mobility of the pelvic joints prepare the pelvis for the passage of the infant. This change can compromise the woman's lower-body strength starting in the second trimester.

The "Middle Ages"

From age 40 to about 65, physiological aging continues. Gray hair, diminished physical abilities, skin wrinkles, and other outward signs of aging are caused by the decreased activity of cells and the lack of their replacement. Most notably, collagen and elastin fibers break down and aren't replaced because there aren't any cells to rebuild them.

For men, reproductive capacity diminishes; for women, it disappears altogether at *menopause,* usually around age 50. Production of some hormones diminishes, triggering anatomical and physiological changes great and small.

Physiologically, menopause essentially reverses the hormonal pathway of adolescence. When a woman enters menopause, her ability to reproduce ends; ovulation stops, and she can no longer become pregnant. The uterus, vagina, and ovaries all shrink; secretions decrease; and libido declines. She may also experience hot flashes and sweat baths if faulty signals from the parasympathetic nervous system disrupt the body's ability to monitor its temperature accurately. A woman's bones also weaken when the breakdown of bone tissue occurs faster than the buildup of bone tissue during bone remodeling, because the protective effect of estrogen is lost. (This is why women are far more prone to osteoporosis.)

REMEMBER

For about six years before menopause, many women experience a stage called *perimenopause,* during which increasingly irregular hormone secretions can cause fluctuation in menstruation, hot flashes, and mood irritability. Around the same age, males may experience *andropause* due to a drop in testosterone, but this drop is gradual and often doesn't completely cease, as estrogen does in women. Some men grow fatigued, irritable, and depressed; some experience a lack of sex drive, while others maintain their drive as well as reproductive functionality (production of sperm) throughout their lives.

As everything seemingly begins to fall apart, the brain continues to develop, cognitively and in many other ways. Recent brain research shows that an older brain thinks better about some things than a younger brain does, including making financial decisions, exercising social judgment (intuitive judgments about whom to trust), and recognizing categories. The older brain is better at seeing the proverbial forest and following the gist of an argument. These years are typically the peak years of professional and occupational achievement. In addition, across all occupations and ethnicities, a sense of well-being peaks as people reach middle age.

An extraordinary aspect of human development is the length of this period. Human females commonly outlive their fertility by three decades — a third of their life span in a stage of life that almost no other animal experiences! Many researchers see a matrix of evolutionary cause and effect conjoining natural selection and cultural evolution in the very existence of your grandma.

Growing creaky

REMEMBER

From age 65 until death is the period of *senescence*. The developments of senescence (growing old) are gradual and widely varied. Body parts don't work quite as well as they used to, and the situation just keeps getting worse. Life expectancy beyond the reproductive years is tied closely to genetics and to everything about a person's life up until that point (diet, exercise, use of sunscreen, and so on). Signs of senescence include

>> Loss of skin elasticity and accompanying sagging or wrinkling

>> Weakened bones and decreasingly mobile joints

>> Weakened muscles

>> Impaired coordination, memory, and intellectual function

>> Cardiovascular problems

>> Reduced immune responses

>> Decreased respiratory function caused by reduced lung elasticity

>> Decreased peristalsis and muscle tone in the digestive and urinary tracts

In particular, the immune system doesn't work as well as it used to, and malformed cells that would have been eliminated immediately at age 35 may now evade immune surveillance and become cancerous.

Age-related changes in the large arteries, as well as cumulative damage to the smaller vessels, can bring about problems with blood pressure.

Inactivity and chronic overconsumption of calories have their worst effects now in all the major systems.

Still, the brain continues to develop. Research has been shown repeatedly that under the right conditions, the brain continues to produce new cells and make new connections among neurons in adults as old as 100.

I hope you don't feel as though you've aged after reading this chapter! Time to test your skills.

Whaddya Know? Chapter 18 Quiz

Quiz time! Complete each question to test your knowledge of the various topics covered in this chapter. You can find the solutions and explanations in the next section.

1 What is the formal term for a 20-month-old child?

(a) Baby

(b) Infant

(c) Toddler

(d) Neonate

(e) Child

2 What are the similarities and differences between menopause and andropause?

_____

_____

_____

_____

For questions 3–9, match the description with the life stage.

(a) Adolescent   (e) Neonate

(b) Child   (f) Old adult

(c) Infant   (g) Young adult

(d) Middle-aged adult

3 _____ The person assumes adult responsibilities, possibly including marriage and a family.

4 _____ Faced with survival, the person must process food, excrete waste, and obtain oxygen.

5 _____ Primary and secondary sex characteristics begin to appear.

6 _____ The person experiences senescence.

7 _____ Deciduous teeth begin to form.

8 _____ Women go through menopause.

9 _____ From 2 years of age to puberty.

10　Brown fat is different from regular adipose tissue in that it (choose all that apply).

(a) stores carbohydrates instead of lipids.

(b) is present only in neonates.

(c) generates heat in addition to insulating.

(d) is found deeper in the tissues.

For questions 11–15, identify the statement as true or false. When the statement is false, identify the error.

11　Infants shiver and sweat to perform thermoregulation.

_____

12　Girls tend to undergo puberty before boys do.

_____

13　The brain finishes development in young adulthood.

_____

14　Teenagers are prone to making poor decisions because their intellectual capabilities outpace their executive functions.

_____

15　Which of the following statements best describes senescence?

(a) The time in life when development is the loss of functions

(b) The loss of cognitive abilities that comes with age

(c) The period when development has ceased

(d) The gradual decline of the body's physical structures

Answers to Chapter 18 Quiz

1. **(b) Infant.** The terms *baby* and *toddler* aren't part of the formal medical lexicon, and a neonate is less than 4 weeks old.

2. **Both are the result of declining sex hormones: estrogen for women and testosterone for men. Those in these phases of life are often tired and more irritable and lack libido. Testosterone production declines slowly over time, though. Estrogen drops off (and sometimes returns, drops off again, and so on), which is why menopause has a much more dramatic effect. The decrease in sperm production makes reproduction less likely but certainly not impossible. Women, however, lose their ability to reproduce in menopause.**

3. **(g) Young adult**

4. **(e) Neonate**

5. **(a) Adolescent**

6. **(f) Old adult**

7. **(c) Infant**

8. **(d) Middle-aged adult**

9. **(b) Child**

10. **(c) generates heat in addition to insulating.** None of the other statements is true. It's fat, so by definition, it stores lipids. Neonates have a great deal of brown fat, but that fat is deposited continually throughout infancy. The final choice doesn't even make sense.

11. **False.** The muscles and sweat glands aren't developed enough yet for those functions. They rely on their brown fat and parental care for thermoregulation.

12. **True.**

13. **True.**

14. **True.** The frontal lobe is the part of the brain responsible for advanced functions. It's also the last part that gets our attention and resources in terms of building capabilities.

15. **(a) The time in life when development is the loss of functions.** Our programmed decline is still considered development.

Index

A

A cells (alpha cells), 244
A-band, 160
abdominal aorta, 376
abdominal cavity, 14
abdominal muscles, 172
abdominopelvic cavity, 13–14
abduction, 137, 175
absorption, 348
accessory organs, 348
accessory structures. *See also* integumentary system
  ceruminous glands, 99
  defined, 96
  hair, 96–97
  mammary glands, 99
  nails, 97
  sebaceous glands, 98
  sweat glands, 98
accommodation, 214
ACE (angiotensin converting enzyme), 385
acetabulum, 131
acetylcholine, 163, 186
Achilles tendon, 175
acidity, 29
acidosis, 384–385
acids, 29
acrosome, 401, 405, 428
ACTH (adrenocorticotropic hormone), 237
actin, 157, 159–161
action potential, 200–201
active labor, 436
active transport, 54
adaptive immunity, 302, 306–309
adduction, 137
adductor longus, 175
adductor magnus, 175
adductors, 175
adenine, 32–33
adenoid tonsil, 299
adenoids, 326
adenosine diphosphate (ADP), 35–36, 39, 154

adenosine triphosphate (ATP), 35–36, 39, 177
ADH (antidiuretic hormone), 237, 240–241, 385
adhesive junctions, 71
adipocytes, 75, 94, 107, 150
adipose tissue, 75, 86
adolescence, 449–450
ADP (adenosine diphosphate), 35–36, 39, 154
adrenal cortex, 241–242
adrenal glands, 234, 237, 241–242. *See also* endocrine
  system
adrenal medulla, 241–242, 255
adrenaline, 242
adrenocorticotropic hormone (ACTH), 237
adsorption, 72
aerobic process, 36
afferent arterioles, 376
afferent lymphatic vessels, 297
afferent nerves, 193, 196
agglutination, 308, 319
air sacs, 329
albumin, 260
alcoholic fermentation, 36
aldosterone, 237, 242, 385, 394
alimentary canal, 347. *See also* digestive system
alkalinity, 29
alkalis, 38
alleles, 399
alpha cells (A cells), 244
alveolar duct, 329
alveolar sacs, 329, 346
alveoli, 329, 346, 349, 411
amino acids
  modified, 235
  overview, 31–32
amniotic fluid, 432
amniotic sac, 432
amphiarthrotic joints, 134
amylase, 351, 357
anabolic reactions, 33–34
anaerobic respiration, 36, 168
anal sphincter, 362
anaphase, 57, 417

B

C

closed double loop, 270

clotting cascade, 262

clotting factors, 262

coccygeal plexus, 197

coccyx, 125

cochlea, 214

cold receptors, 213

cold thermoreceptors, 100

collagen, 31

collagenous fibers, 74

collecting duct, 376

collectins, 305

colon, 361

colostrum, 451

columnar cells, 73

common bile duct, 357

common iliac arteries, 282

common iliac veins, 284

compact (cortical) bone, 113–114

complement cascade, 308

complementary pairs, 32

compounds, 26, 38

concentration, 334

concentration gradient, 52

concentric contraction, 167

conducting fibers, 277

conductivity, 193

condyloid joint, 135

cones, 214

conjunctiva, 213

connective tissues
  defined, 71
  overview, 15
  proper, 74–76
  specialized, 76

control center, 48

coordination, 192

cornea, 213

corona radiata, 400, 413, 428

coronary artery, 271

coronary sinus, 273

corpora cavernosa, 407

corpus albicans, 413

corpus callosum, 207

corpus luteum, 413, 424

corpus spongiosum, 407

cortisol, 237, 242, 249

costal cartilages, 126

countercurrent exchange, 378

covalent bonds, 26, 38

Cowper's glands, 407

coxal bones, 129

cramps, 168

cranial cavity, 14

cranial nerves, 210

cranium, 110, 123–124

creatinine, 387

cribriform plate, 216

cricoid cartilage, 326

cross bridge, 161–162, 187

cross section, 10

crossing-over, 399

crown, 350

crypts, 299

CSF (cerebrospinal fluid), 204, 208, 221

cuboidal cells, 73

culture, 449

cumulus oophorus, 413

cusps, 272

cuticle, 97

cystic duct, 357

cytokines, 308

cytokinesis, 56, 58, 62

cytoplasm, 48, 61

cytosine, 32–33

cytoskeleton, 48

cytosol, 48, 61

cytotoxic T cells, 308

D

D cells (delta cells), 244

Dalton's Law, 334

DCT (distal convoluted tubule), 376

dead space, 334

deep position, 10

defecation, 348

defecation reflex, 362

defensins, 305

deglutition, 348, 351

degranulation, 303, 304

delivery, 436, 437–438

delta cells (D cells), 244

deltoid muscle, 173

F

facial bones, 124
facilitated diffusion, 53, 69
fallopian tubes, 410
false ribs, 126
false vocal cords, 327
fascicle, 165, 196
fast-twitch fibers, 168, 179
feces, 361
female reproductive system
  breasts and, 411–412
  menstrual cycle, 411
  ovaries, 410
  overview, 409
  uterine tubes, 410
  uterus, 410
  vagina, 410–411
  vulva, 411
female urethra, 383
feminization, 242
femur, 112, 131
fenestrations, 262, 267, 377
fertilization, 428
fetal development, 432–433, 439–440
fetal stage, 432
fetus, 430
fever, 305
fibers, 155, 160, 166
fibrin, 260, 262
fibrinogen, 260
fibrocartilage, 76, 113, 150
fibrous joints, 134–135
fibula, 131
fibularis brevis, 175
fibularis longus, 175
filtration, 337
fimbriae, 410
first polar body, 401
flagellum, 48, 49
flatulence, 361
flatus, 361
flexion, 137
flexor carpi radialis, 174
flexor carpi ulnaris, 174
flexor digitorum longus, 175
flexor digitorum muscles, 173
floating ribs, 126
follicle-stimulating hormone (FSH), 412, 449

foramen, 112
foramen magnum, 123, 205
foreskin, 407
formal students, 2
formed elements, 259
fovea centralis, 214
fraternal twins, 413, 446
free nerve endings, 213
frontal (coronal) plane, 10–11
frontal bone, 123
frontal lobe, 206–207
fructose, 31, 38, 358
FSH (follicle-stimulating hormone), 412, 449
fundus, 354
funiculi, 204

G

G_0 phase, 56
G_1 subphase, 57
G_2 subphase, 57
galactose, 358
gallbladder, 357
gametes
  defined, 397
  meiosis, 398–400
  ova, 400–401
  producing, 245, 398–402
  spermatozoa, 401–402
gametocytes, 398
ganglia, 196–197
gap, 57
gap junctions, 71
gas exchange, 334–336
gastric juice, 355
gastrin, 355
gastrocnemius muscle, 175
gastrulation, 430
gene expression, 58
genome, 48, 445
germ cells, 398
germ layers, 430, 439
germinal center, 97
GFR (glomerular filtration rate), 384
GH (growth hormone), 237
gingiva, 349
glands, 234
glans penis, 407

monomers, 30

monosaccharides, 30–31, 38, 358

monozygotic twins, 446

mons pubis, 411

Moore, John T., 25

morula, 428, 439

motility, 406

motor end plate, 163

motor nerve fibers, 210

motor neuron, 163–164

motor neurons, 194

mouth, 349–350

mucin, 355

mucosa, 349

mucous cells, 355

mucus, 325

multinucleated cells, 48, 77, 156

multipolar neurons, 194

multipotent stem cells, 58

muscle contraction, 47, 167

muscle fatigue, 37, 167

muscle fibers, 155, 160, 168

muscle spindles, 156, 168

muscle tissues
    defined, 71
    overview, 15
    types of, 77, 155–156

muscle tone, 167–168

muscular system
    answers to chapter quiz, 186–187
    chapter quiz, 181–185
    functions of, 154
    names of muscles by region, 169–175
    organizational structure of muscle, 165–166
    overview, 153–154
    practice questions, 157–158, 162–163, 164–165, 166, 168–169
    practice questions answers and explanations, 177–180
    relaxation, 164
    roots and word fragments, 8
    sliding-filament theory, 158–162
    stimulation, 163–164
    structure, 155–156

myelin sheath, 196

myelinated neurons, 201

myocardium, 271, 272, 277, 287

myocytes, 77

myofibrils, 157, 159–160

myosin, 157, 159–161

N

nail bed, 97

nail root, 97

nails, 97

naïve B cells, 308

nasal bones, 124

nasal cavities, 325, 326

nasal conchae, 325

nasal mucosa, 325

nasolacrimal ducts, 214

nasopharynx, 214, 326, 349, 351

natural killer cells, 305

neck (tooth), 350

negative feedback, 46, 69

neonates, 448

nephrons, 376, 378

nerve fibers, 195, 210

nerve impulse, 193

nerve tracts, 204

nerves, 100, 193, 195–196

nervous system
    action potential, 200
    answers to chapter quiz, 230–232
    central nervous system, 203–207
    chapter quiz, 223–229
    functions of, 192
    ganglia, 196–197
    glial cells, 195
    impulse transmission, 198–202
    nerves, 195–196
    neurons, 193–195
    overview, 191–192
    parasympathetic, 211
    peripheral nervous system, 210–216
    plexus, 196–197
    practice questions, 197–198, 202–203, 208–209, 215–218
    practice questions answers and explanations, 219–222
    roots and word fragments, 8
    sensory receptors, 212–216
    sympathetic, 211
    synapse, 201–202

nervous tissues
    defined, 71
    overview, 15
    types of, 77–78

neural accommodation, 193

neuroglia, 192, 195

neurolemma, 196

neurons, 77, 100, 192, 193–195, 236

neurotransmitters, 32, 194, 198, 200, 201–202

neutrons, 26

neutrophils, 303

nipples, 411

Nissl bodies, 194

nociceptor, 230

nodes of Ranvier, 201

nonidentical twins, 413

nonsteroid hormones, 235, 248

noradrenaline, 242

norepinephrine, 237, 242

nose, 325–326

nuclear envelope, 48

nuclease, 357

nuclei, 205

nucleic acids, 32–33

nucleolus, 48

nucleotides, 32–33

nucleus, 26, 48

O

oblique muscle, 349

oblique section, 10

occipital bone, 123

occipital lobe, 206–207

olecranon, 130

olfactory bulbs, 215

olfactory receptor cells, 216

oligodendrocyte, 195

oligodendrocytes, 201

ontogeny, 445

oocyte, 428

oogenesis, 400, 410

oogonia, 401

operculum, 451

optic chiasma, 206

optic disc, 214

oral cavity, 349

orbitals, 38

organ systems, 15

organelles, 48–49

organic chemistry, 25

organisms, 14

organogenesis, 427

organs, 15

orgasm, 407

oropharynx, 326, 349, 351

osmolarity, 53

osmosis, 53–54

osmotic pressure, 53, 260, 384

ossein, 111, 140

osseus (bone) tissue
overview, 76
structure of, 116–117
types of, 113–114

ossicles, 214, 222

ossification, 119–120

osteoblasts, 76, 116, 119–120, 139–140

osteoclasts, 116

osteocytes, 116, 117, 119–120

osteons, 76, 116

osteoprogenitors, 116

outer ear, 214

ova, 398, 400–401

ovarian cycle, 412–414

ovarian ligament, 410

ovaries, 237, 245, 410, 413, 451

ovulation, 401, 410, 413

oxytocin, 237, 240–241, 411, 438

P

P wave, 280

Pacinian (lamellated) corpuscles, 101, 107, 213

pain receptors, 213

palatine bones, 124, 349

palatine tonsils, 349

palpebrae (eyelids), 213

pancreas, 237, 244, 357. *See also* digestive system; endocrine system

pancreatic juice, 357

papilla, 96

papillae, 351

papillary muscles, 273

papillary region, 94

parafollicular cells (C cells), 243

paranasal sinuses, 124

parasympathetic nervous system, 211–212

parathyroid, 243–244

parathyroid hormone, 237

reticular formation, 206
reticular region, 94
reticular tissue, 75
retina, 214
rib cage, 126
ribonucleic acid (RNA), 32–33
ribosomes, 48, 56, 69
ribs, 126
right bundle branches, 277
right common carotid artery, 274
right lobe, 356
right lymphatic duct, 296
right subclavian artery, 274
right subclavian vein, 296
RNA (ribonucleic acid), 32–33
rods, 214
root, 350
root canal, 350
roots, 211
rotation, 138
Ruffini's corpuscles, 101, 107
rugae, 354
RV (residual volume), 332

S

S subphase, 57
sacral plexus, 197
sacroiliac joint, 130
sacrum, 125
saddle joint, 135
sagittal plane, 10–11
saliva, 351, 365
salt, 27
saltatory conduction, 201
sarcolemma, 157
sarcomeres, 160
sarcoplasmic reticulum (SR), 157, 164
sartorius muscle, 175
satellite cells, 195
scapula, 112, 129
Schwann cell, 195
Schwann cells, 201
sclera, 213
scrotum, 403, 417
sebaceous glands, 90, 98, 107
sebum, 90, 98, 107
secondary bronchi, 328

secondary follicle, 413
secondary oocyte, 401, 444
secondary oocytes, 399
secondary response, 310
secretin, 355
secretory phase, 414
secretory vesicles, 49
sella turcica, 124, 239
semen, 406–407
semicircular canals, 215
semilunar (SL) valves, 272
semimembranosus, 175
seminal vesicles, 406
seminiferous tubules, 405, 424
semipermeability, 51
semitendinosus, 175
senescence, 453
sensation, 192
sensory nerve fibers, 210
sensory neuron, 194
sensory receptors, 192, 212–213
sensory system, 212–216
septum, 270, 325
serosa, 348
serous membranes, 14
serratus anterior, 169, 173
Sertoli cells, 405
set point, 45
sex cells. See gametes
sex cells (gametes), 397
shaft, 96
sigmoid colon, 361
simple columnar epithelium, 73
simple cuboidal epithelium, 73
simple squamous epithelium, 73
sinoatrial (SA) node, 277–278
sinuses, 124, 325
sister chromatids, 57, 399
skeletal muscle tissue, 77, 156
skeletal system
    answers to chapter quiz, 149–151
    appendicular skeleton, 129–131
    articulations, 134–135
    axial skeleton, 122–126
    chapter quiz, 143–148
    functions of, 110
    general discussion, 109–110
    movements, 137–138

V

vaccine, 310
vagina, 410–411
vas deferens, 406
vascular endothelium, 283
vasoconstriction, 282, 385
vasodilation, 282
vasomotor center, 205
vasopressin, 240–241
vastus intermedius, 175
vastus lateralis, 175
vastus medialis, 175
VC (vital capacity), 332
veins, 282, 284–285. *See also* cardiovascular system
ventilation, 324, 331, 339
ventral cavity, 13–14
ventral position, 10
ventral root, 211
ventricles, 207, 207–208, 273–274
ventricular syncytium, 277
venules, 282
vermis, 206
vertebrae, 125
vertebral (spinal) cavity, 13
vertebral column, 126
vertebral foramen, 125
vesicles, 49, 61
vestibular glands, 407, 411
vestigial organ, 299, 361
villi, 357, 373
visceral membrane, 14
visceral pericardium, 14, 272
visceral peritoneum, 348
visceral pleura, 14, 333
vital capacity (VC), 332
vitamin A, 31
vitamin D, 31
vitamins, 361

vitreous humor, 213, 222
voice box, 327
Volkmann's canals, 117, 139, 149
voltage-gated channels, 53, 200
volume-to-surface ratio, 447
voluntary muscles, 155
vomer bone, 124, 325
vulva, 411

W

water, 27–28, 47
water breaking, 435
water reabsorption, 378
websites
  Cheat Sheet, 3
  personal identification number, 3
  registration, 3
white blood cells (leukocytes), 76, 110, 261, 319
white matter, 204, 206
white pulp, 298
windpipe, 326
womb, 410

X

xiphoid, 126

Y

yellow marrow, 114, 150
yolk sac, 429
young adulthood, 450–451

Z

Z-line, 160
zona pellucida, 400, 413, 428
zygomatic bones, 124
zygotes, 398, 445

About the Author

Erin Odya is a teacher of anatomy & physiology and genetics, a tutor of all things life science, and an all-around phabulous physiologist. She has degrees in biology and education from Indiana University and is currently working to shape young minds at Carmel High School in Carmel, Indiana. This is her third *For Dummies* book.

Author's Acknowledgments

Writing a book while teaching full-time is not an easy feat. My support system runs deep, and I am unabashedly appreciative for everyone's support.

Thank you to Lindsay Lefevere and Vicki Adang for trusting in my vision, to Paul Levesque and Kelsey Baird for keeping me on task, and the rest of the Wiley staff that helped behind the scenes,

To Bear, P, and my Greyhound Family who cheered me on and helped both in my classroom and out.

And to Carey's Crew, who've always encouraged me to take on new challenges, you're the best family a gal could ever ask for.

Dedication

To Gabe and to all of my students — past, present, and future — who challenge my knowledge and inspire my curiosity.

Publisher's Acknowledgments

Acquisitions Editor: Lindsay Lefevere

Senior Project Editor: Paul Levesque

Copy Editor: Keir Simpson

Tech Editor: Frederick Dooley

Production Editor: Mohammed Zafar Ali

Cover Image: © Anatomy Image/Shutterstock